Concrete Technology for a Sustainable Development in the 21st Century

Concrete Technology for a Sustainable Development in the 21st Century

Edited by

Odd E. Gjørv and Koji Sakai

Routledge
Taylor & Francis Group
LONDON AND NEW YORK

First published 2000 by E & FN Spon

2 Park Square, Milton Park, Abingdon, Oxfordshire OX14 4RN
52 Vanderbilt Avenue, New York, NY 10017

Routledge is an imprint of the Taylor & Francis Group, an informa business

First issued in paperback 2019

Publisher's Note

This books has been prepared from camera-ready copy provided by the author.

British Library Cataloguing in Publication Data
A catalogue record for this book is available from the British Library

Library of Congress Cataloging in Publication Data
A catalogue record for this book has been requested

Publisher's Note
The publisher has gone to great lengths to ensure the quality of this reprint but points out that some imperfections in the original may be apparent.

ISBN 978-0-419-25060-9 (hbk)
ISBN 978-0-367-86408-8 (pbk)

CONTENTS

Preface

This book is the result of the International Workshop on "Concrete Technology for a Stustainable Development in the 21st. Century", that was organized in Lofoten, Norway, June 24-26, 1998. The Workshop took place in conjunction with the Second International Conference on Concrete Under Severe Conditions – Environment and Loading" CONSEC'98, in Tromsø, Norway, June 21-24, 1998. Both events were a following up from CONSEC'95 in Sapporo, Japan, August 2-4, 1995, and the International Workshop on "Rational Design of Concrete Structures" in Hakodate, Japan, August 7-9, 1995.

The purpose of the Lofoten Workshop was to focus on both problems and challenges for the concrete industry in the 21st. Century with emphasis on environmental consciousness. The concrete industry is a large consumer of natural resources that has a big impact on the natural environment. Therefore, the concrete industry has a great responsibility both for a more efficient utilization of these resources, and for taking steps to prevent environmental degradation and to create a better harmony with the natural environment.

In addition to the consumption of natural resources for concrete production, the production of portland cement is based on a very energy consuming industrial process. The production of each ton portland cement also releases approximately one ton of carbon dioxide in addition to a number of other polluting constituents to the atmosphere. Both premature corrosion of embedded steel and inadequate structural design for seismic loading are two examples of reduced service life of concrete structures that not only represent technical and economical problems, but also a huge waste of natural resources and hence both a local and global environmental problem.

It is a great challenge for the concrete industry, therefore, to obtain a more controlled service life of concrete structures. In this respect, it is important to further develop and utilize high-performance concrete and to integrate structural and durability design as well as to develop more appropriate systems for operation and preventive maintenance of concrete structures. For seismic loading, it is important to improve the basis for structural design. A more rational planning and flexible design for future needs of the individual concrete structures are also important.

In many countries, large amounts of concrete waste from demolished concrete structures are produced, most of which is simply dumped. Another challenge, therefore, is to develop a basis for recycling as much as possible or for total recycling of the concrete like that for other structural materials such as steel and aluminum.

In addition to the conventional utilization of concrete as a structural material, concrete also has a great potential for a further contribution to the solving of many environmental problems. More industrial wastes and by-products can be utilized for concrete production. Concrete can also be more utilized for containment and embedment of hazardous, industrial wastes. New types of ecological concrete can be further developed and utilized in order to reduce the general burden both on living organisms, animal life and natural vegetation.

As a conclusion from the presentation and discussion of the 33 papers at the workshop, the participants made the following Lofoten Declaration:

> "We concrete experts shall direct concrete technology towards a more sustainable development in the 21st. Century by developing and introducing into practice:
>
> 1. Integrated performance-oriented life cycle design.
>
> 2. More environmental-friendly concrete construction.
>
> 3. Systems for maintenance, repair and reuse of concrete structures.
>
> In addition, we will share information on all these issues with technical groups and the general public".

The Lofoten Workshop was part of a joint research program between Norwegian University of Science and Technology and Kagawa University as part of a bilateral agreement between the Japanese and Norwegian governments.

As editors, we hope that the results from this workshop will contribute to a more sustainable development of concrete technology in the 21st. Century. As editors, we would also like to express our sincere gratitude to all participants for their valuable contributions to the workshop. We are also thankfull to the Norwegian Public Roads Administration, Nordland, for financial support. A special appreciation to Grete Lind, who did an excellent job as the secretary to the first editor.

Trondheim, Norway — Odd E. Gjørv

October, 1998 — Koji Sakai

CONTROLLED SERVICE LIFE OF CONCRETE STRUCTURES AND ENVIRONMENTAL CONSCIOUSNESS

O. E. GJØRV
Norwegian University of Science and Technology, Trondheim, Norway

Abstract
For many years, an uncontrolled service life of concrete structures with an increasing spending of resources on repairs, rehabilitation and premature demolition do not only represent technical and economical problems. This is poor utilization of natural resources and hence also, an environmental problem. In the present paper, current experience on field performance and engineering practice including strategies for a more controlled service life is outlined and discussed.
Keywords: Concrete structures, service life, steel corrosion, condition monitoring, preventive maintenance, life cycle management.

1 Introduction

Both on the basis of weight, volume and money, the construction industry is the largest consumer at materials in our society. From a production point at view, several of these materials have a great impact on both the local and global environment. This is particularly true for concrete as one of the most dominating construction materials. Therefore, increased environmental consiousness in the form of a better utilization of concrete as a construction material is an increasing challenge to the construction industry.

So far, a poor durability and premature service life of concrete structures have created serious problems in many countries. An uncontrolled service life of concrete structures with an increasing spending of resources on repairs, rehabilitation and premature demolition do not only represent technical and economical problems. This is poor utilization of natural resources and hence also, an environmental problem.

In the following, current experience on field performance and engineering practice including strategies for a more controlled service life is outlined and discussed.

Concrete Technology for a Sustainable Development in the 21st Century edited by O.E. Gjørv and K. Sakai.
Published in 2000 by E & FN Spon, 11 New Fetter Lane, London EC4P 4EE, UK. ISBN: 0 419 25060 3

2 Field performance

2.1 General

Although deteriorating processes such as freezing and thawing, expansive alkali reaction and chemical attack also create problems, it is not the disintegration of the concrete itself but rather electrochemical corrosion of embedded steel which poses the most critical and greatest threat to the long-term performance and service life of concrete structures [1]. In particular, this is true for concrete structures exposed to chloride salts. In the following, therefore, the outline of field performance is mainly concentrated to concrete structures in chloride containing environments.

2.2 Bridges

In many countries, deicing salts have created serious problems to corrosion in concrete bridges [2]. Only in the US, it has been estimated that the cost of correcting the corroding bridges is \$ 24 billion, with an annual increase of \$ 500 million [3].

Coastal environments may represent even more severe conditions [4-7]. Recent investigations in Norway have shown that more than 50 % of all the larger bridges along the Norwegian coastline either have a varying extent of steel corrosion or has been repaired due to steel corrosion [8]. Most of these bridges have been built during the last 25 years, of which one was so heavily corroded that it had to be demolished after a service period of 25 years [9].

For all the corroding bridges along the Norwegian coastline, extensive field investigations have revealed that large amounts of chlorides have penetrated the concrete. A typical pattern for the chloride penetration is that those parts of the bridges that are the most exposed to prevailing winds and salt spray have the least chloride penetration compared to that of the more protected parts (Figs. 1 and 2). This is due to the rain intermittently washing off the salt from the most exposed surfaces, while on the more protected surfaces, the salt accumulates. Another pattern observed is that older bridges built before approximately 1970 have performed better than bridges built later on. Based on all this experience, there has over the last few years been an increased focus on improved performance for new bridges.

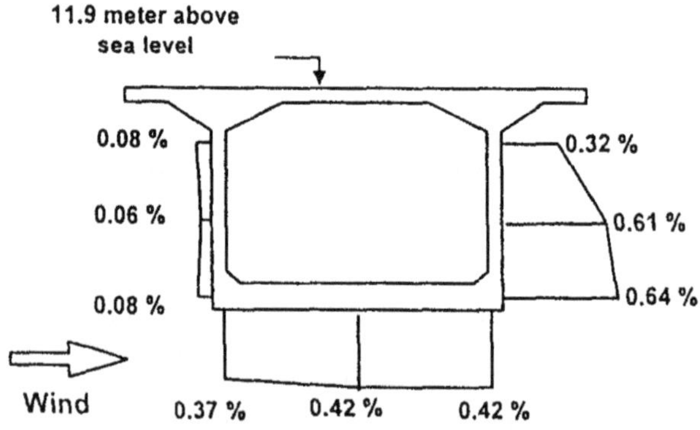

Fig. 1. Typical distribution in surface chloride concentration on a bridge exposed to a Norwegian marine environment [10].

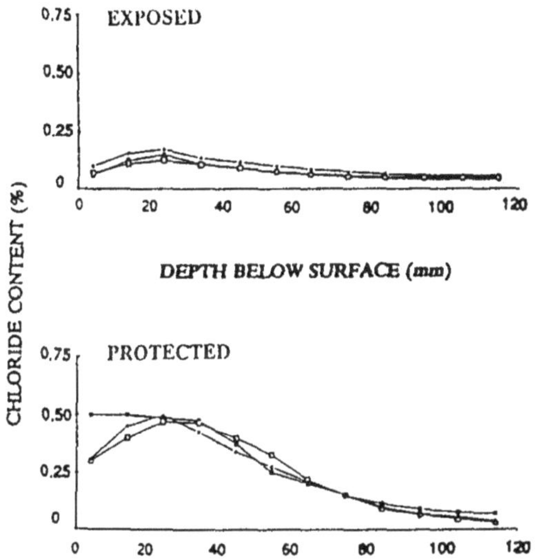

Fig. 2.
Typical chloride penetration into those parts of a concrete coastal bridge that are the most exposed to prevailing winds and salt spray compared to that of the more protected parts [8].

2.3 Harbor structures

Also concrete harbor structures are severely exposed to chlorides. An extensive field investigation carried out on 219 concrete structures in Norwegian harbors during 1962-68 showed that 84 % of all the structures had a varying extent of steel corrosion, and the first visible sign of corrosion typically appeared after a service period of 5-10 years [11]. In addition, 34 % of the structures had been repaired due to steel corrosion.

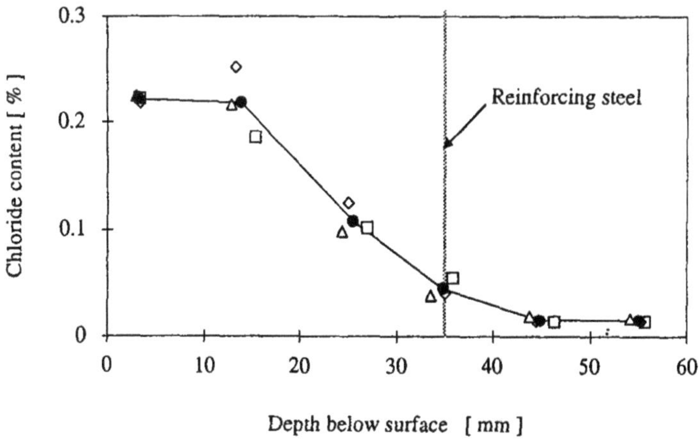

Fig. 3. Penetration of chlorides into a concrete wharf after 4.5 years of exposure [12].

Based on improved technology, many new concrete structures in Norwegian harbors appear to perform much better. However, new extensive field investigations carried out on Norwegian concrete harbor structures show that corrosion problems still occur on several new structures already after a service period of less than 10 years (Fig. 3). For most of these structures, current standard requirements to both concrete quality and thickness of concrete cover have been fulfilled.

For many years the worldwide problem with premature corrosion in concrete harbor structures has been the subject for numerous papers [1].

2.4 Offshore structures

In spite of a very harsh and aggressive environment in the North Sea, where the concrete platforms are completely splashed over by seawater several times a year, no or very little corrosion on embedded steel has been reported so far [13].

For all the installations in the North Sea, extensive surveying programs both above and below water are regularly carried out. So far, the most serious problems reported have been related to accidental loads from ships and falling items, and most concrete surfaces have been observed to be in good condition without any visible sign of deterioration. More detailed inspections of eleven of the oldest platforms installed during 1973-78 and reported in 1982, also showed good condition [14].

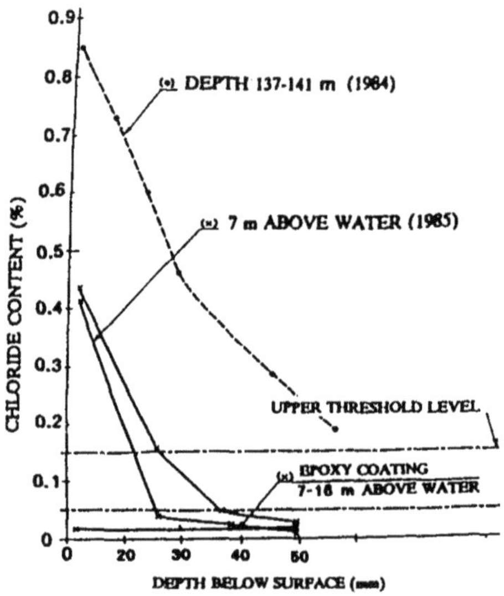

Fig. 4.
Penetration of chlorides into the Statfjord A platform (1977) after 8 years of exposure in the North Sea [15].

However, monitoring of chloride penetration demonstrates that the chlorides also penetrate the high quality concrete in the North Sea installations, only at a slower rate.

Fig. 4 shows the chloride penetration into the Statfjord A platform (1977) after 8 years of exposure, while Figs. 5 and 6 show the chloride penetration into the Brent B (1975) and Brent C (1976) platforms after 20 years of exposure.

For the Oseberg A platform (1988), where the concrete cover has partly been less than prescribed, corrosion has already occurred, and cathodic protection is currently being installed [17].

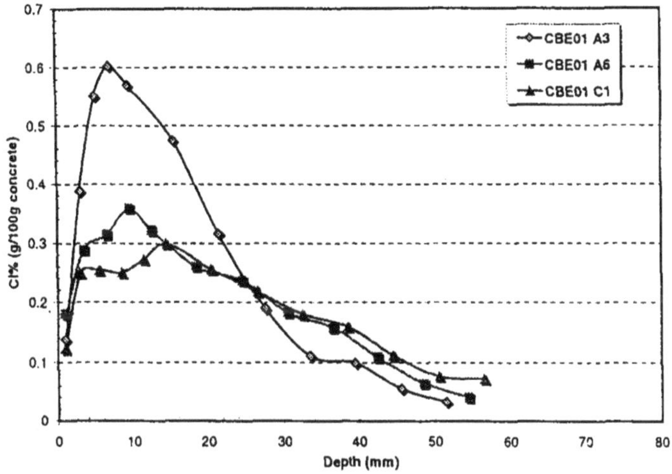

Fig. 5. Penetration of chlorides into the Brent B platform (1975) after 20 years of exposure in the North Sea (14.4 m above sea level [16].

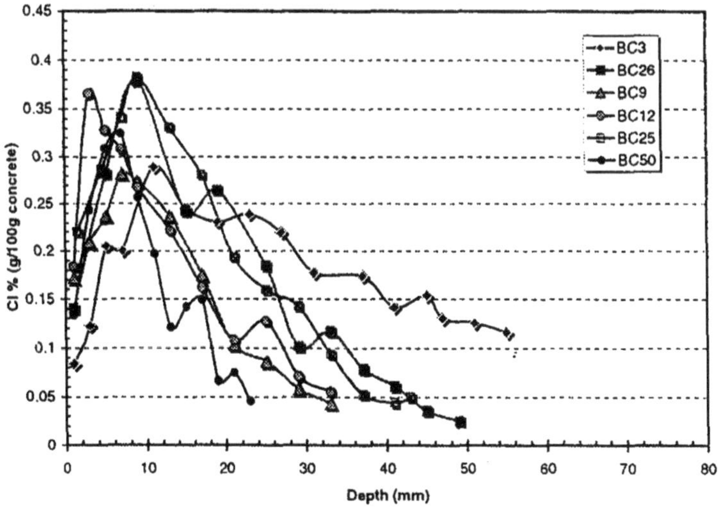

Fig. 6. Penetration of chlorides into the Brent C platform (1976) after 20 years of exposure in the North Sea (11 to 18 m below sea level) [16].

3 Codes and practice

The better performance of offshore concrete structures compared that of the onshore structures observed over the last 25 years, raises the question why the offshore structures in the harsh environment in the North Sea perform so much better than the land-based concrete structures along the Norwegian coastline.

When the first reinforced and prestressed concrete concept for platform construction in the North Sea was introduced in the late 1960's, the offshore technical community showed much scepticism. However, at the same time the results from the comprehensive field investigation of existing concrete structures along Norwegian coastline were published [11]. The overall good performance of these structures even after a service period of 50 to 60 years contributed to convincing the offshore technical community that reinforced and prestressed concrete also could be a reliable construction material for installations in the North Sea.

For the international operators in the North Sea, however, the appearance of reinforcing bar corrosion after a service period of only 5-10 years was not acceptable. Therefore, for the Ekofisk Tank (1973) which was the first concrete platform in the North Sea, both increased concrete quality and increased concrete cover to the steel were required. At the same time, very strict quality assurance and quality control programs were introduced on the construction site.

When the first FIP publication "Recommendations for the Design and Construction of Concrete Sea Structures" was introduced in 1973 [18], also these durability requirements were based on the results from the comprehensive field investigations in Norwegian waters. Already in 1977, the third revised edition of the FIP Recommendations were published, and meanwhile, a number of new both national and international guidelines and codes had been introduced [9]. Since the Ekofisk Tank was installed in 1973, the concrete quality requirements steadily increased. From 1973 to 1986, concrete with a water-cement ratio of less than 0.40-0.45, a cement content of more than 400 kg/m^3 and a concrete permeability of less than 10^{-13} kg/Pa.m.s. was required. In the same period, the compressive strength increased from 45 to 70 MPa (Fig. 7), while the minimum concrete cover was 50-75 mm. If the observed concrete cover did not meet the specified requirements, additional precautions in the form of surface coatings were required.

The good performance of the concrete platforms in the North Sea demonstrates that already from the early 1970's, it was possible to design and produce quite durable concrete structures even for the most aggressive and harsh marine environment. However, proper utilization of existing knowledge and experience was essential.

For the land-based concrete structures, the traditional requirement to compressive strength was almost the only concrete quality requirement until the late 1980's. During the booming construction activities in the 1970's and 1980's, both harbor structures and coastal bridges were typically produced with a compressive strength of 30 to 40 MPa and a minimum concrete cover of 25 to 50 mm.

That other structures built before approximately 1970 performed better than structures built later on, may be related to the lack of general knowledge in the construction industry in combination with the rapid development in concrete technology that started in the early 1970's. Thus, until the late 1960's, a 30 MPa type of concrete in Norway would typically be produced with a cement content of 400 to 500 kg/m^3. During the 1970's and 1980's, however, new organic and mineral admixtures were introduced, and new and more efficient cements became available. At the same time, both aggregate production and concrete mixing were optimised. Therefore, it successively became possible to meet the specified requirement to compressive strength with successively less cement. Also, an increased production track often resulted in more poor workmanship and curing of the concrete. As a consequence, the

C45 - C50

Cement (SP30)	430 kg
Sand, 0 - 10 mm	900 kg
Coarse aggreg.,	
10 - 32 mm	900 kg
Betokem LP	4 ltr.
Slump	120 mm
W/C ratio	0,41
Obtained m.s.	55 MPa

Beryl A (1973 - 75)

C55 - C60

Cement (SP30 - 4A)	380 kg
Sand, 0 - 5 mm	1030 kg
Coarse aggreg.,	
5 - 20 mm	845 kg
Betokem Pa (B)	4 ltr.
Slump	220 mm
W/C ratio	0,42
Obtained m.s.	67 MPa

Statfjord C (1981 - 83)

C65 - C70

Cement	
(SP30 - 4AMod)	380 kg
Sand, 0 - 5 mm	940 kg
Coarse aggreg.	
5 - 20 mm •	945 kg
Betokem Pa(B)	6 ltr.
Slump	240 mm
W/C ratio	0,43
Obtained m.s.	79 MPa

Gullfaks C (1986 - 87)

C80 - C90

Cement (SP30 - 4 AMod)	420 kg
Micro silica	30 kg
Sand, 0 - 5 mm	875 kg
Coarse aggreg., 5 - 20 mm	945 kg
Betokem Pa (B)	6 ltr.
Betokem P	2,5 ltr.
Slump	260 mm
W/C + S	0,34
Obtained m.s.	98 MPa

Test (1983 - 84)

Fig. 7. Typical concrete mixtures and compressive strenght for concrete platforms in the North Sea [20].

ability of the concrete to resist chloride penetration and to protect embedded steel from corrosion got successively impaired.

Poor workmanship in combination with lack of proper quality control on the construction site often also resulted in very small concrete covers. This situation is typically illustrated by the concrete covers observed in a Norwegian bridge from 1978-81 shown in Fig. 8, where the graph is based on 2028 single observations.

The situation represents a general problem in many countries as demonstrated in Fig. 9, where the observations from the Norwegian bridge (N) are shown together with similar observations from a Japanese bridge [27] (J) and more than 100 concrete structures from the Arabian Gulf [23] (AG).

Although both codes and job specifications for land-based concrete structures have been changed several times during the last 30 years, the requirements to concrete durability are still almost exclusively based on traditional requirements to concrete composition, procedures as well as curing conditions, and where the compressive

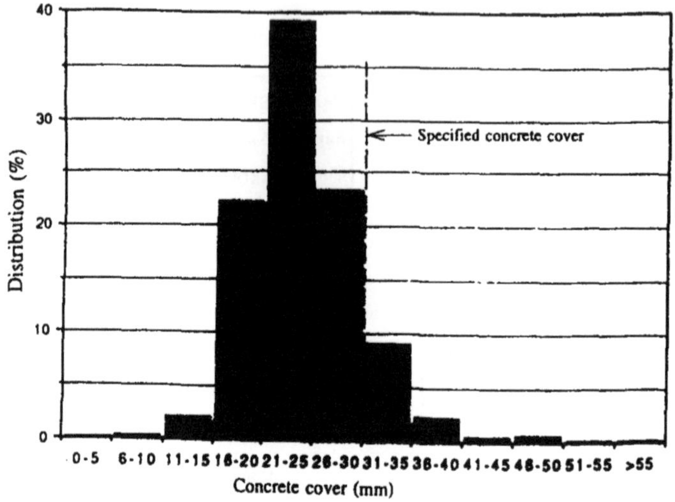

Fig. 8. Concrete covers observed in Gimsøystrømmen brigde (1978-81).

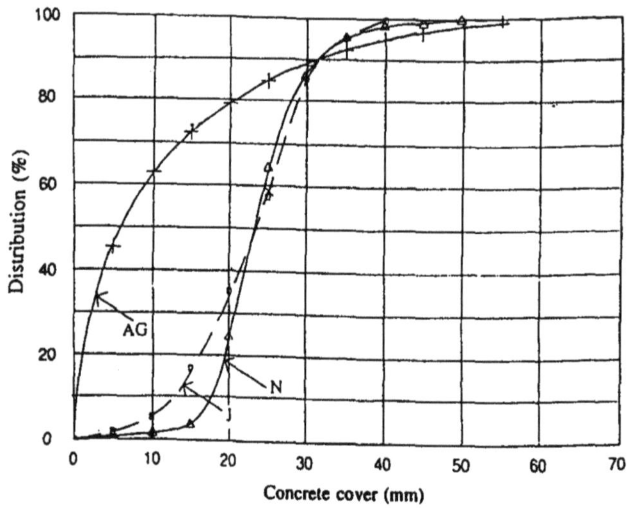

Fig. 9. Concrete covers in Gimsøystraumen bridge compared to that in a Japanese bridge (J) and more than 100 concrete structures from the Arabian Gulf (AG) [21].

strength is the only performance criteria. For several reasons, this descriptive approach has shown to yield insufficient and unsatisfactory results.

For many new concrete structures, however, there has been an increasing focus on additional quality parameters based on water suction and chloride diffusivity [24]. Such concrete quality parameters are used both for job specification and control of in situ quality. They are partly also used as input parameters in models for service life design.

4 Service life design

In recent years, there has been an increasing focus and interest for development of models for service life design of concrete structures. As a basis for such design, however, it is very important to have a proper calculation model for each mechanism involved in the deterioration process. Also, it is very important to be able to properly describe the environmental exposure conditions.

As far as chloride penetration into concrete is concerned, a model based on Fick's 2. law was introduced already in the early 1970's [25, 26]. Although several modifications of this model have been introduced later on [27-29], the mechanism for chloride penetration is not only controlled by diffusion. Even for submerged conditions, the chloride penetration is the result of several very complex mechanisms [30]. Above water, which represents the most severe conditions, the effect of water suction also complicates the penetrating mechanisms. In addition, the environmental exposure conditions become very complex and difficult to describe as already discussed for bridges (Figs. 1 and 2).

In addition to all the deterministic models for chloride penetration proposed in recent years, there has been an increasing focus on the development of models based on a probabilistic approach [31]. Combinations of deterministic and probabilistic models have also been proposed.

Also for a probabilistic approach, however, a proper mathematical model for the chloride penetration, taking into account the variability both in the exposure conditions and the resistance of the concrete against chloride penetration, is essential. Relevant and sufficient data on the variability are also essential, but nor such data are easily available.

So far, current standards and codes of practice have given no or very little assistance in design for service life, but work with new guidelines is under way, a presentation of which has recently been given by Cather and Marsh [32]. These authors also point out, however, that much of the research work that has been carried out in this area can not readily be translated into practical guidance. As an example, reference is made to the time from the onset of corrosion to the point when surface cracking is clearly visible, which may be the point chosen for defining the end of the service life. This adds to the already extensive uncertainty that exists to the lack of accuracy involved in predicting the rate of chloride penetration, and the level of chloride concentration necessary at the reinforcement for initiation of corrosion. Therefore, although several numerical models for chloride penetration are available, it appears that these models have not yet any satisfactory and practical validation.

Another question is how the end of the service life is defined. Even if it would be possible to predict a reliable time in service until onset of corrosion or visible sign of corrosion, it will not necessarily be so that the concrete structure will be abandoned at that time. As soon as the structure becomes chloride contaminated and corrosion starts, however, it will both be difficult and very expensive to carry out a repair. It may be better, therefore, to have a strategy for a more controlled prevention of chloride penetration based on an appropriate program for life cycle management [1].

5 Life cycle management

By use of special technology, research has shown that it is possible to produce concrete that will resist almost any kind of chloride penetration [33, 34]. Based on practical experience from concrete structures in high chloride containing environments, however, it appears to be just a question of time before detrimental amounts of chlorides will reach embedded steel even through 50–70 mm covers of high quality concrete. In order to obtain a more controlled chloride penetration, therefore, protective surface treatments or coatings should be applied, either during the early suction period immediately upon the formwork removal or at a later stage. For those offshore concrete platforms that during construction got a solid epoxy coating on the exposed concrete in the splash zone, no chloride penetration has been observed later on, even after 15 years of severe exposure [35].

For most other construction materials, it is generally accepted that a regular preventive maintenance is needed. This is the reason why several steel bridges still exist in good condition even after more than 100 years of service. The only difference between steel and concrete structures from a performance point of view, should be that concrete structures should need much less preventive maintenance than steel structures, and the more durable the concrete structure are produced, the less preventive maintenance should be needed.

For all offshore concrete structures in the North Sea, a regular assessment of the current condition in combination with preventive maintenance make up a very important part of the established life cycle management. Also for structures and installations in most other industrial areas, regular programs for condition assessment and preventive maintenance are well established. The typical situation for most land-based concrete structures, however, is still that nothing is done until an advanced stage of visual damage has developed. For a chloride-induced corrosion, repairs are then both technically difficult and disproportionately expensive compared to that of a regular monitoring of chloride penetration and the application of a protective coating at a proper stage.

For concrete structures in chloride containing environments, in appears that a regular monitoring of the chloride penetration into the most exposed parts of the structure should provide a most reliable basis for the preventive maintenance and hence, for a more controlled service life. In principle, the monitoring of the chloride penetration can be based on a regular coring or drilling of samples for chloride analysis (Fig. 6). For new structures, however, the monitoring can also be based on embedded probes [36, 39], that can either be part of a semi or fully automatic monitoring system. Even for existing concrete structures, probes for monitoring of chloride penetration can now be installed [38]. Often, a combination of different monitoring techniques or different preventive measures should be part of the corrosion protection strategy [39].

6 Conclusions

For many years, an uncontrolled service life of concrete structures with an increasing spending of resources on repairs, rehabilitation and premature demolition do not only represent technical and economical problems. This is poor utilization of natural resources and hence also, an environmental problem.

It is not the disintegration of the concrete itself but rather electrochemical corrosion of embedded steel which poses the most critical and greatest threat to the service life of concrete structures. This is particularly true for concrete structures exposed to chloride containing environments. In such environments, it appears to be just a question of time before detrimental amounts of chlorides will reach embedded steel even through 50-70 mm covers of high quality concrete.

For all concrete structures of a certain importance, appropriate programs for life cycle management should be established. As a basis for such programs, a regular assessment of current condition in combination with preventive maintenance are very important.

For chloride containing environments, a regular monitoring of the chloride penetration should provide a most reliable basis for the preventive maintenance and hence also, for a more controlled service life. Often, a combination of different monitoring techniques or different preventive measures should be part of the corrosion protection strategy.

7 References

1. Gjørv, O.E. (1996) Performance and Serviceability of Concrete Structures in the Marine Environment, Proceedings, Odd E. Gjørv Symposium on Concrete for Marine Structures, St. Andrew By-The-Sea, ed. by P.K. Metha, CANMET/ACI, pp. 259-279.
2. Camptroller General of the United States (1979) Solving Corrosion Problems of Bridge Surfaces Could Save Billions, United States Accounting Office PSAD – 79-10.
3. American Association of State Highway and Transportation Officials (1986) Strategic Highway Research Program Research Plans, Transportation Research Board.
4. Nilsson, I. (1991) Repairs of the Ølands Bridge, Experience and Results of the First Six Columns Carried Out in 1990, Report, NCC, Malmø, Sweden (In Swedish).
5. Stoltzner, E. and Sørensen, B. (1994) Investigations of Chloride Penetration into the Farø Bridges, Dansk Beton, Vol 11, No. 1, pp. 16-18 (In Danish).
6. Beslac, J., Hranilovic, M., Maric, Z. and Sesar, P. (1997) The Krk Bridge: Chloride Corrosion and Protection, Proceedings, International Conference on Repair of Concrete Structures – From Theory to Practice in a Marine Environment, ed. by A. Blankvoll, Norwegian Public Roads Administration, Oslo, pp. 501 – 506.
7. Wood, J.G.M. and Crerar, J. (1997) Tay Road Bridge: Analysis of Chloride Ingress Variability & Prediction of Long Term Deteroration (1997) Construction and Building Materials, Vol 11, No. 4, pp. 249-254.
8. Østmoen, T., Liestøl, G., Grefstad, K.A., Sand, B.T. and Farstad, T. (1993) Chloride Durability of Coastal Bridges in Concrete. Report, Norwegian Public Roads Administration, Oslo (In Norwegian).
9. Hasselø, J.A., (1997) Ullasundet Bridge – The Life Cycle of a Concrete Structure, EEU Course on Life Cycle Management of Concrete Structures, The Norwegian University of Science and Technology, Trondheim, 15 p. (In Norwegian).

10. Fluge, F. (1997) Enviromental Loads on Coastal Bridges, Proceedings, International Conference on Repair of Concrete Structures – From Theory to Practice in a Marine Environment, ed. by A. Blankvoll, Norwegian Public Roads Administration, Oslo, pp. 89-98.
11. Gjørv, O. E. (1968) Durability of Reinforced Concrete Wharves in Norwegian Harbours, Ingeniørforlaget AS, Oslo, 208 p.
12. Lahus, O. and Gjørv, O.E. (1998) Condition, Operation and Maintenance of Norwegian Concrete Wharves, Betongindustrien, Vol. 30, No. 1, pp. 10-12 (In Norwegian).
13. Durability of Concrete Structures in the North Sea (1994) State-of-the-Art-Report. FIP Concrete Sea Structures Commission, London.
14. Fjeld, S. and Røland, B. (1982) Experience from In-Service Inspection and Monitoring of 11 North Sea Structures, Offshore Technology Conference, Paper No. 4358, Houston, USA, 9 p.
15. Sandvik, M. and Wick, S.O. (1993) Chloride Penetration into Concrete Platforms in the North Sea. Proceedings, Workshop on Chloride Penetration into Concrete Structures. Division of Building Materials, Chalmers University of Technolgy, Gothenburg, ed. by L.O. Nilsson, 7 p.
16. Brent Concrete Core Testing, Department of Building Materials, The Norwegian University of Science and Technology, Trondheim (Unpublished data).
17. Østmoen, T. (1998) Field Tests with Cathodic Protection on the Oseberg A Platform, Ingeniørnytt,Vol. 34, No. 6, pp. 16-17 (In Norwegian).
18. Recommendations for the Design and Construction of Concrete Sea Structures (1973), FIP, London.
19. Marshall, A.L. (1990) Marine Concrete, Blackie and Son Ltd, London.
20. Norwegian Contractors A.S. (Private communication).
21. Kompen, R. (1994) New Specifications for Securing the Concrete Cover, Betongindustrien, Vol 26, No. 2, pp. 28-32 (In Norwegian).
22. Ohta, T., Sakai, K., Obi, M. and Ono, S. (1992) Deterioration in a Rehabilitated Prestressed Concrete Bridge. ACI Materials Journal, Vol. 89, pp. 328-336.
23. Matta, Z.G. (1993) Deterioration of Concrete Structures in the Arabian Gulf, Concrete International, Vol. 15, pp. 33-36.
24. Gjørv, O. E. (1994) Important Test Methods for Durability of Reinforced Concrete. Proceedings, V. Mohan Malhotra Symposium on Concrete in the 21st Century, San Franscisco, ACI SP-144, ed. by P.K. Metha, pp. 545-574.
25. Collepardi, M., Maricialis, A. and Turiziani, R. (1970) The Kinetics of Chloride Ions Penetration in Concrete, Il Cemento, Vol 67, No. 4, pp. 157-164 (In Italian).
26. Collepardi, M., Marcialis, A. and Turriziani, R. (1972) Penetration of Chloride Ions into Cement Pastes and Concretes, Journal of the American Ceramic Society, Vol. 55, No. 10, pp. 534-535.
27. Takewaka, K. and Mastumoto, S. (1988) Quality and Cover Thickness of Concrete Based on the Estimation of Chloride Penetration in Marine Environment. ACI SP-109, ed. by V.M. Malhotra, pp. 381-400.
28. Maage, M., Helland, S. and Carlsen, J.E. (1997) Service Life Prediction of Concrete in Marine Environment, Proceedings, International Conference on Repair of Concrete Structures – From Theory to Practice in a Marine Environment, ed. by A. Blankvoll, Norwegian Public Roads Administration, Oslo, pp. 177-187.

29. Poulsen, E. (1997) Four-Parametric Description of Marine Exposure and Concretes Response to its Chloride Intensity, Proceedings, International Conference on Repair of Concrete Structures – From Theory to Practice in a Marine Environment, ed. by A. Blankvoll, Norwegian Public Roads Administration, Oslo, pp. 189-189.
30. Zhang, T. and Gjørv, O.E. (1986) Diffusion Behavior of Chloride Ions in Concrete, Cement and Concrete Research, Vol. 26, No. 6, pp. 907-917.
31. Bamforth, P. (1997) Probabilistic Performance Based Durability Design of Concrete Structures for Long-Term Serviceability, ed. by E.A. Byars and T. McNulty, Thomas Telford Publishing, London, pp. 33-44.
32. Cather, B. and Marsh, B. (1997) Service Life Design of Concrete Structures – Current and Emerging Approaches, Management of Concrete Structures for Long-Term Serviceability, ed. by E.A. Byars and T. McNulty, Thomas Telford publishing, London, pp. 21-32.
33. Gjørv, O.E., Ngo, M.H., Sakai, K., and Watanabe, H. (1998) Resistance Against Chloride Penetration of Low-Heat, High-Strength Concrete, Proceedings, Second International Conference on Concrete under Severe Conditions – Environment and Loading, ed. by O.E. Gjørv, K. Sakai and N. Banthia, E & FN Spon, London (In print).
34. Nilsson, L., Ngo, M.H. and Gjørv, O.E. (1998) High-Performance Repair Materials for Concrete Structures in the Port of Gothenburg (1998) Proceedings, Second International Conference on Concrete under Severe Conditions – Environment and Loading, ed. by O.E. Gjørv, K. Sakai and N. Banthia, E & FN Spon, London (In print).
35. Aarstein, R., Rindarøy, O.E., Liodden, O. and Jenssen, B.W. (1988). Effect of Coatings on Chloride Penetration into Offshore Concrete structures, Proceedings , Second International Conference on Concrete under Severe Conditions – Environment and Loading, ed. by O.E. Gjørv, K. Sakai and N. Banthia, E & FN Spon, London (In print).
36. Gjørv, O.E. and Vennesland, Ø. (1982) A New Probe for Monitoring Steel Corrosion in Offshore Concrete Platforms, Materials Performance, Vol. 21, pp. 33-35.
37. Eri. J., Vælitalo, S., Gjørv, O.E. and Pruckner, F. (1998) Automatic Monitoring for Control of Steel Corrosion in Concrete Structures, Proceedings, Second International Conference on Concrete under Severe Conditions – Environment and Loading, ed. by O.E. Gjørv, K. Sakai and N. Banthia, E & FN Spon, London (In print).
38. Schiessl, P. and Raupach, M. (1998) A New Sensor System for Monitoring the Corrosion Risk in Existing Structures, Proceedings, Second International Conference on Concrete under Severe Conditions – Environment and Loading, ed. by O.E. Gjørv, K. Sakai and N. Banthia, E & FN Spon, London (In print).
39. Tromposch, E.W., Dunaszegi, L., Gjørv, O.E. and Langley, W.S. (1996) Corrosion Protection Philosohpy – Northumberland Strait Bridge Project, Proceedings, Third CANMET/ACI International Conference on Performance of Concrete in Marine Environment, St. Andrews By-The-Sea, Supplementary papers, pp. 327-341.

INTEGRATED DESIGN OF CONCRETE STRUCTURES AND TECHNOLOGY DEVELOPMENT

K. SAKAI
Kagawa University, Takamatsu, Japan
N. BANTHIA
University of British Columbia, Vancouver, Canada

Abstract
In design of concrete structures, it is important to consider not only structural safety but also durability. However, durability problems have not properly been considered in design codes. This has caused serious deterioration in concrete structures. There is no doubt that we have to establish a reasonable design method for concrete structures. This paper critically reviews the current status of design concerning durability and proposes integrated design methods which include structural design and durability design. The direction of technology development is also described in the scope of integrated design.
Keywords: Concrete technology, durability design, integrated design, limit states, performance-based design, structural design.

1 Introduction

In recent years, it has been pointed out that not only structural safety but also durability is important when designing concrete structures. However, in traditional structural design the degradation of structures over long periods of time is not regarded as a serious problem; the destruction of structures has a much greater impact on people, as exemplified by the Great Hanshin Awaji Earthquake. In Japan, it is estimated that the maintenance and renovation costs for infrastructure will exceed 70% of total public investment in 2010, assuming that the growth rate of new construction and improvement costs remain at 0% [1]. Restraint in public investment and reductions in costs have also been demanded. In the United States, where public investment began earlier than in Japan, exhaustion of infrastructure has become an even more serious problem [2]. The limitation of resources should be considered from the perspective of

Concrete Technology for a Sustainable Development in the 21st Century edited by O.E. Gjørv and K. Sakai.
Published in 2000 by E & FN Spon, 11 New Fetter Lane, London EC4P 4EE, UK. ISBN: 0 419 25060 3

global environmental concerns as well.

Taking these social factors into consideration, it is obvious that we should not consider the problem of concrete durability as a problem of the past. It is necessary to develop rational design methods and the technology to support these methods to meet various demands [3]. However, there is no method available today which can be truly called a design method for durability. The fact that the enormous amounts of research data on durability have not developed through to the design level is unfortunate. Recently, the importance of clarifying the relationship between the performance and cost of a structure, and of the prediction of the lifetime of the structure have been recognized. Thus, establishment of durability design has become a matter of social importance. In addition, if a design method for durability is systematically established, the technological points needed to achieve it will become clearer.

This paper reviews the current status of design concerning durability of concrete structures and proposes integrated design methods, including structural design and durability design. The direction of technology development in the scope of integrated design is also described.

2 Changes in design methods

It was after the establishment of the allowable stress design method as a structural design method that civil engineering structures began to be built according to a certain engineering rationality. In Japan, this idea for civil engineering structures was first adopted in the "Detailed Regulations for Road Structures (Draft) [4]," prepared by the Ministry of Home Affairs in 1926. The basics of this design method are still used in the present "Specification for Bridges[5]." In this design method, safety is achieved by defining limit values for the stress acting upon cross-sections of structural members, but the stress limit values are not necessarily clear. However, it is known from long-term experience that the performance of a structure is maintained at a level by this design method, although limit values have been altered. The problem with this method is that the performance which the allowable stress design method ensures is obscure in its relationship with the actual behavior of a structure. This is not rational from a "cost-performance" point of view.

In recent years, the design method for concrete structures internationally has been the limit state design method. In this method, dimensions concerning structural members and cross sections are determined so that they satisfy required performance for all the relevant limit states. The limit state design method has been created to solve the problems of the allowable stress design method, and was established based on the accumulation of a vast amount of research data on the behavior of concrete structures. In other words, as the behavior of concrete structures became clearer, this rational design method emerged.

Can we then declare that the current limit state design method is a perfect method? The answer is no. The framework of the limit state design method is said to be roughly systematic, with each detail being insufficient.

Both the allowable stress design method and the limit state design method are structural designs, which in principle do not include durability design. Although it is needless to mention again that durability is important in concrete structures, there are no designs which can be truly called durability designs. This is partially due to the fact that historically, the primary purpose of design was to avoid collapse of a structure, and thus the design system itself has been created within such a framework. It cannot be denied that room for incorporating the results of such research into design codes was not provided, and consequently the stance of such research eventually became inattentive to the application to design, even though a number of research studies have been conducted on concrete durability.

Given these circumstances, the performance verification design method (hereafter called the performance-based design method) is receiving attention these days. As the history of the performance-based design method was described in detail in another paper [6], its details are omitted here. This method is said to have received attention as an effective means of minimizing friction generated by the globalization of various activities. That is to say, this idea is convenient in constructing a system of which people can generally approve. At the same time, in the design of a concrete structure, structural design and durability design can both be treated by the same concept. Thus, it can be said that the framework to achieve durability design has been provided.

The limit state design method is said to be basically a performance-based design method. The volume for seismic design of the Standard Specification for Design and Construction of Concrete Structures recently published by the Japan Society of Civil Engineers [7] is the first standard which clearly defines the concept of performance-based design.

3 Current state of durability design

The items or provisions concerning durability in the Standard Specification for Design and Construction of Concrete Structures by the Japan Society of Civil Engineers [8, 9] were investigated. Design codes and standards in Japan and other countries can be considered as being basically the same.

3.1 Design lifetime

The determination of design lifetime of a structure is the most critical step in design. Up until now, it has been believed impossible to determine this on reasonable grounds, and the Standard Specification only states "consider the period of service, the environmental conditions and durability performance." The determination of design lifetime is the essence of durability design, and therefore the measures needed to achieve a required durability periods should be clearly indicated. In other words, to realize durability design, the behavior of a structure needs to be understood as a function of time.

3.2 Material factor

In structural design, various uncertainties are generally treated by introducing various safety factors. It is also clear that material deteriorates over time, and it is important to evaluate deterioration. The Standard Specification considers the phenomenon with a material factor. However, the fluctuation in material strength, differences of material properties between specimens and those used in the structure, as well as the influence of material property on limit state, are also taken into consideration in the material factor. Thus, the practical significance of the material factor is not clear. Even if this could be dealt with structurally, the deterioration of material causes a problem of durability and, in turn influences the structural behavior. As is apparent from this, structural and durability problems should be treated within a unified system.

3.3 Crack width

To prevent the corrosion of reinforcing steel and to maintain watertightness, cracks which occur in concrete must be controlled. Regarding the corrosion of steel, the Standard Specification classifies environments into three categories: general environment, corrosive environment and extremely corrosive environment, which stipulates the allowable width of cracks according to the type of steel and as a function of concrete cover, and at the same time provide an equation to calculate the width of flexural cracks which occur in a structural members. However, the relationship between the width of cracks and the corrosion of steel has not been entirely defined. In particular, the relationship between the corrosion of steel due to cracks, crack width and the corrosion of steel due to the formation of micro-cells caused either by the penetration of chloride ions, which is dependent largely on the quality of the concrete, or caused by the carbonation has not been clarified.

To prevent steel from corroding, an evaluation which rationally takes these influences into consideration is needed.

Regarding watertightness, allowable crack width is specified depending on required watertightness and the predominant cross-sectional force.

3.4 Concrete cover

Concrete cover is the most important factor in ensuring adequate durability of concrete structures. As a structural detail, the Standard Specification defines the minimum value of concrete cover, which is obtained by multiplying the basic cover value for each structural member and for each of the aforementioned environmental conditions by the coefficient according to the design strength of concrete.

Concrete cover should be evaluated according to the required performance of a structure, based on the rate of movement and accumulation of chloride ions in concrete or of carbonation, in other words by taking the influence of time into consideration. However, because of many factors and the lack of reliable data over a long time period, only rough and possibly even conservative values emerge.

To realize durability design of concrete structures, it is essential to rationally evaluate the corrosion of steel based on the mechanism of the movement of deleterious agents in concrete in relation to design lifetime or to the required performance. Concrete cover, by nature, should not be treated as merely a structural detail in the specifications.

3.5 Water-cement ratio

In general, the quality of concrete can be evaluated by its strength. In other words, strength can be said to be an effective index indicating the "comprehensive performance" of concrete. Because the strength of concrete is greatly influenced by its water-cement ratio, the water-cement ratio has been used as an index in specifying the durability of concrete.

For example, the maximum water-cement ratio needed to ensure frost resistance of concrete is specified depending on the exposure conditions of the structure, the meteorological conditions and the types of water supplied. Furthermore, regarding off-shore concrete, the maximum water-cement ratio to be determined according to required durability is specified by classifying the environmental conditions into "in the atmosphere above the sea," "seawater spray area" and "submarine." These water-cement ratios are determined by considering the corrosion of steel, frost damage and chemical erosion. However, it is unreasonable to treat phenomena with different deterioration mechanisms using single values. They should each be specified based on performance considering different actions.

There seem to be two problems in using the water-cement ratio as a durability index. One is the fact that the influence of the type of cement is not considered. It is well known that the movement of substances in concrete, frost damage and chemical erosion differ greatly according to the type of cement used. The other problem is that the index cannot be selected with relation to the required performance.

Thus, the current provisions for durability are in a state in which the precise relationship between environmental actions and performance as a function of time cannot be evaluated.

4 Performance-based design

In recent years, the concept of performance-based design has been attracting attention. The development of this concept was prompted by the unification of Europe. The unification resulted in demands for consistency in all systems. Matters concerning the construction industry were no exception, and efforts have been made to unify various standards and specifications.

The preparation of European Codes (EC) 2 for concrete began in 1979. At that time, it was based mainly on the CEB/FIP Model Code 1978, and the results were eventually published as the "European Pre-Standard ENV 1992-1-1" by CEN/TC 250.

Since European countries naturally had designed structures according to their own standards, the unified codes had to be lenient. Namely, the strategies of CEN to gradually promote EC2 were to: (1) prepare its own effective ENV and announce its existence and to (2) allow each country's own standards to remain effective in parallel with ENV until the final decision to shift to EN from ENV, if the national standards were incompatible with ENV. Thus, to make the unified codes function, it is most important to have contents with which each country can basically agree. This is the only choice available.

The design method in EC2 is basically a limit state design. In other words, it is a type of performance-based design in which it is verified whether required performance is satisfied using proper indices. The provisions for durability in EC2 are the so-called item specifications which provide such values as the maximum water-cement ratio and the minimum water-cement ratio according to comparatively detailed and classified environmental conditions, but are not based on the true performance of the concrete structure.

On the other hand, the activity of the Technical Committee on concrete of the International Standardization Organization (ISO) was re-started under the leadership of the United States in 1995, after an interruption for the preparation of EC2 in 1987.

The fundamental idea behind the restarting was to prepare the "Umbrella Performance-Based Structural Concrete Standard" (Level 1), which possesses unified standards and consists of broad fundamental principles which all countries in the world can accept and adopt. According to the Standard, each country or region should prepare more detailed standards according to their respective conditions. If these standards are in compliance with the concept of ISO Level 1, they may be adopted as ISO standards. Thus, the strategy for achieving the establishment of unified codes is basically the same as that for EC2. More practical manuals and guides are defined as Level 3.

Thus, performance-based design can be thought of as having wide applications as design methods for concrete structures. The problem is what kind of design system is to be established.

5 Integrated design

The concept of performance-based design is said to be extremely useful in dealing with the design of concrete structures in a unified manner. That is, the concept enables landscape design and constructability, as well as structural and durability design to be considered in the same way. Such a design system is referred to as integrated design. The framework of the integrated design system is described below.

A structure must fulfill its purpose of use, be safe and durable, and at the same time must be designed taking constructability, maintenance and management, load on the environment and landscape into consideration. The design lifetime should be determined by taking into consideration the service period required for the structure, environmental conditions, durability performance of the structure and its maintenance and management plan.

A structure should be designed so that the required performance is attained during construction, as well as during the design lifetime. In principle, the required performance of a structure should be verified by a direct analysis of time-dependent behaviors of a structure under the assumed load and environmental actions. However, it should be also allowed to consider appropriate limit states which indirectly guaranttee the required performance. In general, the limit performance can be considered as follows:

(i) Landscape limit states
(ii) Constructability limit states
(iii) Ultimate limit states
  (a) Static performance
    • Bending and axial force
    • Shear
    • Torsion
  (b) Seismic performance
    • Seismic performance 1: The structure is functionally normal even after an earthquake and is usable without repairs.
    • Seismic performance 2: The function of the structure can be restored in a short time and no reinforcement is required.
    • Seismic performance 3: The entire structural system does not collapse due to an earthquake.

(iv) Serviceability limit states
(v) Durability limit states

- Durability performance 1: Performance level at the beginning of service is maintained throughout the design lifetime.
- Durability performance 2: Although some deterioration of structural material occurs during the design lifetime, the structure can be used without strengthening.
- Durability performance 3: Large-scale repair and/or strengthening in a structure are assumed at a certain time.

The essence of performance-based design methods lies in verification techniques to ensure required performance. Generally speaking, technologies continually develop, thus making it impossible to classify technical levels. It is, however, possible to classify verification levels of present design techniques into macro, meso and micro.

The macro-level is a very simple verification level. The allowable stress design method, for example, is thought to be included at this level. Some structural details may also be included in this category. The relationship between actual performance of structures and verification techniques at this level is however not necessarily clear.

The meso-level is a level at which verifications are conducted based on indexes which are theoretically or experimentally quantified. For example, it can be considered that verifications using the general shear capacity formula are included in this level. Many of our present techniques are included in this category. It seems that degree of accuracy for each verification index differs greatly.

At the micro-level, the entire performance of a structure is verified based on modeling of materials and structural characteristics. The results therefore greatly dependd upon the reliability of models and sophistication of analytical techniques. Recent development of analytical techniques, including the finite element method, has been remarkable. Many problems, however, such as the calculation of shear failure or the post-peak performance of complicated structures remains unresolved.

The verification of various limit performances is conducted using the following equation:

$$R(t_d) > S(t_d)$$

where $R\ (t_d)$ is either the landscape performance, constructability performance, mechanical performance or durability performance which a structure or structural member possesses during the construction or during its design lifetime (t_d), and $S(t_d)$ is the index value determined under the conditions of load and environmental actions or any other given condition.

To attain necessary performances in integrated design, development of verification techniques to guarantee such performances is crucial. In general, verification is conducted for a certain limit state. The limit state is determined from a certain state of the "performance" of a structure. It is therefore important to fully understand performances in response to all environment and load actions. The level of understanding of these performances relates directly to the level of design. Fig. 1 shows a flow of performance-based integrated design which can treat structural safety and durability of concrete structures in an integrated manner. Two extreme cases can be considered. One is the method to grasp the behavior of a structure, based on a

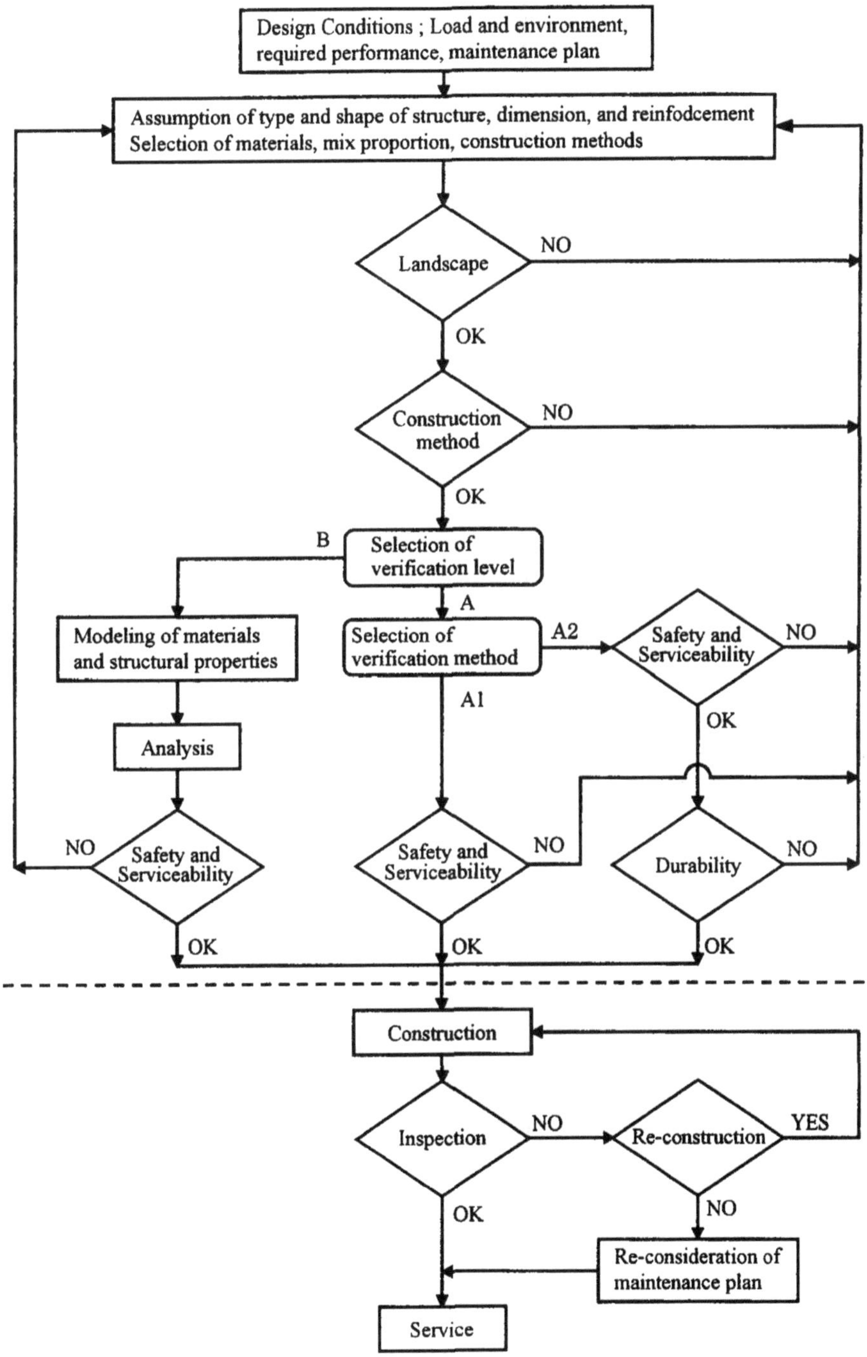

Fig.1 Flow of integrated design

model in terms of the material properties as a function of time and their relationship with the mechanical properties of a cross section, an element or a structural member and to verify whether they fulfill the required performances (Flow B). This method would be difficult at present and should be established in the future. The other case is to examine mechanical and durability aspects separately, and to fulfill the required performance for each aspect (Flow A2). An intermediate method between the two will be also possible (Flow A1). Although the examination for mechanical behavior is not necessarily at a satisfactory technological level, it can be regarded as usable as a whole, as seen in the limit state design. However, there is no doubt that durability design is at an extremely primitive design level.

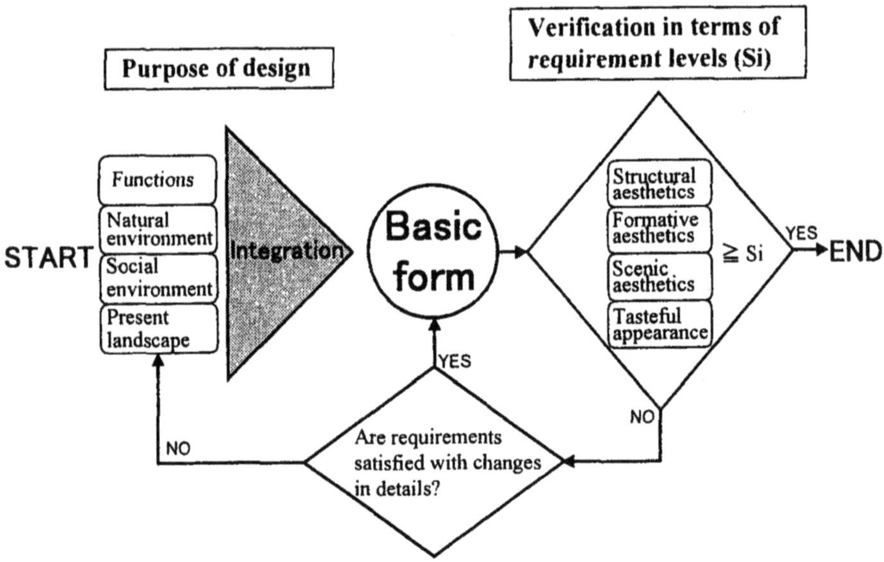

Fig.2 Integrated landscape design

The verification for landscaping are conducted as follows. When construction of a structure is planned for a given location and landscape design is deemed to be necessary, the purpose of design must first be clarified. Then, in connection with the design purpose, functions, the surrounding natural environment, the social environment and the present landscape are respectively evaluated and integrated to create a basic form. Finally, the basic form is verified using appropriate indexes such as structural, formative, scenic aesthetics and tasteful appearance to determine whether the form meets requirements shown in the design purpose. Fig. 2 shows the flow of design and its verification.

Regarding the verification of construction method, it is important to ensure if the construction methods planned can be applied and if the methods can guarantee a predetermined quality.

The verification of ultimate limit states can be conducted to:

(1) static performance

(a) bending and axial force

- cross-sectional capacity

(b) shear
- shear capacity

(c) torsion
- torsional capacity

(2) seismic performance [7]

(a) Seismic performance 1
- Verification of seismic performance 1 is to be conducted for a Level 1 earthquake.
- The stress which occurs on reinforcing bars and concrete during earthquakes should be verified as being lower than the design strength of the material.

(b) Seismic performances 2 and 3
- Verification of seismic performances 2 and 3 is to be conducted for a Level 2 earthquake.
- Using a proper analysis model, cross-sectional force and displacement should be calculated.
- Verification of seismic performance 2 is conducted by confirming that response displacement or residual displacement is within limit values. When the failure mode is not incorporated into the analysis model, determination of failure mode should be conducted and then safety concerning ductility and shear is checked.
- Verification of seismic performance 3 should be conducted by confirming that a structure does not collapse.

Regarding the verification of serviceability limit state, deflection or crack widths should be checked using appropriate methods.

The verification of durability limit states can be conducted to:

(1) Durability performance 1
- Confirm that the values obtained by the equations for bending, shear and tortional cracks are below limit values.
- Confirm that cracks and/or scaling are not caused by actions such as freezing and thawing, drying and wetting or by detrimental chemicals.
- Confirm that abrasion is not caused by abrasive action on concrete.
- Confirm that chloride ions on reinforcing bars are less than the limit value for the occurrence of corrosion.
- Confirm that the depth of the progress of carbonation is lower than the concrete cover.

(2) Durability performance 2 and 3
- Confirm that any cracks in, and/or scaling of the concrete, caused by such actions as freezing and thawing, drying and wetting or by harmful chemicals, are below limit values.
- Confirm that abrasion of concrete caused by abrasive action does not exceed the limit value.
- The degrees of corrosion of reinforcing bars caused by chloride ion penetration and carbonation and the occurrences of cracks, as well as the load capacity of a structure or structural member, are calculated using proper analytical models.

6 Research for the future

Figure 3 shows the relationship between the amount of corrosion of reinforcing bars and time. The figure illustrates that reinforcing bars corrode as the result of the movement and accumulation of chloride ions in concrete or due to carbonation and cracks occur in concrete at time t . For example, in a case in which the presence of chloride ions is a problem, the accumulation of chloride ions over time should be evaluated to verify that durability performance 1 is being fulfilled. In general, Fick's diffusion equation is used for the evaluation, but it cannot be said to be at a satisfactory level at present, because there are many influencing factors, including environmental conditions, quality of concrete and cracks. Regarding durability performances 1 and 2, there is little information on the conditions in which cracks occur, the morphology of cracks, bond property, the cross-sectional loss of reinforcing bars and changes in failure mode. Much further research is needed.

The most significant phenomenon concerning the durability of concrete is the occurrence of cracks. These occur in a variety of morphologies and sizes. One of the major tasks in concrete engineering has been how to control cracks. However, how various cracks grow from microscopic to macroscopic sizes and influence the durability of concrete or reinforcing bars is not clear, with the exception of those on the limitations on mechanical crack widths. In the framework of integrated design, the understanding of these factors is very important to ensure each durability performance. In other words, only when these are clarified, integrated design is realized.

As a method to control cracks, fibers are used. Fibers are also used to improve the toughness of concrete. There is a great deal of information about the latter and the use of fiber in such a way has reached a practice level although the actual range of application is still narrow. On the other hand, there is not necessarily sufficient information about the former. In particular, the effects of fiber reinforcement to control

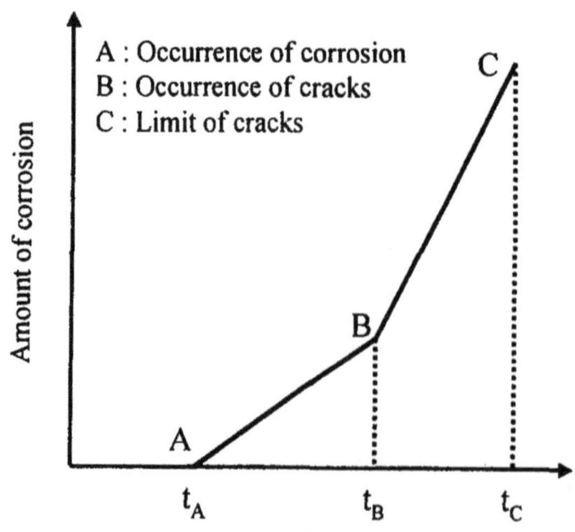

Fig.3 Relationships between the amount of steel corrosion and time

micro- and macro-cracks in concrete, that is, the improvement of durability performance, has hardly been examined. As one of the technologies to attain durability performances 1 and 2, it is significant to pursue the possibility of fiber reinforcement.

The global environment has been the focus of discussion in recent years. Investigations and research on that are being conducted in the field of construction. It is obvious that the environment must be taken into consideration in all fields in the future and it is important to accordingly include environmental assessments in the design of structures ("Environmental Design"). Lifecycle cost is a possible method for evaluating the relationship between the design of structures and the global environment. If the performance of structures as a function of time, can be evaluated as it is, it will be possible to design structures in terms of lifecycle cost. It will be needed to study long-term behavior of structures in this perspective. It is also essential to promote the diversification of performance requirements to allow and promote concrete recycling.

The design levels that structure owners will or should demand will also be important. Performance requirements are thought to differ depending on differences in customer's awareness on technology levels. It is therefore desirable to consider basic ideas on these factors in developing performance verification techniques.

7 Concluding remarks

Technology is the reflection of a social system. That is to say technology develops in the framework of the social system of an era, therefore, the direction of development of technology changes when the value standards of people change along with their social environment. The situation surrounding concrete structures has also changed greatly. First, the myth that concrete is maintenance-free was proven untrue. This was a natural course, because there was no rational design method for durability. Second, because of the serious economic situation, a tight control over construction costs and guarantees of longer life have been strongly demanded. This means that a structure must be designed taking cost-performance into consideration. These demands stand as gigantic challenges in the field of concrete technology. However, there will be limited scope in concrete technology unless we tackle these problems seriously. The direction in which we have to proceed is obvious. It is hoped that the integrated design described in this paper will be a catalyst for such a movement.

8 References

1. Economic Planning Agency. (1988) Japan Living with the World - *A Five Year Plan for Eco-nomic Management*, Report of the Social Capital Sub-committee, Economic Council.
2. Dunker, K.F. and Rabbat, B.G. (1993) Why America's Bridges are Crumbling. *Scientific American*, No.3, pp. 66-72.
3. Sakai, K. (editor). (1995) *Integrated Design and Environmental Issues in Concrete Technology*, E & FN SPON.
4. Ministry of Home Affairs. (1926) *Detailed Regulations on Road Structure* (Draft).
5. Japan Road Association. (1996) Specification for Highway Bridges. *Concrete*

Bridge.
6. Sakai, K. (1997) Performance-Based Design. *Bridge and Foundation Engineering*, Vol.31, No.8, pp.73 -83.
7. Japan Society of Civil Engineers. (1996) Standard Specification for Design and Construction of Concrete Structures. *Seismic Design.*.
8. Japan Society of Civil Engineers. (1986) Standard Specification for Design and Construction of Concrete Structures. *PART 1 Design* .
9. Japan Society of Civil Engineers. (1996) Standard Specification for Design and Construction of Concrete Structures. *PART 2 Construction* .

INTEGRATED LIFE CYCLE DESIGN OF CONCRETE STRUCTURES

A. SARJA
Technical Research Centre of Finland, VTT, VTT Building Technology, Espoo, Finland

Abstract
Integrated life cycle design is an important tool towards more sustainable construction. Sustainable building is defined as a technology and practice which meets the multiple requirements of the people and society in an optimal way during the life cycle of the build facility. The most important sustainability factors in performance for structures with long target service life can therefore generally be defined as flexibility towards functional changes of the facility and high durability, while in the case of the structures with moderate or short target service life changeability and recycleability are dominating. Building envelope has to have a good hygrothermal behaviour. The competitiveness in sustainability of the comparison between materials and structures is focusing on the question, which materials and structures are able to produce, designed and manufactured with skill and at the same cost the best sustainability of the building facility.
Keywords: Design, life cycle, concrete structures, sustanability, service life, durability, recycling

1 Introduction

Sustainability is related to multiple requirements of people, societies and nature. Referring to Fig. 1 we could give a technical definition for sustainable building as follows (Sarja, 1997) /2/: "Sustainable building is a technology and practice which meets the multiple requirements of the people and society in an optimal way during the life cycle of the build facility."

Ecology can be interpreted as the economy of the nature (Sarja, 1995) /3, 4/. The term gives us a quite concrete starting point for the application of this aspect to materials and structural engineering and it can be concretised in the life cycle methodology in design, manufacturing, construction and management. The expences of the natural economy are the consumption of non-renewable natural resources, the production of pollutants into air, soil and water and the qualitative goals regarding non-calculative effects like biodiversity and noise.

Sustainability must always be treated according to the life cycle principle — in other words with the application of life cycle methodology to design, manufacture, construction, maintenance, and the management of building projects, companies and other organisations in building. Through the principle of sustainability, resistance design will be expanded into durability design to include time as a new dimension in the design calculations. Health aspects are generally related to the control of moisture and thermal conditions and to special subjects like hazardous emissions from material. Design for recycling is an important tool for saving natural non-renewable resources and for reducing the environmental impact.

Concrete Technology for a Sustainable Development in the 21st Century edited by O.E. Gjørv and K. Sakai.
Published in 2000 by E & FN Spon, 11 New Fetter Lane, London EC4P 4EE, UK. ISBN: 0 419 25060 3

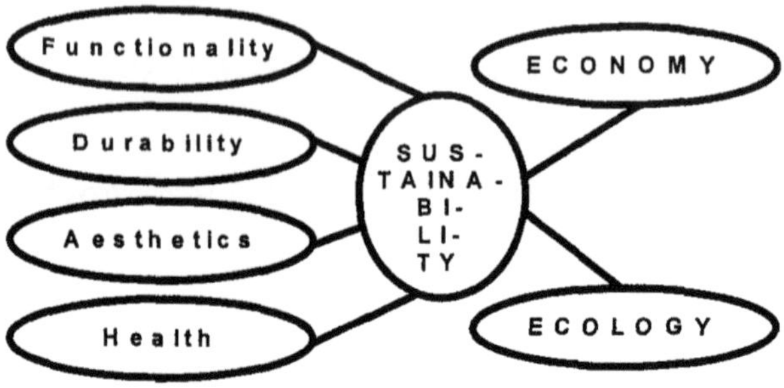

Fig. 1. Multiple requirements for sustainable buildings /2, 8/.

2 Integrated Life Cycle Design

2.1 General framework, process and procedure

The overall scheme of the integrated structural design is presented in Fig. 2.

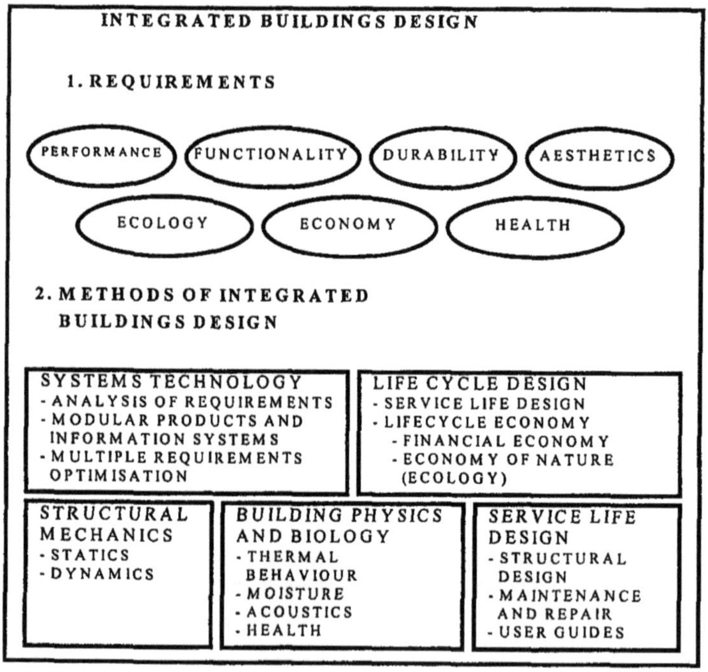

Fig. 2. Framework of integrated structural design (Sarja, 1996) /2, 6, 7/.

The process of integrated design consists of the following five design phases:

Phase 1: Analysis of functional requirements and specification of technical performance requirements.
Phase 2: Creation and schetching of alternative structural systems.
Phase 3: Environmental design: Analysis of the environmental expenditures, checking environmental requirements and criteria, development of the structural system and renewed analysis.
Phase 4: Multiple criteria decision making and selection of the optimal alternative: Definition of performance parameters, transfer of values of environmental expenditures from the analysis results, definition of the weighting factors for performance and expenditure factors, and multiple criteria selection between the design alternatives.
Phase 5: Final integrating design: Mechanical design, Physical design, Durability design.

3 Design methods

3.1 Environmental design

The results of comparisons regarding all types of the main sustainability requirements shown in Fig. 1, generally speaking, lead to the conclusion that differences between different materials and structural solutions of the construction phase are quite small. On the contrary, recarding to the monetary and nature expences during the period of use, we can find quite large differences between existing buildings or between other facilities. Those differences are caused by differences in the basic factors of sustainability, which are the flexibility in design, the changeability during the use, the durability in comparison to the design service life and the recyclability of those components, which have a quite short service life. In order to focus on these factors we have to take a separate look at the the different types of the building facilities.

In buildings the energy consumption is economically important and mostly dictating the environmental properties in the life cycle, the differences in environmental economy between different structural systems being otherwise quite small. For this reason, beside a well-controlled heating, ventilating and heat recovery, the thermal insulation of the envelope is important. The bearing frame is the most massive and long lasting part of the building, and the durability and flexibility in view of the functional changes, spaces and service systems are very important. The envelope shall be durable and, as mentioned above, have an effective thermal insulation and a safe static and hygrothermal behaviour. The internal walls have a more moderate length of service life, but they have a requirement of coping with relatively high degrees of change, and must therefore possess good changeability and recyclability. An additional property of an environmentally effective structural system is a good and flexible compatibility with the building service system, as the latter is the most frequently changed part of the building. In the production phase it is important to ensure the effective recycling of the production wastes in factories and on site. Finally, the requirement is to recycle the components and materials after demolition.

Engineering structures like bridges, dams, towers, cooling towers etc. often are very massive and their target service life is long. Therefore environmental efficiency is tied to selection of environmentally friendly local raw materials, high durability and easy maintainability of the structures during use, minimising and recycling of construction wastes, and finally recycling of the components and materials after demolition. Some parts of the engineering structures like water proofing membranes and railings, have a short or moderate service life and therefore the aspects of easy reassembly and recycling are most important.

The key issues from the environmental viewpoint are the life cycle design and the service life design. With the aid of life cycle design we can optimise the monetary (financial) economy and the economy of the nature (ecology). Through the service life (performance and durability) design, the targeted service life can be guaranteed. The conceptual, creative design phase is very decisive in order to utilise the potential benefits of integrated design process effectively. In this phase, the design is made on the system and module level. The modular systematics help the rational design, because the structural system typically owns different parts or modules with different requirements, e.g. regarding to durability and service life requirements.

It is also critical to introduce a time factor into the detailed structural design. This includes service life design to guarantee the durability and performance of the structures to corresponding target values /4/.

Health aspects can be identified from medical knowledge and applied as criteria into structural design. These are mainly related to building physics and biology, including several hygrothermal and acoustical methods.

3.2 Calculation of life cycle monetary and natural expences

The life cycle costs are calculated with the equation:

$$E_{tot}(t_d) = E(0) + \Sigma\,[N(t) \bullet E(t)], \qquad (1)$$

where E_{tot} is the design life cycle cost
$E(0)$ construction cost
$N(t)$ coefficient for calculation of the current value of the cost at the time t after construction
$E(t)$ cost to be born at time t after construction.

In the case of annual cost, the current value can be calculated with the equation

$$E_{tot}(t_d) = E(0) + N_s(t_d) \bullet E_a \qquad (2)$$

where $E(0)$ is the construction cost
$N_s(t_d)$ the multiplication coefficient for the annual cost (e.g. 50, 100 or 150 a)
E_a annual cost.

The coefficients $N(t)$ and N_s can either be calculated with known equations or taken from the general financial calculation tables. Equation (1) is suited for costs of repair, renewal, demolition, recycling and wasting, while equation (2) is suited for maintenance, energy and other annual costs. The life cycle costs can be utilised both for optimisation and as a selection criteria between design alternatives and products.

For the ecological calculations the expenses are, as presented above, the components of the ecological profile, e.g. consumption of non-renewable raw materials and energy, and the production of pollutants into air, soil and water, including CO_2, CO, SO_2, NO_x, dust, solid wastes and noise. There are also aspects, like biodiversity, which can not be calculated numerically. They shall be only evaluated qualitatively.

3.3 Recycling in the life cycle design

The components of environmental profile of the basic materials already include the recycling efficiency, which means the environmental expenses in recycling. It is important to realise that in design, the recycling possibities of building components, modules and even technical systems shall be additionally considered. The higher the hierarchical level of recycling is, the higher is the ecological and economical efficiency of recycling will be. Typical examples of rough estimates of recycling efficiency are the following:

1. Recycling of materials:
   - consumption of raw materils 0,5-0,9
   - consumption of energy 0,0-0,5
   - production of pollutants 0,1- 0,7
2. Recycling of structural or installation components:
   - consumption of raw materils 0,7-0,9
   - consumption of energy 0,7-0,9
   - production of pollutants 0,8-0,9

3.4 Multiple requirements optimisation

In the optimisation, the following types of variables are used: Calculated numerical monetary and ecological expenses and values, classified expenses and values, estimated qualitative expenses and values, normative expenses and values, ranking of the importance of expenses and values, or intuitive valuation.

3.5 Durability design of concrete structures

3.5.1 Theoretical background

Statistical service life

The simplest mathematical model for describing the event "failure" comprises a load variable S and a response variable R. In principle the variables S and R can be any quantities and expressed in any units. The only requirement is that they are commensurable. Thus, for example, S can be a weathering effect and can be the capability of the surface to resist the weathering effect without too large visual damages or loss of the concrete cover of reinforcement.

If R and S are independent of time, the event "failure" can be expressed as follows /15/:

$$\{\text{failure}\} = \{R < S\} \quad (3)$$

The failure probability P_f is now defined as the probability of that "failure":

$$P_f = P\{R<S\} \quad (4)$$

Either the resistance R or the load S or both can be time dependent quantities. Thus the failure probability is also a time dependent quantity. Considering $R(\tau)$ and $S(\tau)$ are instantaneous physical values of the resistance and the load at the moment τ the failure probability in a lifetime t could be defined as:

$$P_f(t) = P\{R(\tau)<S(\tau)\} \quad \text{for all } \tau \le t \tag{5a}$$

The determination of the function $P_f(t)$ according to the formula 3a is mathematically difficult. That is why R and S are considered to be stochastic quantities with time dependent or constant density distributions. By this means the failure probability can usually be defined as:

$$P_f(t) = P\{R(t)<S(t)\} \tag{5b}$$

According to the definition 3b the failure probability increases continuously with time as schematically presented in Figure 3.

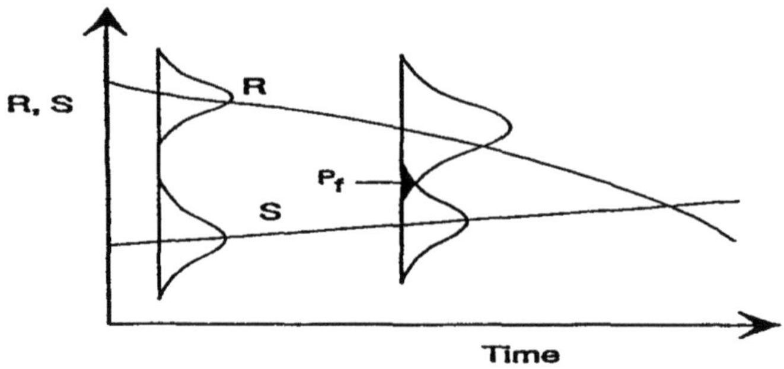

Fig. 3. The increase of failure probability. Illustrative presentation.

At a certain moment of time the probability of failure can be determined as the sum of products of two probabilities: 1) the probability that R < S, at S = s, and 2) the probability that S = s, extended for the whole range of S:

$$P_f = \sum_{\forall s} P\{R < S|S = s\}P\{S = s\} \tag{6a}$$

Considering continuous distributions the failure probability P_f at a certain moment of time can be determined using the convolution integral:

$$P_f = \int_{-\infty}^{\infty} F_R(s) f_S(s) ds \tag{6b}$$

where $F_R(s)$ is the distribution function of R,
$f_S(s)$ the probability density function of S, and
s the common quantity or measure of R and S.

The integral can be solved by approximative numerical methods.

Lifetime safety factor method

In practice it is reasonable to apply the lifetime safety factor method when the formulation of the design procedure returns to the deterministic form. The design service life is determined by multiplying the target service life by a lifetime safety factor:

$$t_d = \gamma_t \cdot t_g \tag{7}$$

where t_d is the design service life,
γ_t the lifetime safety factor and
t_g target service life.

The design formulae can then be written by applying either the performance principle or the service life principle as follows:

$$R(t_d) - S(t_d) \geq 0 \tag{8}$$

$$t_L - t_d > 0 \tag{9}$$

Figure 4 shows a distribution of service life and the relationships between the target service life, failure probability and mean service life. The lifetime safety factor is the relation of mean service life to the target service life.

$$\gamma_t = \frac{\mu(t_L)}{t_g} \tag{10}$$

where γ_t is the central lifetime safety factor,
$\mu(t_L)$ the mean service life and
t_g the target service life.

Using the lifetime safety factor, the requirement of target service life (corresponding to a maximum allowable failure probability) is converted to the requirement of mean service life.

The mean service life is approximated by service life models which show the crossing point of the degradation curve with the limit state of durability (Fig. 4). The mean service life evaluated by the service life model must be greater than or equal to the *design service life*, which is the product of the lifetime safety factor and target service life.

$$\mu(t_L) \geq t_d \tag{11}$$

$$t_d = \gamma_t \, t_g \tag{12}$$

where t_d is the design service life.

The lifetime safety factor depends on the maximum allowable failure probability. The lifetime safety factor also depends on the form of service life distribution. Figure 5 illustrates the meaning of lifetime safety factor when the design is done according to the performance principle. The function R(t) - S is called the safety margin.

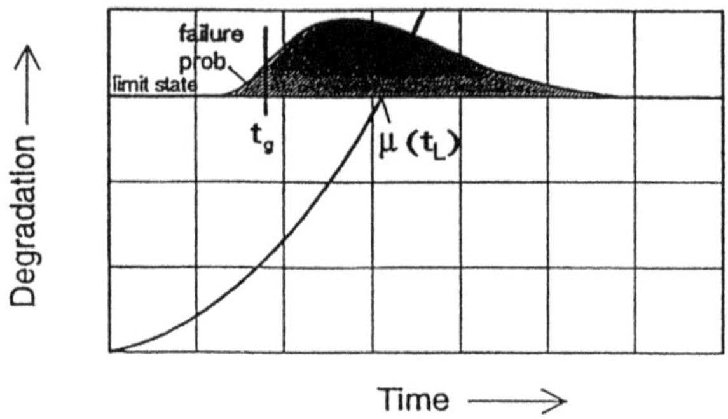

Fig. 4. Relationship between mean service life and target service life.

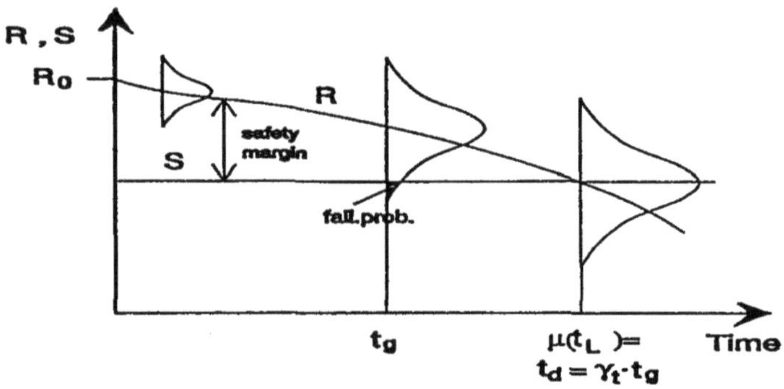

Fig. 5. The meaning of lifetime safety factor in a performance problem.

Performance behaviour can always be translated into degradation behaviour. By definition, degradation is a decrease in performance. The transformation is performed by the following substitutions:

$$R_0 - R(t) = D(t) \quad (13a)$$

$$R_0 - S = D_{max} \quad (14a)$$

or

$$R_0 - R_{min} = D_{max} \tag{14b}$$

Figure 6 shows the principle of design in a degradation problem. D(t) is the degrading effect of environmental loading on the performance of the structure. The range D_{max} - D(t) is the safety margin.

Let us consider that the degradation function is of the following form:

$$\mu(D(t)) = a \cdot t^n \tag{15}$$

where $\mu(D(t))$ is the mean of degradation,
a the constant coefficient,
t time, and
n the exponent.

The exponent n may in principle vary between $-\infty$ and $+\infty$.

The coefficient a is fixed when the mean service life is known:

$$a = \frac{D_{max}}{\mu(t_L)^n} \tag{16}$$

Degradation is assumed to be normally distributed around the mean. It is also assumed that the standard deviation of D is proportional to the mean degradation, the coefficient of variation being constant, v_D.

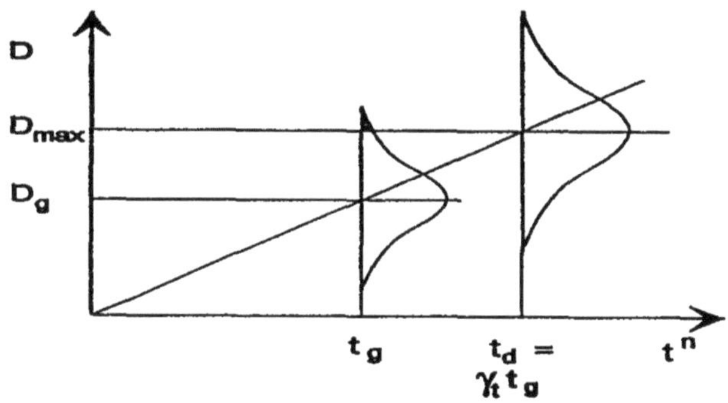

Fig. 6. Determination of lifetime safety factors with normally distributed D.

Figure 6 shows the degradation as a function of t^n. The value of γ_t can be determined as follows: The index β of standard normal distribution at t_g is:

$$\beta = \frac{D_{max} - D_g}{v_D D_g} = \frac{1}{v_D} \cdot \left(\frac{D_{max}}{D_g} - 1 \right) \tag{17}$$

where D_{max} is the maximum allowable degradation,
D_g the mean degradation at t_g, and
v_D the coefficient of variation of degradation.

From Figure 6 we get:

$$\frac{D_{max}}{D_g} = \frac{(\gamma_t t_g)^n}{(t_g)^n} = \gamma_t^n \tag{18}$$

By assigning this to Formula 15 we obtain:

$$\gamma_t = (\beta \cdot v_D + 1)^{1/n} \tag{19}$$

The lifetime safety factor depends on β (respective to the maximum allowable failure probability at t_g), the coefficient of variation of D and the exponent n. Thus the lifetime safety factor is not directly dependent on t_g.

Durability design method

Degradation factors and their structural effects

The following degradation factors are most commonly treated:

1. Corrosion due to chloride penetration
2. Corrosion due to carbonation
3. Mechanical abrasion
4. Salt weathering
5. Surface deterioration
6. Frost attack

Additionally there exist some internal degradation processes, such as alkaline-aggregate reaction, are not treated here as they can be solved by a proper selection of raw materials and an appropriate design of concrete mix.

Degradation factors affect either the concrete or the steel or both. Usually degradation takes place on the surface zone of concrete or steel, gradually destroying the material.

The main structural effects of degradation in concrete and steel are the following:

1. Loss of concrete leading to reduced cross-sectional area of the concrete.
2. Corrosion of reinforcement leading to reduced cross-sectional area of steel bars. Corrosion may occur
   a) at cracks
   b) at all steel surfaces, assuming that the corrosion products are able to leach out through the pores of the concrete (general corrosion in wet conditions).

3. Splitting and spalling of the concrete cover due to general corrosion of reinforcement, leading to a reduced cross-sectional area of the concrete, to a reduced bond between concrete and reinforcement and to visual unfitness.

Design procedure
The design procedure includes the phases presented in Fig. 7.
Ordinary mechanical design is performed using conventional design methods.

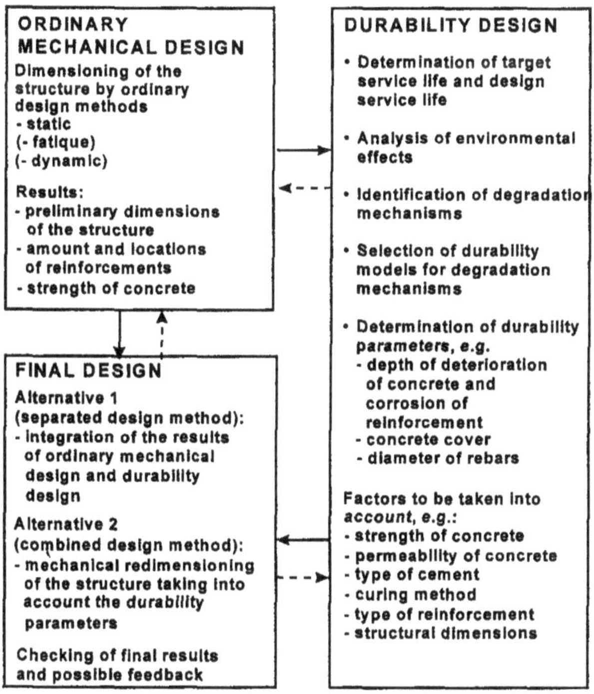

Fig. 7. Flow chart of the durability design procedure.

The durability design procedure is the following:

1. Specification of the target service life and design service life
2. Analysis of environmental effects
3. Identification of durability factors and degradation mechanisms

4. Selection of a durability calculation model for each degradation mechanism
5. Calculation of durability parameters using available calculation models
6. Possible updating of the calculations of the ordinary mechanical design
7. Transfer of the durability parameters into the final design

3.5.4 Description of the phases of the durability design

Phase1: Specification of target service life and design service life
The target service life is defined corresponding to the requirements given in common regulations, codes and standards in addition to possible special requirements of the client. Typical classes of service life are 10, 25, 50, 75, 100 etc. years. The safety classification of durability design is presented in Table 1.

The design service life is determined by formula:

$$t_d = \gamma_t \cdot t_g \qquad (20)$$

where t_d is the design service life,
γ_t the lifetime safety factor, and
t_g the target service life.

Table 1. Safety classification of durability design and the corresponding safety factors.

Limit state	Safety class of durability design	Separated design		Combined design 1)	
		Lifetime safety factor γ_t 2)	Load and material safety factors	Lifetime safety factor γ_t 3)	Load and material safety factors
Ultimate limit state	1. Serious social, economic or ecological consequences of a mechanical failure.	3.3	normal 4)	2.5	normal 4)
	2. Consequences of a mechanical failure are not serious.	2.9	normal 4)	2.2	γ_g 1.3 5) γ_p 1.38 γ_c 1.4 γ_s 1.13
Service-ability limit state	1. Noticeable consequences and considerable repair costs.	2.5	-	1.9	-
	2. Non-noticeable consequences and repair costs.	1.9	-	1.5	-

1) Check for m (relative reduction of the safety margin during $0 \rightarrow t_d$) ≤ 0.7 is required.
2) Cf. Chapter 5.3.2 Table 7.
3) Cf. Chapter 5.3.3.4 Tables 8 - 11.
4) Load and material safety factors specified for ordinary mechanical design are used. Reduced values of load and material safety factors may be used in the durability design (cf. Chapter 5.3.4). However, the safety at the start of service life (t = 0) must be at least the same as that required in ordinary design.

Phase 2: Analysis of environmental effects
The analysis of environmental effects includes identification of the climatic conditions such as temperature and moisture variations, rain, condensation of moisture, freezing, solar radiation and aerial pollution, and the identification of geological conditions such as the location of ground water, possible contact with sea water, contamination of the soil by aggressive agents like sulphates and chlorides. Man-made actions such as salting of roads, abrasion by traffic etc. must also be identified.

Phase 3: Identification of degradation factors and degradation mechanisms
Based on the environmental effect analysis the designer identifies the degradation factors to which the structure will most likely be subjected. Some kind of degradation process is usually assumed to take place in both the concrete and the reinforcement.

Phase 4: Selection of durability models for each degradation mechanism
A designer must determine which degradation factors are decisive for service life.. The models presented in the report may be applied in these evaluations. In *concrete structures exposed to normal outdoor conditions* the effects of degradation mechanisms can be classified into the following structural deterioration mechanisms:

1. Corrosion of reinforcement at cracks, causing a reduction in the cross-sectional area of steel bars.
2. Surface deterioration or frost attack, causing a reduction in the cross-sectional area of concrete.

Phase5: Calculation of durability parameters through calculation models
The damages are determined using the design service life, t_d, as time. Selected calculation models are presented in the appendix of the TC 130-CSL report /9/.

Phase 6: Possible updating of calculations in ordinary mechanical design

Some durability parameters may influence the mechanical design. An increase in concrete dimensions, increases the dead load, thus increasing the load effects on both the horizontal and vertical structures.

Phase 7: Transfer of durability parameters to the final design
The parameters of the durability design are listed and transferred to the final design phase for use in the final dimensioning of the structure.

Phase 8: Final design
Separated design method

In the separated design method the mechanical design and the durability design are separated. The ordinary structural design (phase1) produces the mechanical safety and serviceability parameters whereas the durability design (phase 2) produces the durability parameters. Both of these groups of parameters are then combined in the final design of the structure.

Combined design method

In the combined design method the mechanical design is newly performed, taking into account the results of the durability design and the required safety at the end of service life. The combined method is especially suited to degradation mechanisms which directly affect the bearing capacity or the mechanical serviceability of structures. Examples of the design are presented in the report /9/.

4 References

1. Sarja, Asko, Environmental Design Methods in Materials and Structural Engineering. CIB Information, Nr. 4/96. Rotterdam, October/November 1996. pp. 23 - 25.
2. Sarja, Asko, A vision of sustainable materials and structural engineering. In: Selected research studies from Scandinavia. Report TVBM-3078. Lund University, Lund Institute of Technology. Lund, Sweden, 1997. Pp. 153-163.
3. Sarja, Asko. Development towards the ecological and recyclabe building materials technology. Second International Conference on Materials Engineering for Resources, Akita, October 19-22, 1994, pp. 139 - 145. Research Institute of Natural Resources attached to Mining College, Akita University, Japan.
4. Sarja, Asko, Methods and methodology for the environmental design of structures. RILEM Workshop on environmental aspects of building materials and structures. Technical Research Centre of Finland, Espoo 1995. 5 p.
5. Sarja, A., Integrated structural design. Nordic Concrete Research, Proceedings: Nordic Concrete Research Meeting, Espoo, Finland, pp. 218 - 219. Publisher: Norsk Betongförening, Postboks 2312, N-0201 Oslo, Norway.
6. Sarja, Asko, Some Principles of Integrated Structural Design. Structural Engineering International (SEI), Nr. 1/1997. IABSE, Zürich. pp. 59-60.
7. Sarja, Asko, Framework and methods of life cycle design of buildings. Symposium; Recovery, Recycling, Reintegration, R'97, Geneva, Switzerland, February 4-7, 1997. EMPA, 1997, Volume VI, pp. 100-105.
8. Sarja, Asko, Prefabrication in relation to sustainable building. Prefabrication - facing the new century. International symposium, 1-3 October, 1997. Concrete Association of Finland, Helsinki, Finland, pp. 143-146
9. Sarja, Asko & Vesikari, Erkki (Editors), Durability design of concrete structures. RILEM Report Series 14. E&FN Spon, Chapman & Hall, 1996. 165 pp.
10. Sarja, Asko & Hannus, Matti, Modular Systematics for the Industrialized Building. VTT Publications 238, Technical Research Centre of Finland, Espoo 1995, 216 p.
11. Sarja, Asko, A vision of sustainable materials and structural engineering. Selected Research Studies from Scandinavia. Lund University, Report TVBM-3078, 1997, pp. 153-163.
12. Sarja, Asko, Prefabrication in relation to sustainable building. Symposium: Prefabrication - facing the new century. Concrete Association of Finland, October 1997, pp. 143 - 146.

A HOLISTIC APPROACH TO STRUCTURAL DURABILITY DESIGN

Holistic durability design

G. SOMERVILLE
Engineering Division, British Cement Association, Crowthorne, UK

Abstract

The structural design is perceived as an essentially numerical procedure, often confused with analysis. Initial attempts at durability design have adopted a similar approach, largely based on reliability methods. In this paper, an overall perspective is presented, which argues the case for an holistic approach embracing design material and construction issues. Many of the essentials which contribute to good durability are not amenable to a pure numerical approach.

Quality of construction, for concrete structures in corrosive environments, is perceived as a key issue - to be addressed by a system rather than a technical approach. The quality of detailing, and the treatment of water on exposed surfaces is also significant. Above all, there is a need for clarity in defining performance requirements, in practical whole life terms, including the future operation of the structure, and its management and maintenance.

Keywords: Holistic durability design; design concept; materials, detailing and construction quality; performance requirements; nature of aggressive actions.

1 Introduction

Good performance-in-service-with time is a sustainability issue. Premature repair, upgrading or replacement involves the unforseen use of scarce resources. Improving performance to meet functional needs over a prescribed period, at minium total cost, is therefore highly desirable. It is increasingly accepted that a structured design approach to durability is the best means of achieving that, but, as yet, there is no consistent pattern in the approaches being put forward, and there are some major difficulties to be overcome.

Concrete Technology for a Sustainable Development in the 21st Century edited by O.E. Gjørv and K. Sakai.
Published in 2000 by E & FN Spon, 11 New Fetter Lane, London EC4P 4EE, UK. ISBN: 0 419 25060 3

The CEB-FIP Model Code [1] gives the basic requirements for adequate service life, in the following terms:-

'Concrete structures shall be designed, constructed and operated in such a way that, under the expected environmental influences, they maintain their safety, serviceability and appearance during an explicit or implicit period of time, without requiring unforeseen high costs for maintenance and repair'.

Few would quarrel with that essential requirement. The real question is how best to achieve it. We now have an underlying science base and considerable understanding of the issues involved (through research and via feedback). How can we use that? How far can we go, in terms of developing a numerate approach, comparable to conventional structural design? Is an overall generic approach feasible? What should our strategy be, as we move towards a routine practical methodology for the future - with confidence?

The main purpose of this paper is to try and answer these questions. In doing so, the word 'holistic' has been included in the title. The dictionary definition is 'tending, as in nature, to form wholes that are more than the sum of the parts, by ordered grouping'. In the present context that means:-

- ensuring that all the factors which contribute to good in-service performance are identified, evaluated, and properly taken into account;
- looking at the complete picture, and not just durability by itself, e.g. the relationship with structural design, or the need to improve productivity and quality in the complete supply chain.

2 The case for a holistic approach

While our generic thinking develops, there is a need to solve on-going problems, as best as we can, based on imperfect knowledge. Two such examples are presented here, to illustrate the trend towards holistic methods.

In September 1992, a ban was introduced in the UK on the use of grouted post-tensioned construction for highway bridges. The prime reason was serious concern regarding corrosion caused by the use of de-icing salts. A concerted effort by all sectors of the industry (owners, designers, contractors, researchers, material and component suppliers) led to new recommendations for standards and practices [2], and the ban was removed in September 1996.

The philosophy adopted was that of multi-layer protection, particularly for the vulnerable tendons and anchorages, but also for the structural elements and the bridge as a whole. It was recognised, with current methods and knowledge, that any single method of protection might fail or wear out, and there was a need for redundancy in the design system.

The recommended protective layers were a mix, which could be broadly categorised under the headings:-

- design (concept and detailing)
- materials (specifications)

• construction (quality, certification, testing, etc.)

Combinations used in a particular case had to be compatible, and had to be chosen from each of the 3 categories - a holistic approach, within a clear overall philosophy.

This example addressed a particular (if somewhat narrow) technology and form of construction, while integrating design, material and construction issues in a unique way, to confidently satisfy defined performance requirements. Better specifications would not have solved the problem alone - nor would rigorous certification - the approach had to be holistic. An important lesson, learned in doing this, was the inadequacy of broadly-based Codes and Standards. Of necessity, these are very general, and often drafted in isolation. This lack of cohesion (and of detail) is an issue which must be addressed, in moving towards a holistic approach.

The second example comes from the construction of two major structures, in two different continents, in broadly the same aggressive environments: these are the Oresund crossing between Sweden and Denmark [3], and the Confederation Bridge in Canada [4]. A key element for each of these major jobs was the concrete specification; a comparison is made in Table 1.

Table 1. A comparison of the specifications for the Oresund Tunnel and the Confederation Bridge

	Oresund Tunnel [3] (Denmark/Sweden)	Confederation Bridge [4] (Canada)
Service life	100 years	100 years
Concrete strength	Not given	$60 N/mm^2$
Maximum W/C	0.40	0.34
Minimum cement content	Not specified	$450 kg/m^3$
Cement types	PC or slag ($\ngtr$66%)	PC+SF(7.5%)+pfa ($\ngtr$10%)
Air entrainment	As required	As required
Admixtures	Superplasticisers	Superplasticisers. Water reducers
Minimum cover	75mm	75mm
Special emphasis, in developing specifications.	Feedback from in service performance. Modelling, and control of early age cracking.	Modelling chloride profiles. Concrete resistivity. Risk analysis.

Both had the same service life requirement, and there are common features in the two specifications (e.g. cover), although the emphasis in developing these specifications was different. This is encouraging in itself, but there are additional common points to be noted; these are:-

• both are major structures, permitting the integration of design, materials and

construction issues, from the very beginning;
- significant pre-planning, and development testing, took place;
- both specifications were closely related to the type of structure, and proposed construction method;
- the need to specify both the material quality and the standards of execution, compatible with the construction method, was recognised;
- both groups drew on existing knowledge and data banks, in different ways: this resulted in some harmony, but also in differing ideas of limits, and of what was important.

The above two examples relate to different situations. Example 1 addresses a particular form of construction; example 2 relates to major structures in an aggressive environment - yet the need for a compatible holistic approach is clearly recognised in each case. An obvious follow-up question is:- can that holism be extended in a general way, or will the scale and nature of the problem be such that we must break our approach down in some way, as we move towards practical solutions at an operational level. This will be explored next.

3 Durability and design - where are we now?

A broad picture of the causes of deterioration for different types of structure is given in Figure 1 (taken from reference 13); this shows the dominance of corrosion (from various sources), but indicates the need to consider other mechanisms as well.

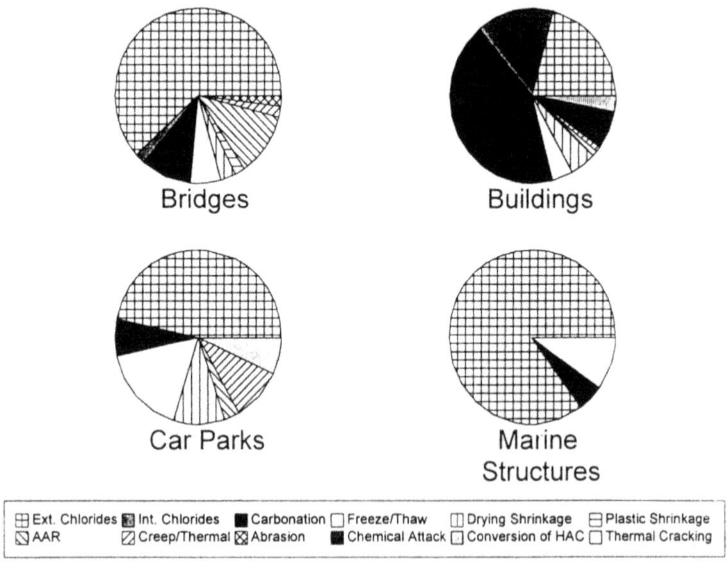

Figure 1. Reported Deterioration Mechanisms for Bridges, Buildings, Car Parks and Marine Structures.

The traditional approach to durability is largely materials-based. For some forms of aggressive action - mostly those which directly attack the concrete - this works well. Specifications for sulfate attack and alkali-silica reaction do exist, while being constantly improved. The biggest problem here, is where more than one action occurs simultaneously, and there is a need to specify for these, while still producing a concrete which can be properly placed under real site conditions. Probably, it is sensible to persevere with this approach, for this type of action.

Most codes also contain general requirements for concrete quality and cover thickness, specified against loosely-defined exposure classifications. This is the traditional defence against corrosion. As awareness of the corrosion problem increased, the natural reaction was to increase either the concrete quality or cover (or both) - while advocating the introduction of materials such as slag, pfa or silica fume. This was substantially backed by laboratory data, supported by increasingly sophisticated predictive models. For the more extreme exposures, additional protective measures entered the arena. These included coatings, sealers, corrosion inhibitors, penetrants, etc. - either individually or on a 'belt and braces' approach, without any clear idea of how long they would last, or, indeed, whether they were truly additive.

To a large extent, this is still the current position, for everyday practice. It largely ignores the strong influence of construction quality and of the benefits of good design concept and detailing.

Fortunately, that situation is now changing. State-of-the-art reports are becoming more definitive, e.g.[5], [6], [7], [8]. More significantly, papers and reports are emerging, which put a strong emphasis on design, while recognising, to varying degrees, the importance of a holistic approach. Some key documents here are listed as references [9] - [14]: all demonstrate the development of numerate design concepts for durability, but differ in how they attempt to deal with reliability, safety and margins. A study of these different approaches, reveals four areas of major difficulty, which need to be addressed before real progress can be made. These are:-

1. The establishment of performance requirements, in terms of design life and minimum technical performance.
2. How to deal with relevant environments, in meaningful design terms.
3. Variations in construction quality, as influenced by design detail, construction method and workmanship.
4. The precise make up of the design framework, which simultaneously embraces all relevant factors and provides the required reliability.

Each will now be looked at briefly, while making suggestions for future development.

4 Design life, and minimum performance criteria

The move towards a performance-based approach requires the establishment of limiting criteria. This has to be at two levels:-

1. A requirement, which deals with the in-service time factor - possibly a target, somehow expressed in terms of 'life'.
2. A requirement, which directly relates to a minimum acceptable technical performance, over that life.

The 'pure' approach to 1. [9], [10] is to use a reference period of, say, 50 years for buildings and relate it to a reliability index of 3.8 for the ultimate limit state (as given in ENV 1991-1) and of 1.5 for serviceability. This sets a mean service life, over which the normal imposed leads and load combinations are to be carried. This can lead to service lives which are well beyond our experience. It also begs the question of what constitutes minimum technical performance, due to effects caused by deterioration mechanisms, and how these might affect normal action effects such as bending and shear.

In considering how to move forward here, a broader view is necessary. Most cities, in most countries, have buildings dating from the last century; few will be exactly the same, due to changes in use, upgrading, new cladding, even some replacements in whole or in part. Obsolescence, and change in use, is therefore an issue. To some extent, the situation is similar, for civil engineering structures, especially for highway structures on principal routes.

In addition, based on experience with modern infrastructure, owners are now more conscious of the need for regular inspection and good maintenance. Further, if structural changes have to be made for whatever reason, they wish the disruption to the operation of the structure to be kept to a minimum. In developing this type of management and maintenance strategy, it is important to remember that a structure is not a single unit; it is made up of numerous elements, services, furniture and fittings - all of which have different lives. The trend here is to plan for this in a systematic and economic way, using performance plans, shown simplistically in Table 2 for a building. Doing this properly involves whole life costing methods, with the objective of minimising total cost and maximising the operational use of the structure.

This type of thinking (obsolescence, systematic performance plans) promotes the idea that design life is not a single absolute value, but will vary, depending on the type of structure, on its use and on the maintenance/replacement plan (Table 2). Actual values will be established, in individual cases, at the conceptual design stage, between the client and the designer. The designer must then have available to him:-

1. A simple classification system for distinguishing between the life time requirements for different elements in the structure as a whole.
2. A design procedure which ensures that the performance requirements are met for each classification.

One possible classification is:-

A. *Life-long* The element is so important that failure would cause cessation of function, and/or major disruption or cost, during remedial work. Highly critical.
B. *Repairable* Efficiency of operation would be reduced, but replacement/remedial work could be done economically. Medium criticality.

C. *Replaceable* Routine maintenance or replacement could be done at minimum cost and little inconvenience. Low criticality.

Table 2. Basics of a performance profile plan for building

System	Criticality	Target life before replacement, years						Capital Costs*	Cost in use target: x years at y price**	Maint-enance Requir-ements
		>5	5-10	10-20	20-40	40-100	> 100			
Foundations	A	••••••	••••••	•••••••	•••••••	••••••	•••	—	—	
Structure	A	••••••	••••••	••••••	••••••	•••		—	—	
External Cladding	B	••••••	••••••	••••••	••••••			—	—	Clear specif-ication for each system
Glazing	B	••••••	••••••	••••••	•••			—	—	
Partitions	B	••••••	••••••	••••••	•••			—	—	
Heating & Ventilation	B	••••••	••••••	••••				—	—	
Water, Public Health Service	A/B	••••••	••••••	••••••	••••••			—	—	
Electrical	A	••••••	••••••	••••••	••••••			—	—	
Decorating	C	••••••	••••••					—	—	

* e.g., 100 units
** e.g., 1500 units over x years

A British Standard [15] exists, based on this principle, which simultaneously encourages the development of materials and components on a targetted service life basis; this is also under development at an European level via CIB W60 and RILEM PSL 140.

One possible way of relating this type of classification to life is illustrated simply in Figure 2, developed from a CEB concept into a 'required performance' cube. The front face of the cube establishes performance in broad conceptual terms; the third axis (performance profile rating) creates an opportunity to establish minimum technical performance requirements for each of the zones on the front - how to do that is explained in reference 14, and guidance on how to select a zone in Figure 2 is given in a more detailed report [16].

The Figure 2 approach has the advantage of directly relating performance to future function and management. The chances are that the chosen life will be less than that arrived at via notional reference lives and applying probability indices. It is therefore attractive to owners, who, currently, see designers using increasingly severe (and expensive) prescriptive measures, without any quantification regarding reduced maintenance in the future.

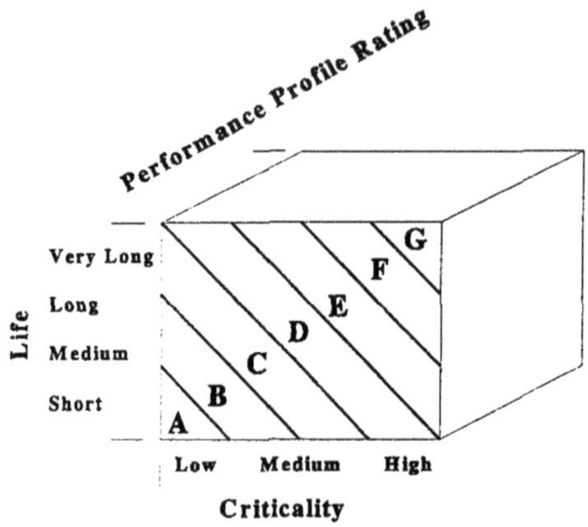

Figure 2. The required performance cube.

The lack of definition of practical performance requirements is probably the biggest obstacle to the future development of durability design - both in terms of better in service performance and of the development of more durable materials and components.

5 Aggressive and environmental loads, and design

A knowledge of environmental 'loads' is much more significant than for imposed loads in structural design, particularly for the major aggressive action - corrosion. It is not difficult to identify the major causes - chlorides and CO_2 - but the effects of these, in penetrating the protective concrete cover, are much influenced by general and local climatic conditions. This involves an almost infinite combination of moisture, wind, and temperature. Further, there is a strong interaction between the overall climatic conditions and the outer fabric of the structure itself. Any planes of weakness (joints, cracks, etc) will quickly be found out - and, therefore, there is a strong relationship between durability and structural design and detailing.

It is difficult to see a unique universal solution to this problem, and it may have to be addressed in different ways. For example, for non-aggressive or mild conditions, the present exposure classifications in most Codes may suffice for routine structures. At the other extreme - major structures in very severe conditions (e.g. references 3 and 4) - detailed investigations and modelling may be required. Certainly, predictive models now exist to permit that, although it has been shown [9] that attempts to use these for general application can be very complex.

The output from the above exercises, whether treated simply or in detail, will generally be a specification for concrete quality and cover, perhaps augmented by some additional protection, e.g. coatings. There is an additional factor, which, in the future, should be integrated into general design and detailing procedures - a positive approach to the treatment of water.

All structural engineers now consider wind and temperature in their design (although neither is fully treated in Codes). They would be concerned with safety, and with stress limits, and movement. It has not always been so. Recently, the author [17] reviewed 15 examples of major bridges where some form of strengthening had been required; mostly, this was due to temperature gradients or a failure to take proper account of long term deformations - neither subject was treated in Codes at the time of design, but this situation has now been rectified.

The treatment of water should be elevated alongside wind and temperature, in general design procedures. Removing, or keeping out, water is fundamental to improved durability. It is a transport mechanism for aggressive actions and, acting with wind and temperature, can produce detrimental physical and chemical effects. This line of thought leads to a 'design-out' option, which is perhaps most appropriate for buildings, where the designer is dealing with an enclosed space, with the attractive possibility of keeping the structural concrete 'in-doors', and providing an enveloping cladding system. Doing this effectively raises some building physics problems; this is illustrated in Figure 3, in terms of mechanisms for rain penetration - careful detailing between cladding and structure is all-important here.

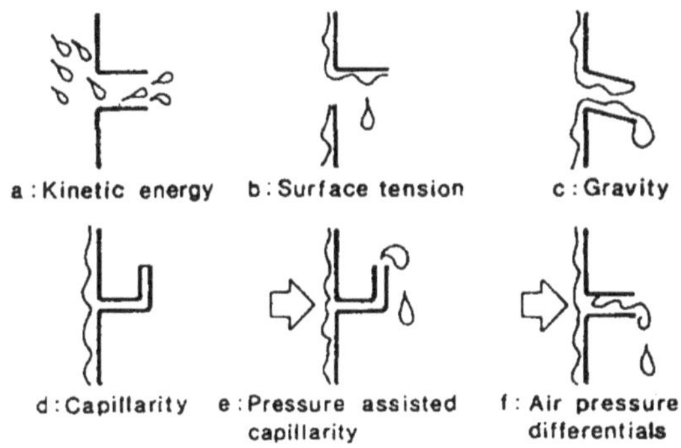

Figure 3. Possible mechanisms by which rainwater leaks through joints.

This type of design-out option is also available - perhaps less effectively - for many types of highway structure. The prevention of leaks can be enhanced by opting for integral or indeterminate bridges. The removal or avoidance of critical moisture can be achieved by good drainage and the shaping of elevations. Having said that, there will always be situations where the concrete has to directly resist aggressive actions. Marine structures are one example: however, much is now known about this i.e. how to design for the critical splash and tidal zones.

6 The influence of construction quality

There is no doubt that the issue of construction quality and variability must be firmly grasped, before any rational approach to durability design can be fully developed. While there is clear evidence [18], [19] that good concrete with the right cover can be placed, and survive under the most severe conditions, (mostly precast concrete, with good control), there is overwhelming evidence of great variability in in-situ construction. This is a key factor in itself, but can also negate any potential benefits from improved material specifications or better design.

Probably the best known and documented quality issue is the failure to meet specified cover requirements. In Germany, a Code [20] has been introduced to improve this situation, and experiences in monitoring its use have been reported [21], [22]. The problem is well recognised, and some positive steps taken to improve matters, without yet reaching the stage of integrating different quality levels into a holistic design approach. A comprehensive and significant recent paper [23] identified the main causes of this chronic problem. Quite apart from matters such as spacers, tolerances, bar shapes and percentage reinforcement, skill levels, etc, the paper identified deficiencies in the entire design-supply-contractual-chain, plus a lack of awareness among designers (using CAD packages) of the practicalities of reinforcement supply and fixing. To a large extent, this is a human communication problem, unlikely to be solved by purely technical methods, or greater rigour involving certification systems. A review of the entire supply system is required, to benchmark best practice, and identify quality within defined confidence limits. Failure to do this effectively rules out any rational design approach, leaving only the expensive belt-and-braces option of multi-layer protection [2].

Achieving the right cover with confidence is not the only quality issue. The quality of the concrete itself can vary. Figure 4 shows concrete strength contours for a vertically-cast wall. This variability is due in part to w/c ratio gradients and to variations in compaction.

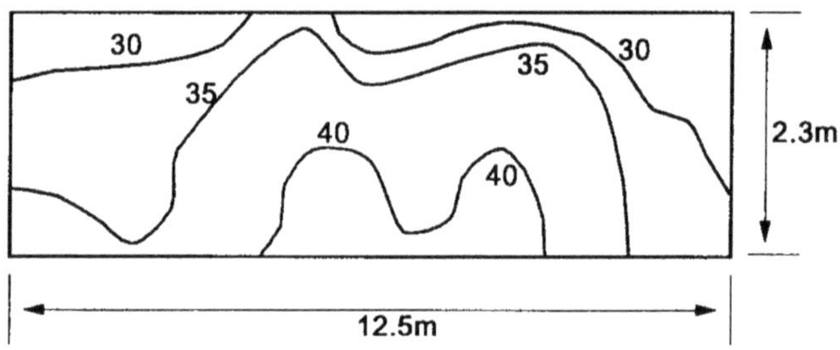

Figure 4. Distribution of concrete strength (N/mm2) in a 300mm thick wall.

The variation shown here in strength terms, probably translates into a factor of 4 in corrosion resistance terms. In the presence of reinforcement, variability in compaction can be even more significant. Figure 5 illustrates this for a beam, as measured by upv methods; clearly, over part of its length, the concrete has been compacted onto the reinforcement, rather than through it; with the trend towards higher steel percentages, this is an important issue.

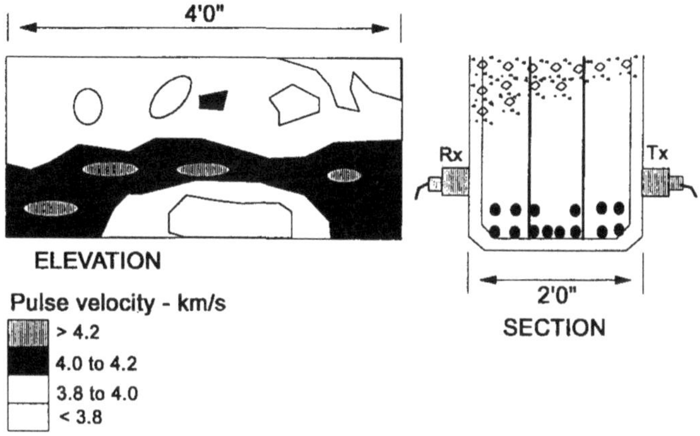

Figure 5. Variation in u.p.v. readings in a badly compacted concrete beam.

Again, there is growing awareness of this problem, and different proposals made to deal with it. The CEB approach [9] is to identify zones where compaction is likely to be less perfect, and to introduce a more rigorous specification for the concrete or additional protection to the concrete surfaces. The recent series of HETEK reports [24], [25], [26], (27) take a different approach. The entire thrust of the detailing reports is to encourage designers to use certain shapes and reinforcement arrangements, which permit practical formwork arrangements and the easy placing of both reinforcement and concrete. The author is very supportive of this approach, since it is his belief that it is the fundamental reason for the successful use of standard bridge beams in the UK, in the most adverse of corrosive conditions for more than 40 years, with no hint of deterioration. [18]

Significant space has been devoted to Quality in this paper, since it is a key issue in the development of a holistic approach to durability design. It is also an issue where the solution lies not just with the contractor but also with the designer. That solution is not dependent on greater understanding or analytical rigour, but on identifying what is already known about best practice, and ensuring that it is applied, via good communications, at an operational level. Since it is a 'systems' issue, it is somewhat difficult to integrate it into traditional numerical design procedures; nevertheless, methods have to be found to deal with it.

7 Elements in a holistic design framework

Traditional structural design can be conveniently considered in 2 phases:-

1. Initial conceptual design, associated with preliminary sizing of elements.
2. Final detailed design, essentially a numerical process, involving structural analysis, section design, and satisfying prescribed limits states.

For the purposes of this paper, phase 2 can be divided into:-

2a The numerate process associated with limit states.
2b Detailing, either in terms of reinforcement arrangements, or the final shaping of cross-sections or elevations, or integrating the base structure into the artefact as a whole (eg. cladding, services, finishes, etc).

The resulting calculations and drawings are then supported by material specifications, backed by QA or certification schemes.

Is it then feasible to produce a parallel compatible system for durability design? The answer is probably 'Yes', but we are not yet ready to do that fully. In addition, there must be a significant change in emphasis between the different design phases; some of these are noted below:-

7.1 Phase 1: conceptual design

This would require significantly more effort, including:-

- a definition of performance requirements, in terms of life and minimum technical criteria, while considering the structure as a whole (Figure 2 and Table 2 concepts).
- a management and maintenance strategy, targeted at minimum whole life costs and disruption to future operations.
- a buildability and construction quality plan.
- decisions on the basic defence strategy in dealing with aggressive actions, ie, a design out solution (e.g. by using a protective envelop for buildings), or the use of one or more protective barriers.
- a strategy for dealing with the effects of wind, water and temperature.

7.2 Phase 2b: detailing

As defined earlier, this includes many items, which follow directly from phase 1, and which are not amenable to numerical calculation. To prepare for this, therefore, it is necessary to build up a portfolio of good practical details (based on feedback) covering two essential themes:-

- the improvement of construction quality (and therefore linked to the specification).
- features, making up part of the basic defence strategy (as illustrated typically in CEB Bulletins 148, 152, 166 and 182).

7.3 Phase 2a: limit states and numerical calculations

To understand the potential role for this, it is necessary to make a list of the possible resistance options, in designing for corrosion protection. A possible list is given in Table 3, under the sub-headings Materials; Design; Construction (with design involving some options from each group).

To apply this, we need to have:-

- detailed knowledge of the loads, pertaining to the particular environment.
- a model which predicts the penetration mechanism of the aggressive action.
- agreed limiting performance criteria, eg. the reinforcement should not be depassivated for x years.
- agreed values for the partial safety factors.

Much work has been done on developing suitable models; little work has been done, on developing the other features in the design condition - and this requires attention. Following on from that, and still referring to Material options in Table 3, work is required on the non-concrete options, in evaluating their real performance in whole life cost-benefit terms.

8 Concluding remarks

In this paper, arguments are presented for a holistic approach to durability design, embracing material, design and construction matters. While much progress has been made, particular areas requiring attention are:-

- the establishment of performance requirements, in practical whole life terms, including the future operation of the structure, and its management and maintenance.
- how to deal with relevant environments, in a design context.
- the need to design for buildability, and hence improve construction quality.

Until progress is made on all these issues, the development of durability design will be handicapped; there is much variability which requires better control and some unknowns which require definition. In moving forward, design must lead, but in a broader context than for structural design. Both concept and detailing (e.g. dealing with water) are more significant, within a holistic approach, and there is a stronger need for a clear design strategy in individual cases. For example, the balance between Materials, Design, and Construction issues (Table 3) may be different for different situations, e.g. the design out option for buildings, mentioned earlier. Traditionally, engineers look for numerical solutions; these do have a role to play in durability design, but in a more limited way, within a holistic approach.

At this time, these ideas are put forward for discussion and debate. A holistic approach does not yet exist in general terms - and it is important that a focus is given to future developments, which must involve all parties in the construction industry including the owners of structures.

9 References

1. CEB, Comité Euro-International de Beton (1993). *CEB-FIP Model Code 1990,*

Table 3. Shopping list of resistance options for corrosion

A Materials	Concrete quality	-	mix proportions
		-	mix ingredients
	Cover	-	minimum
		-	tolerances
	Permeable formwork		
	Concrete protection	-	sealers
		-	coatings
		-	penetrants
		-	layers
	Rebar protection	-	epoxy coating
	Special steels		
	Non-corrodible reinforcement		
	Cathodic protection		
B Design	Design concept		
	Structural detailing		
	Cladding, services, fittings, finishes		
	Articulation, joints, movement		
	Treatment of water, drainage		
	Control of the environment, barriers		
	Provision for inspection, maintenance, replacement		
	Accurate assessment of effects of deterioration mechanisms		
C Construction	Construction methods		
	Quality control		
	Certification		
	Testing		
	Rationalisation, standardisation, simplification		

'Most of the Design and Construction options fall into Phases 1 and 2b above - and are largely not numerate. However, material-based options would also be part of a multi-barrier approach, and, initially, the designer might wish to evaluate combinations of concrete quality and cover. In that context, numerical modelling could have a role to play. If that is related to conventional structural design, we have to satisfy the usual design condition.

$$S_K \cdot \gamma_S \leq \frac{R_K}{\gamma_R} \tag{1}$$

where

S_K is the characteristics value of the aggressive action
γ_S is a partial safety factor for this loading
R_K is the characteristic resistance
γ_R is a partial safety factor for that resistance

Design Code. Thomas Telford, London.
2. Concrete Society (1996). *Durable bonded post-tensioned bridges*. Technical Report 47. The Concrete Society, Slough, UK. p64.
3. Munch-Petersen, C. et al (1997). *Concrete strategy for the Oresund Crossing*. Betonwerk & Fertigteil - Technik. Bft 11. November 1997. pp 44-54.
4. Dunaszegi, l. (1996). *High performance concrete a key component in the design of the Confederation Bridge*. Newsletter, Beton Concrete Canada. Vol 1, No. 3. December 1996. pp 1-2.
5. Kropp, J. and Hilsdorf H.K. (1995). *Performance criteria for concrete durability*. RILEM Report 12. E&FN Spon, London. p327.
6. Cementa AB (1996) *Durability of concrete in saline environment*. Cementa AB, Danderyd, Sweden. May 1996. p 206.
7. Broomfield, J.P. (1977) *Corrosion of steel in concrete: understanding, investigation and repair*. E&FN Spon, London. p240.
8. Reinhardt, H.W. (1997). *Penetration and permeability of concrete*. RILEM Report 16. E&FN Spon, London. p331.
9. CEB, Comité Euro-International du Beton (1997). *New approach to durability design. An example for carbonation induced corrosion*. Bulletin 238. CEB, Lausanne. May 1997. p138.
10. Sarja, A. and Vesikari, E. (1996). *Durability design of concrete structures*. RILEM Report 14. E&FN Spon, London. p165.
11. Sakai, K. (1996). *Integrated design and environmental issues in concrete technology*. E&FN Spon, London. p308.
12. Japan Society of Civil Engineers (1989). *Proposed recommendation on durability design of concrete structures*. JSCE, Tokyo. Concrete Library International No. 14.
13. British Cement Association (1997). *Development of an holistic approach to ensure the durability of new concrete construction*. Final report to the Department of the Environment on Project 38/13/21 (cc 1031). BCA, Crowthorne. October 1997. p81.
14. Somerville, G. (1997). *Engineering design and service life - a framework for the future. Prediction of concrete durability*. Proceedings of STATS 21st Anniversary Conference. E&FN Spon, London. pp58-76.
15. British Standards Institution (1992). Guide to: *Durability of buildings and building elements products and components*. BS7543: 1992. BSI, London.
16. British Cement Association (1996). *Durability by Design*. Final report to the Department of the Environment, on project CI.38/13/12 (cc 716). BCA, Crowthorne. p47.
17. Somerville, G. (1998). *Concrete balanced cantilever bridges - historical feedback from performance in service*. To be published, but available from British Cement Association, Crowthorne, UK.
18. Somerville, G. (1994). *Concrete bridges for motorway widening*. Proceedings of a conference on Bridge Modifications. The Institution of Civil Engineers, London. 23 March 1994. Thomas Telford Ltd.
19. Bernander, K.G. & Oberg, M. (1996). *Field surveys on carbonation depths in precast concrete elements*. Betonwerk & Fertigal - Technik. Vol 11/1996. pp 65-72.
20. Deutscher Beton-Verein E.V. und Fachvereinigung Deutscher Betonfertigteilbau

EV. *Merkblatt Betondeckung, Sieherung der Betondeckung beim Entwerfen, Herstellen and Einbauen der Bewehtung sowie des Betons* (Fassung Marz 1991). Beton-Kalender, 1992 Teil II, S. 190-195.

21. Dobbelmann, H. et al (1996). *Einhaltung der Mindestbeton-deckung-Empfehlungen für Bauplanung and Bauausführung.* Beton-and Stahlbetonbau 91 (1996) Heft 8. pp176-182. (English translation available from British Cement Association, Crowthorne, UK).
22. Hosser, D. & Gensel, B (1996). *Einflusse auf die Betondeckung von Stahlbetonbauteilen-statistische Analyse von Messungen an Wanden, Stützen and Unterzugen.* Beton and Stahlbetonbau 91 (1996) Heft 10. (English translation available from British Cement Association, Crowthorne, UK.)
23. Clark, L.A. et al (1997). *How can we get the cover we need?* The Structural Engineer, Vol 75, No. 17; 2 September 1997, pp289-296. The Institution of Structural Engineers, London.
24. Danish Road Directorate (1997). *Structural design and workmanship - state of the art.* HETEK Report No. 85. Copenhagen. p24.
25. Danish Road Directorate (1997). *Guidelines for structural detailing to improve concreting.* HETEK Report No. 96. Copenhagen. p36.
26. Danish Road Directorate (1997). *Control of early age cracking in concrete guidelines.* HETEK Report 120. Copenhagen. p81.
27. Danish Road Directorate (1997). *Curing guideline.* HETEK Report No. 121. Copenhagen. p86.

UNIFICATION OF THERMO-PHYSICS OF MATERIALS AND MECHANICS OF STRUCTURES
- TOWARD A LIFE SPAN SIMULATOR OF STRUCTURAL CONCRETE

K. MAEKAWA and T.ISHIDA
University of Tokyo, Tokyo, Japan

Abstract
The performance based design obligates quantitative assessment of required performances by means of transparent and objective science. In this design scheme, simulation of both macro and micro-scale structural behaviors is thought to be a powerful tool. This paper proposes a way how to unify the structural safety and serviceability check method and durability assessment of RC structures. Though component chemical-physical processes are crudely assumed, system dynamics of micro-scale pore structure formation and macro-scale defects and deformation of structures was shown as a possible and promising approach in future.

The authors understand that the unification of structural and durability design has just started. For further progress and development, predictive tool of structural behaviors from birth to death of concrete under any specified environment and load serves as an essential technicality.
Keywords: Performance based design, Life-span Simulation, Structural performance, Durability.

1. Introduction

The structural serviceability under permanent loads and daily live actions is not ever-lasting but does change in general under combined long-term environmental and repeated mechanical actions. The safety potential against extremely large actions such as earthquake, explosion and accidental forces during the life of structures might be gradually degraded due to change of quality and material performances in long ranged time. Engineers' concern is to find out engineering solutions to satisfy multi-oriented requirements with respect to structural serviceability and safety from birth to death of

Concrete Technology for a Sustainable Development in the 21st Century edited by O.E. Gjørv and K. Sakai.
Published in 2000 by E & FN Spon, 11 New Fetter Lane, London EC4P 4EE, UK. ISBN: 0 419 25060 3

concrete and concrete structures that continue to serve as a part of social stocks.

For meeting the challenge to make such a design possible, a performance based design scheme is thought to be most promising in the worldwide public of the next millenium. Clear and transparent description of required performances are to be presented to practitioners. Designers are expected to create their design for materials, structural dimensioning & details and construction systems so that their proposals might pass the requirements with reasonable reliance and acceptance by clients and societies. Here, assurance of expected performances on serviceability and safety in the whole span of life is exactly requested and in fact, transparently and objectively to be shown to the sectors which issue design tasks.

It can be said that the limit state design scheme has been functional for examining structural safety in the past decades. Here, the material quality change and reform of micro and macro-scale structures during the design life are assumed not to be predominant. In fact, limit states have been conservatively selected so that the designs which do not attain them would certainly fulfill the safety and serviceability within the span of life. Both material based limit states (for example, carbonation depth, chloride penetration, corrosion limit) and capacity based ones (i.e., flexural and shear capacities, ductility) have been defined basically with no mutual obvious interaction. In principle, both structural and material based limit states defined within the limit state design scheme have a mutual linkage. But, the interaction can be ignored provided that the durability oriented limit states in terms of material characteristics be defined conservatively. For example, if the corrosion of reinforcing bars is not allowed at any location, we can apply nonlinear structural analysis on the basis of complete stability of constituent materials. If the crack width is set within a reasonably small limit, we can apply mass transport theorem and corrosion prediction by assuming that crack width is unchanging.

However, the structural deformation and capacity are really linked with both micro-pore based deterioration and large-scale mechanical defects represented by

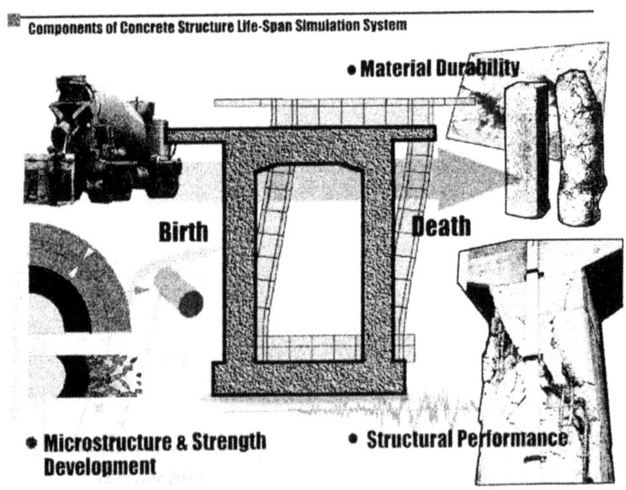

Fig.1 Life-span simulation of structural concrete from birth to death and from micro-pore to macro-structures.

cracking and yielding of materials with some control volume. In turn, the progress in macro-scale material damage and defects are also dependent on both the structural deformation and environmental boundary conditions. Here, nonlinearly accelerated change of material and structural performances can take place. For example, corrosion and associated volume expansion induces additional cracks and defects which also accelerate the migration of moisture and ions. The authors understand that the unified approach of mechanics which governs stress and strain fields and thermo-hygro physics ruling mass and energy transport associated with thermo-dynamic state equilibrium would serve as a technicality of ensuring total performances of concrete structures as well as structural concrete performance over the life span of concrete structures.

In this paper, the unification of mechanics and thermo-dynamics of materials and structures will be tackled for showing the possibility and future direction of research and development. The unified system will lead to overall damage assessment of structural concrete from micro scale geometry to macro defects related to cracking and continuum plasticity. The behavioral simulation of both macro structure and micro pore geometry is the authors' destination in civil engineering through concrete as shown in Fig.1.

2. Thermo-hygro physics for concrete performance -- DuCOM sub-system --

The hydration of cement in concrete volume, corrosion of steel, moisture migration and associated transport of solved ion ingress and micro-pore structural formation exhibit strong mutual inter-link. The authors have been developing 3D finite element analysis program for simulating these transient phenomena in space and time and a research fruit is in public as a computer code named DuCOM partially available through internet (http://concrete.t.u-tokyo.ac.jp/) [1]. Since the details were presented in literatures, just core points are reviewed hereafter in line with overall performance assessment.

The hydration of both constituent minerals of cement and Pozzolans is formed by simultaneous differential equations based on the simple chemical reaction law [5][6].

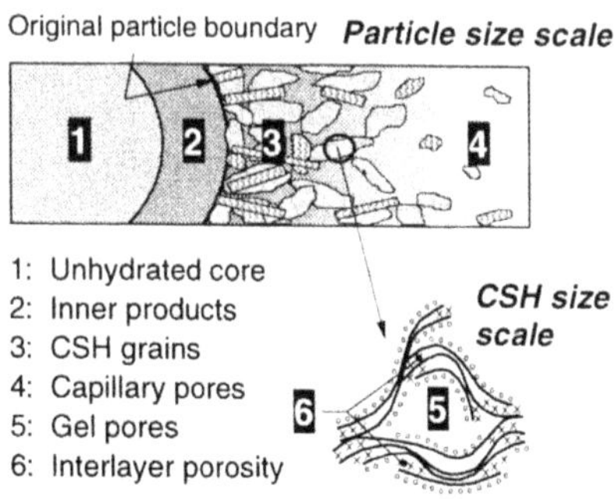

Fig.2 Hydrated and non-hydrated regions around dispersed cement particles

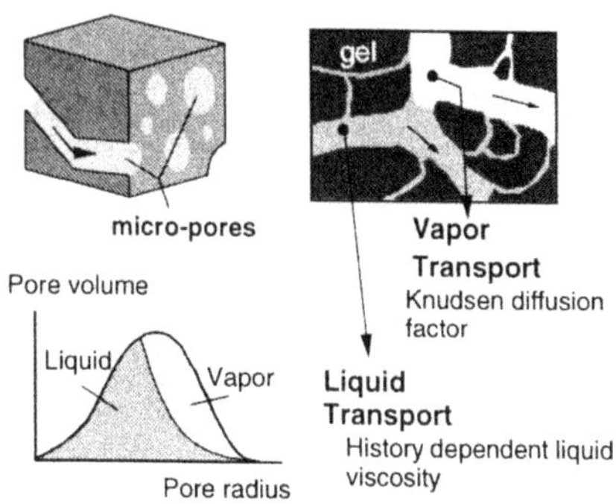

Fig.3 Permeability modeling – water passing through fictitious pipes made of micro-pores -

The rate of hydration is mathematically specified in terms of temperature, free water content in capillary pores, degree of hydration and associated cluster thickness of C-S-H gel layers precipitated around non-reacted cement particles (Fig.2). The pore solution of $Ca(OH)_2$ is simultaneously computed and free water loss trapped by hydration process of cement minerals is simulated. Then, the chemical process and its interaction among minerals and additive Pozzolans are considered by sharing common variables associated with pore solution, water and temperature. The released heat energy computed is conveyed to the entarpy equilibrium equation as discussed below.

During the hydration process, mass and heat energy conservation have to be satisfied with respect to moisture, ion solution and temperature. At the same time, moisture migration in terms of vapor and liquid movable free water and heat flux are incorporated in the conservation conditions of the second law of thermo-dynamics. The equilibrium conditions are simultaneously to be solved together, and the mass and energy transport resistance denoted by permeability and conductivity has to be formulated. Here, transport of chloride ion solution is defined in the mode of diffusion and bulk one accompanying the liquid water in motion [11]. The oxygen gas, carbon dioxide and related carbonation can be overlaid on the moisture transport with the same manner. In this framework, it is assumed that the moisture migration is not influenced by the existence of solved ion mass but perfectly independent. These material performances have explicit and direct relation with micro-pore structure of hardened cement paste with non-uniform pore size distribution.

The permeability of vapor and liquid water is mathematically formulated based on the micro-pore size distribution as demonstrated in Fig.3 [7][8]. The path of moisture in cement past is thought to be assembly of small sized fictitious pipes and its integration results in the macroscopic permeability. Tortuousity on parcolation and the thermo-dynamic activation of surface energy onto the micro-scale viscosity of pore water are taken into account. It is to be noted that the simple micro-mechanical

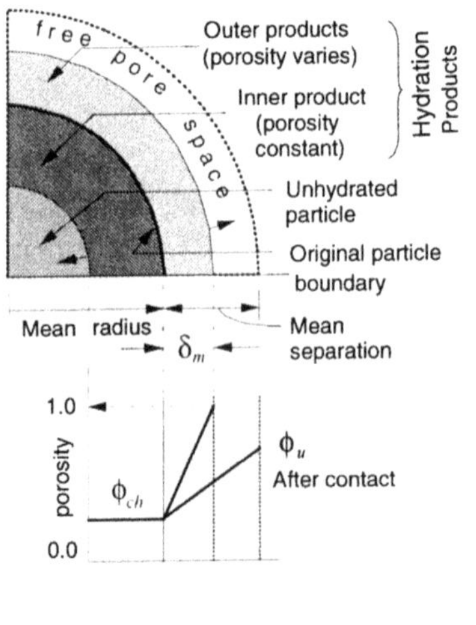

Fig.4 Micro-pore structure formation around cement particles

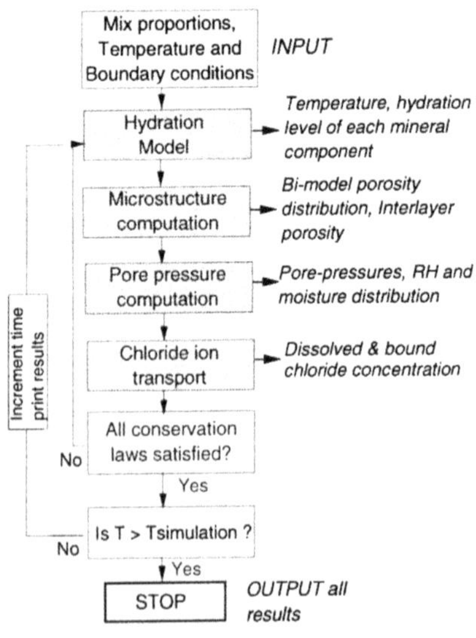

Fig.5 Framework of DuCOM thermo-hygro physics

modeling is applied without any variable fitting. These internal factors are thought to be associated with creep deformability of green cement paste at early ages. The authors are discussing the general time dependent constitutive law for cement hardened paste with this information of material evolution [19]. In this study, the linear convolution type creep law fairly used in practice is tentatively utilized for demonstrating how the entire system be operational for performance assessment in durability and safety [18].

As a natural way, the pore structure formation model as illustrated in Fig.4 is added in the system dynamics of transient concrete performance modeling [1]. The statistical approach to the micro pore structural geometry of hardened cement paste having interlayer, C-S-H gel and capillary pores is made in use. The hydration degree of each mineral compound in concrete specifies pore structure evolution. The porosity distribution of hydrated and non-hydrated compounds around referential cement particles is calculated and the surface area of micro-pores is estimated mathematically. By assuming statistical distribution function with regard to the pore sizes, the authors extend the geometrical description of micro pores. The connective mode of each pore volume is also defined with simple provability on the basis of which the path-dependency of isotherm of moisture is successfully described [9].

The structure formation model makes the entire system complete and consistent with hydration model and mass and energy transport as shown in Fig.5 [1]. In order to mathematically make clear the interrelation of each physical component, a set of governing simultaneous equations solved in finite element scheme are lined up in appendix. Here, it must be noted that the mechanical damage concerning cracking and compressive fracturing of localization is also influential on the mass transport of moisture and solved ion migration [16][17]. In fact, comparatively large-scale defects like cracks are not represented by the pore size distribution on the basis of C-S-H hydrate scales, and this damage can only be predicted by solving solid stress and deformational fields with mechanical constitutive law of concrete. Then, the simultaneous solution with mechanics of structures is further needed in view of life-span simulation of concrete inside the structure. If we would work within the closed field of thermo-hygro physics, we can predict concrete material performance alone, but cannot reach structural concrete quality with presence of cracks and mechanical defects.

3. Continuum mechanics of materials and structures -- COM3 and sub-system of mechanics --

For simulating structural behaviors expressed by displacement, deformation, stresses and macro-defects of materials in view of continuum plasticity, fracturing and cracking, well established continuum mechanics can be used as illustrated in Fig.6. The compatibility condition, equilibrium and constitutive modeling of material mechanics are the basis and the spatial averaging of overall defects in control volume of finite element is incorporated into the constitutive model of quasi-continuum as summarized in the appendix, too. The authors adopted a 3D finite element computer code named COM3 for structural dynamics, which has been also developed at the University of Tokyo for static as well as dynamic ultimate limit states [2][4].

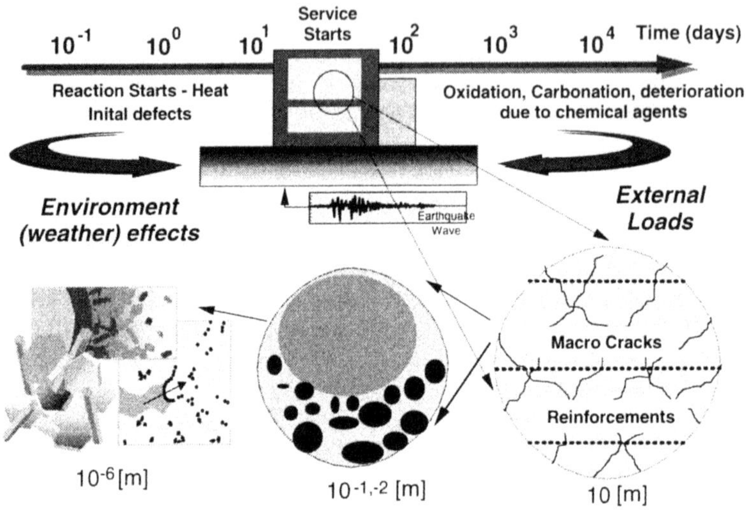

Fig. 6 Macro-scale defects and micro-scale pore structures.

This frame of structural mechanics has an inter-link with thermo-hygro physics in terms of mechanical performances of materials through the constitutive modeling in both space and time. In this study, the instantaneous stiffness, short-term strengths of concrete in tension and compression, free volumetric contraction rooted in coupled water loss and self-desiccation caused by varying pore sizes are considered in the creep constitutive modeling of liner convolution integral as summarized in the appendix. The volumetric change provoked by the hydration in progress and water loss is physically tied with surface tension force developing inside the micro-capillary pores [10]. Of course, the micro-pore size distribution and moisture balance of thermo-dynamic equilibrium are given from the code DuCOM at each time step.

The cracking is the most important damage index associated with mass transport inside the targeted structures. Cracks are assumed to be induced normal to the maximum principal stress direction in 3D extent when the tensile principal stress exceeds the tensile strength of concrete. As stated before, the strength is numerically evaluated from the degree of micro structural formation [10]. In reality, the explicit relation of the specific strength and formed porosity with intrinsic sizes is adopted in this study. After crack initiation, the tension softening on progressive crack planes is taken into account in the form of fracture mechanics. In the reinforced concrete zone in which bond stress transfer is expected being effective, the tension stiffness model is brought together. Since the external load level with which the environmental action be coupled in design is rather lower than the ultimate limit states, compression induced damage accompanying dispersed micro-cracking is disregarded in this study.

In COM3, the authors furnished with reinforced concrete smeared crack modeling being developed in the past decade [2][4]. The constitutive law can handle multi-directional cracking in 3D space by decomposing stress bearing mechanism into orthogonal 2D sub-planes on which 2D in-plane constitutive model of reinforced or plain concrete as illustrated in Fig. 7.

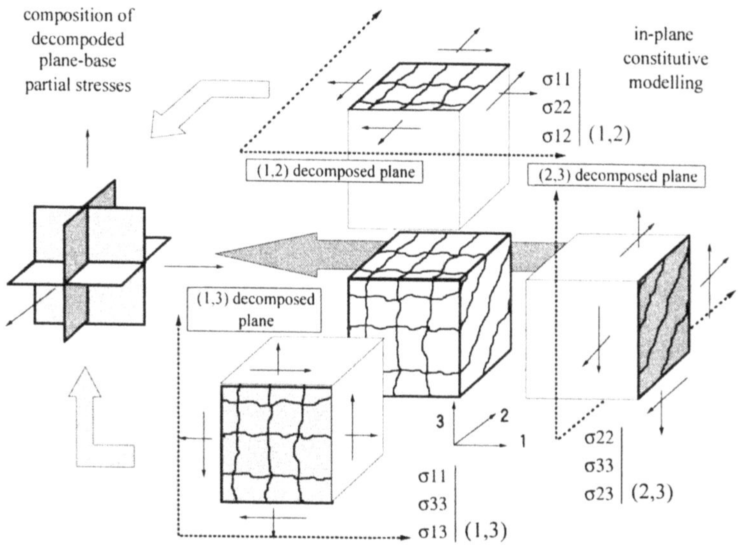

Fig. 7 3D smeared crack approach for reinforced concrete with stress induced cracking of multi-directions.

4. Dual parallel processing of thermo-physics and structural mechanics

For solving two sub-systems as for micro structural by DuCOM and macro structural performances by COM3 with mutual interaction, complete amalgamation of these dual systems in a single processing is one of the possible measures for unification of material and structural performance simulators. The authors tried this unification with a complete finite element discretization based on weighted residual function method. It has to be reported that the theoretical formulation does not matter, but the numerical stability is hardly maintained due to the matrix processing mixed with different dimensional diagonal terms of a great deal gap of algebraic orders. In fact, the stiffness terms of equilibrium and diffusivity ones for mass balance have quite different values which are hardly solved numerically without any cracked accuracy.

The reasonable and simple way of coupling two sub-systems is proposed by the authors for establishing more general umbrella system as shown in Fig.8. That is the parallel processing on the UNIX and/or Windows environments. In the multi-task operation system, one processor unit or a fraction of the common processor is assigned to each sub-system, that is to say, DuCOM and COM3 finite element programs. The common data storage is arranged in a memory device. Through the common area, two sub-systems are connected by high speed signal bus or networks so as to share the common data information mutually. In the common area, intrinsic variables which are conjointly bestowed as control variables of both sub-systems are renewed time by time and referred by both sub-systems. In this parallel processing, each sub-system can manage the same dimensional variables in a matrix form and the cross terms between

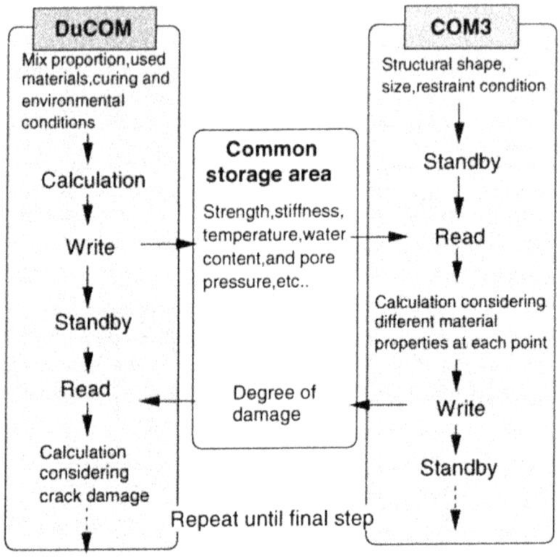

Fig. 8 Parallel processing of DuCOM and COM3

two sub-systems are not explicitly included in the matrix processing, but implicitly taken into account by communicating in between thermo-hygro and structural mechanics systems. In order to unify the system clock of the virtual world, sea-saw switching of two processing ruled by two processors is performed with small time intervals as shown in Fig.8.

The chief advantage of unifying material and structural analysis in this manner is numerical stability of explicit scheme. Furthermore, this coupling method under multi-task operation enables engineers easily to link independently developed computer codes even if being written by different computer languages and algorithms. As a matter of fact, slight modification for data exchange with the common memory space through high speed bus is needed with a short system manager program alone. Therefore, finite element structural analysis program and Poisson type equation solver such as heat conduction and diffusion can serve as a platform of assessing durability and structural safety/serviceability performances in the life span of concrete structures.

5. Integration of sub-systems on the world-wide network

In Chapter 4, a proposal of integrating material physics and structural mechanics was made on the parallel processing environment in a hardware. Here, the data on 3D geometry and time is exchanged between two sub-systems. If the two systems are simultaneously loaded on the single core memory, the data exchange is physically made within a CPU of extremely high speed. When the coupled systems are loaded on the separated processors or the computers, data transfer between the thermo-physicals and the mechanicals is performed through the high speed data bus. On this line, the sub-systems are not necessarily mounted on the single hardware, but they can be worldwide dispersed and linked with each other by the network.

The authors picture the near future environment of integrating material and structural knowledge that are possessed by several sectors but are inter-connected through WWW as shown in Fig.9. Here, what we should do is to make a consensus on the common data storage protocol as discussed in Fig.8. Let us consider the case where the hygro-physics and hydration model, chloride diffusion, corrosion, large scale defects rooted in cracking and carbonation model are steered by independent sectors and/or researchers. Provided that each knowledge would be open in the network, it is easy to sequentially call each sub-systems through the network as shown in Fig.5. Substantially, this sequential processing can be made by users on their personal computer environment if their computational environment be connected to the common network. In this mutual linkage of research and development, a practitioner feels that all softwares are ready on his PC.

It may be the time to discuss the possibility of constructing protocol for integrated system of structural mechanics and chemo-physics of concrete. For that, it is indispensable for each developers to mount their software that can be used by others on the commonly accessible network. As stated in Chapter 2, DuCOM is open to anyone who wants to use it and can get the computed data downloadable. As a matter of fact, the integration of DuCOM with other software for computing thermal stresses was performed by a private sector for solving engineering problems related to durability of massive concrete. Each sub-system exists for the whole but is totally independent. The authors understand that the centralization of technology is not the principle but the lateral and multiple linkage of knowledge and technology with "dispersed" media will be the standard for "integration" of durability and structural safety design.

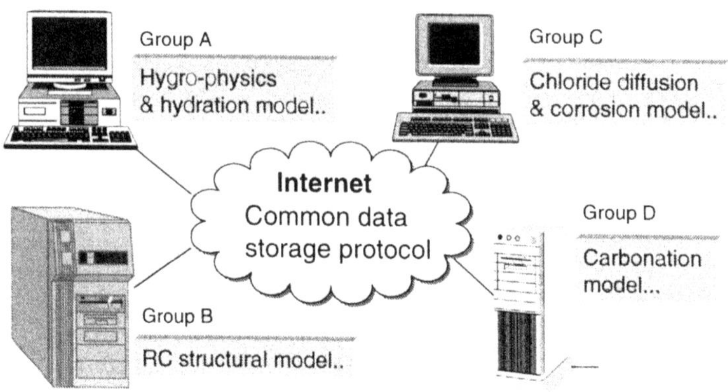

Fig.9 Mutual linkage of each sub-system on chemo-physics and mechanics.

6. Numerical simulations

6.1 Moisture distribution in cracked concrete

Using the proposed computational system, numerical simulations about the behaviors of moisture loss in cracked concrete were done. It has been reported that there would be close relationship between the moisture conductivity and the damage level of cracked concrete, that is, moisture conductivity would be dependent on the crack width, or the continuity of each cracking. The proposed system, in which the information can be shared between thermo-hygro and structural mechanics system, can describe this aspect quantitatively by considering the inter-relationship between the moisture conductivity and properties of cracking. However, in this example, after cracking the both of liquid and vapor conductivity was assumed becoming to ten times the moisture conductivity before cracking, for the sake of checking the validity of proposed system itself.

The target structure in this analysis is a concrete slab, which has 30% water to powder ratio using medium heat cement. The volume of aggregate was 70%. After 3 days of sealed curing, the specimen was exposed to 50%RH. Fig.10 shows the mesh layout and the restraint condition used for this analysis.

Fig.10 shows the cracked elements, the distribution of moisture, and normalized tensile stress at each point from the boundary surface exposed to drying condition. Moisture distribution calculated without stress analysis is also shown in fig.10. As shown in the results, the crack occurs from the element near the surface, and the crack progresses internally with the progress of drying. It is also shown that the amount of moisture loss becomes large due to cracking.

6.2 Ingress of chloride ion in RC beam damaged by external load

The second case is the numerical simulation about the ingress of chloride ion in RC beam damaged by external load. Fig 11 shows the size of the beam, layout of FE mesh and load condition used in this analysis. The reinforcement ratio is 0.96%. For FE analysis of RC structures, AN proposed the model which combines the nonlinearity of

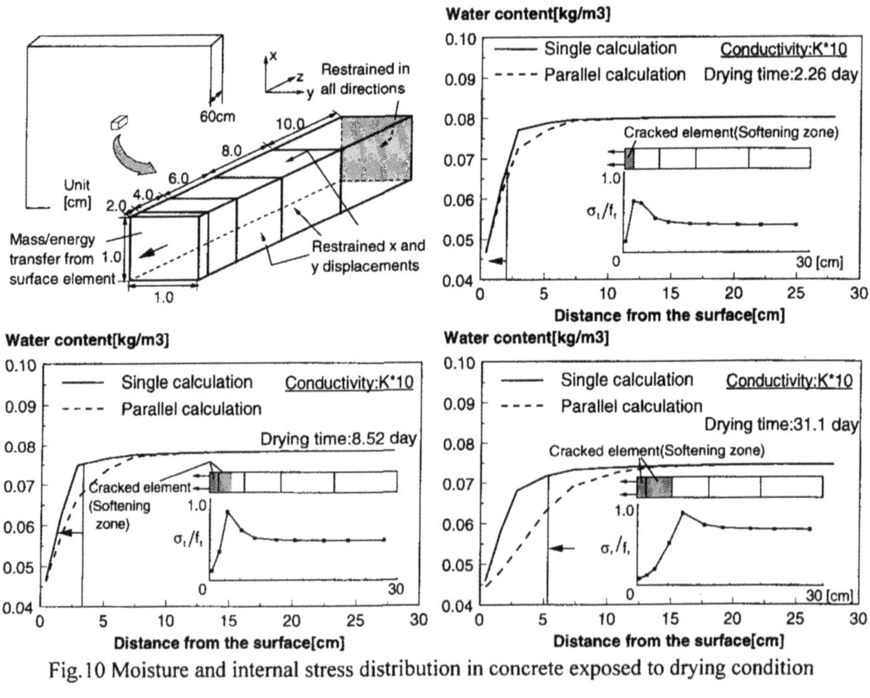

Fig.10 Moisture and internal stress distribution in concrete exposed to drying condition

cracked concrete in RC zone and plain concrete zone (PL zone) [15]. In this analysis also, we considered two different zones in RC beams to take into account the difference of concrete mechanics near or far from reinforcing bars (Fig.11). As for the mix proportion given to DuCOM, water to cement ratio is 45%, and the volume of aggregate is 65%. After 7days of sealed curing, load is applied with displacement control. Fig.12 shows the load-deflection relationship and cracked elements due to bending.

After loading, behaviors of chloride transport into damaged RC beam were simulated. The bottom surface of the beam is exposed to the concentration of chloride ion 1.4 [mol/l] under alternate drying (7days) and wetting (7days) cycles. Wetting is simulated by an environmental relative humidity of 95%, whereas drying condition is given as 50%RH. Similar to previous analysis, the moisture conductivity of cracked area is roughly assumed to be 10 times before cracking. Fig.12 shows each distribution of chloride ion at point a, b and c. The parallel simulations clearly shows deeper ingress of chloride ion within 100 days, compared to the results without consideration

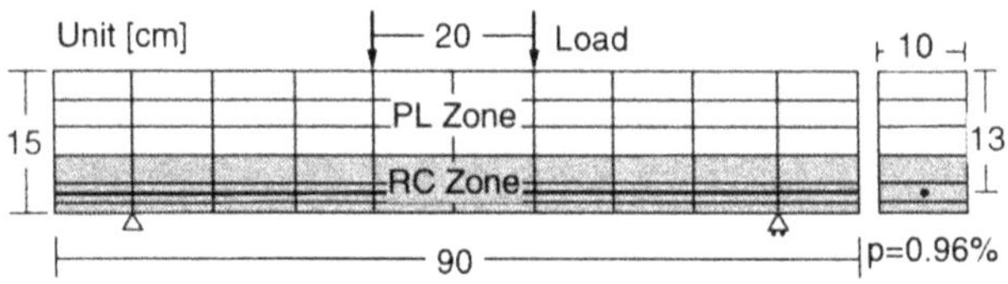

Fig.11 Mesh layout and load condition used in FE analysis

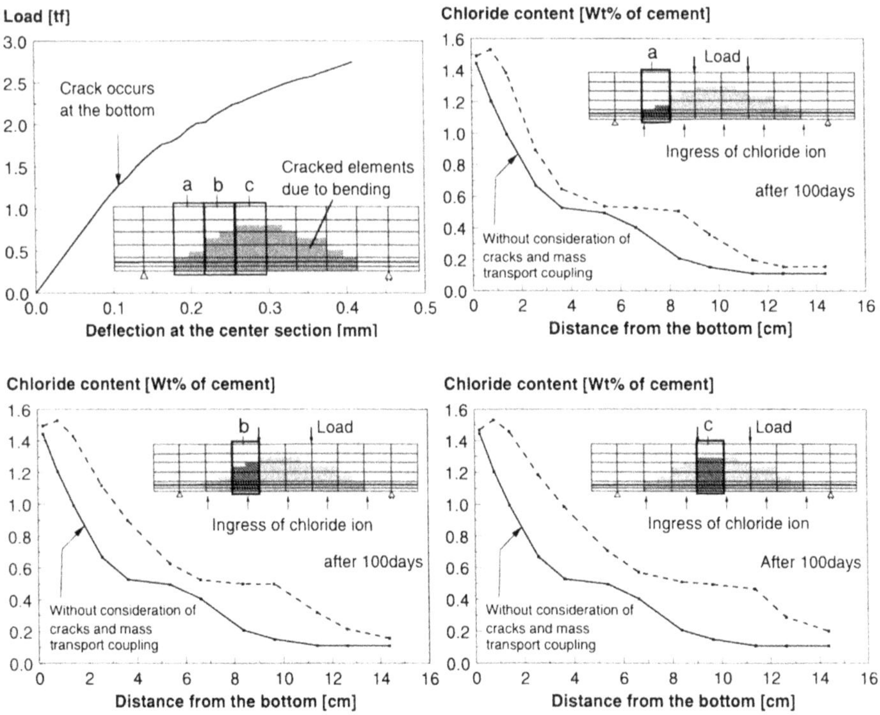

Fig.12 Distribution of chloride ion in damaged RC beam due to external load.

cracks and mass transport coupling. It can be also seen that the amount of ingress of chloride ion increases near the center section, since in cracked element the bulk movement of chloride ion in pore water can easily take place.

7. Concluding remarks

The performance based design obligates quantitative assessment of required performances by means of transparent and objective science. For that purpose, simulation of both macro and micro-scale structural behaviors is thought to be a powerful tool even though plenty of articles have to be investigated. In other words, if we may simulate cyber materials and structural behaviors in a virtual world under artificial environments and mechanical actions, we could see what happens within shortened period. This paper proposes a way how to unify the structural safety and serviceability check method and durability assessment in this regard. Though component chemical-physical processes are crudely assumed, system dynamics of micro-scale pore structure formation and macro-scale defects and deformation of structures was shown as a possible and promising approach in future.

For accomplishing entire system, migration of oxygen and carbon dioxide has to be added, and rate of corrosion and carbonation is required to be installed with hydration process of cement in concrete. For versatility of ion transport, magneto-electrical field governed by Maxwell's principle should be overlaid on the mass, momentum and energy conservation conditions. Thus, the authors understand that the unification of structural and durability design has just started. For further progress and development, predictive tool of structural behaviors from birth to death of concrete under any specified environment and load serves as an essential technicality.

8. Appendix

Governing equations solved by the proposed system are shown in Fig. a, b, and c. Fig. a shows mass and energy conservation laws which govern thermo-physics of materials, and balance equation which governs mechanics of structures. These conservation laws should be satisfied in all material systems as well as in the field of concrete materials. Fig. a also shows mass, energy and momentum transport, and sink term in above equations. These governing equations are derived from the specific characteristics of concrete materials. Fig. b and c show state, compatibility and constitutive law of mass, energy and concrete solid. These equations are also modeled by considering specific characteristics of concrete.

Table. I shows input values needed for calculation. In addition to the list of Table. I, x, y, and z-coordinate of each node, element type/shape of analytical target should be required. These input value determine the initial value at time=0, and also determine some of the material constants shown in Table. II. Other material constants are given through the experimental results. By solving above governing equations under given material constants and initial conditions, variables shown in Table. III can be obtained with 3-D space and time domain.

Conservation laws (Mass, energy and momentum conservation)

Conservation of mass and energy

$$\alpha\frac{\partial X}{\partial t} + div\mathbf{J}_X - Q_X = 0$$

Moisture[1]: $\left[\phi_{tot}S_{tot}\frac{\partial\rho_l}{\partial P_l} + \rho_l\frac{\partial(\phi_{tot}S_{tot})}{\partial P_l}\right]\frac{\partial P_l}{\partial t}$

Chloride ion [11]: $\frac{\partial}{\partial t}\left[(\phi_{gl}+\phi_{cp})S_{cg}C_{Cl}\right]$ Heat energy[14]: $\rho c\frac{\partial T}{\partial t}$

Balance equation

$$\frac{\partial\sigma_{ij}}{\partial x_j} + X_i = 0$$

Mass, energy and momentum transport term

Moisture [7][8]

$$J_w = -D_w\nabla P_l \quad D_w = K_l + K_v$$

$$K_l = \frac{\rho_l\phi_{tot}^2}{50\eta}\left(\int_0^{r_c} r dV\right)^2 \quad K_v = \frac{\rho_v\phi_{tot}D_0}{\Omega}\int_{r_c}^{\infty}\frac{dV}{1+N_k} \quad N_k = \frac{l_m}{2(r-t_a)}$$

$$\eta = \eta_l\exp(G_e/RT) \quad G_e = G_{\max}H_d \quad \ddot{H}_d + \left(\frac{1+\ddot{\eta}_e}{\eta_e}\right)H_d = \frac{RH}{\eta_e}$$

$$\eta_e = a(1+bH_d^c) \quad a = \left[1.59\left(\frac{\phi_{lr}}{\phi_{cp}+\phi_{gl}}\right)+0.7\right]^5 \quad b = 2.5a \quad c = 2.0$$

After cracking

$$D_w = 10\cdot(K_l+K_v)\ ,or\ = 50\cdot(K_l+K_v)$$

Chloride ion [11]

$$J_{Cl} = (\phi_{gl}+\phi_{cp})S_{cg}\cdot(-D_{Cl}\nabla C_{Cl} + \tau u_w C_{Cl})$$

$$u_w = J_l/\rho_l = -(K_l\nabla P_l)/\rho_l$$

Heat energy [14]

$$J_H = -K_H\nabla T$$

Momentum transport with mass transport is neglected

Mass, energy and momentum sink term

Moisture [1]

$$Q_p = Q_{pd} + Q_{hyd} = \rho_l\frac{\partial(\phi_{tot}S_{tot})}{\partial t} - W_{pow}\frac{\partial\beta}{\partial t}$$

$$\beta = \sum p_i\cdot\frac{\overline{Q}_i}{\overline{Q}_{i,\infty}}\cdot w_i$$

Chloride ion [11]

$$Q_{Cl} = \frac{\partial C_{bound}}{\partial t}$$

Heat energy [5][6]

$$Q_H = W_{pow}\sum p_i\overline{H}_i$$

$$\overline{H}_i = \gamma_i\cdot\beta_i\cdot\mu_i\cdot H_{i,T_o}\exp\left[-\frac{E_i(\overline{Q}_i)}{R}\left(\frac{1}{T}-\frac{1}{T_o}\right)\right]$$

$$\overline{Q}_i \equiv \int\overline{H}_i dt$$

$$\beta_i = f(\omega_{free},\eta_i) \quad \eta_i = 1-\left(1-\frac{\overline{Q}_i}{\overline{Q}_{i,\infty}}\right)^{\frac{1}{3}}$$

Momentum generation by product/loss of mass is neglected

Fig.a Conservation laws, mass/energy transport and sink term

State, compatibility and constitutive law of mass,energy and solid

Strength, elastic modulus, and deformability against pore pressure[10][13]

$$f'_c = a\exp(-b\cdot V_{pore}) \quad a,b: \text{constant}$$

$$V_{pore} = \phi_{cp}\cdot\exp(-B_{cp}\cdot r_{50}) \quad r_{50} = 50\times10^{-9}[m]$$

$$E = 40000 f_c'^{1/3} \quad f_t = 0.58 f_c'^{2/3}$$

Unrestrained strain due to temperature [14]

$$\varepsilon_T = \alpha_c\Delta T$$

Unrestrained shrinkage strain due to pore pressure [10]

$$\varepsilon_{sh} = \phi_{tot}S_{tot}f_{pow}\frac{P_l}{E_{sh}}$$

$$E_{sh} = E/3$$

Compatibility equation

$$\varepsilon_{xx} = \frac{\partial u}{\partial x} \quad \varepsilon_{xy} = \frac{\partial v}{\partial x} + \frac{\partial u}{\partial y}$$

$$\varepsilon_{yy} = \frac{\partial v}{\partial y} \quad \varepsilon_{yz} = \frac{\partial w}{\partial y} + \frac{\partial v}{\partial z}$$

$$\varepsilon_{zz} = \frac{\partial w}{\partial z} \quad \varepsilon_{zx} = \frac{\partial u}{\partial z} + \frac{\partial w}{\partial x}$$

Constitutive law of concrete solid [14][18]

$$\varepsilon_{V}(t) = \int_0^t\frac{1+\varphi(t,t')}{3K(t')}\frac{d\sigma_V}{dt'}dt' + \varepsilon_T + \varepsilon_{sh}$$

$$\varepsilon_V \equiv \frac{1}{3}(\varepsilon_{xx}+\varepsilon_{yy}+\varepsilon_{zz}) \quad \sigma_V \equiv \frac{1}{3}(\sigma_{xx}+\sigma_{yy}+\sigma_{zz})$$

$$e_{ij}(t) = \int_0^t\frac{1+\varphi(t,t')}{2G(t')}\frac{dS_{ij}}{dt'}dt'$$

$$e_{ij} \equiv \varepsilon_{ij} - \delta_{ij}\cdot\varepsilon_V \quad S_{ij} \equiv \sigma_{ij} - \delta_{ij}\cdot\sigma_V$$

$$K = \frac{E}{3(1-2\nu)} \quad G = \frac{E}{2(1+\nu)}$$

$$\varphi(t,t') = \varphi^T(t_e^{-m}+\chi)(t-t')^{\eta_T}$$

$\varphi^T = f(\text{mix proportion}, T)$

$t_e = f(\text{Temp.history})$

$m = f(\text{Comp. strength})$

$\chi = f(\text{W/P})$

$\eta^T = f(\text{Temp. history})$

Fig. b State and compatibility law of mass, energy and solid (No.1)

State, compatibility and constitutive law of mass,energy and solid

Pore structure formation [1]

$$V_s = \frac{\alpha W_{pow}}{1-\phi_{ch}}\left(\frac{1}{\rho_p} + \frac{1}{\rho_w}\cdot\frac{\beta}{W_{pow}}\right) \qquad \phi_{lr} = (t_w s_l \rho_g)/2$$

$$\alpha = \sum p_i \cdot (\overline{Q}_i/\overline{Q}_{i,\infty}) \qquad \phi_{gl} = V_s\phi_{ch} - \phi_{lr}$$

$$\phi_{cp} = 1 - V_s - (1-\alpha)\frac{W_{pow}}{\rho_p} \qquad \phi_{tot} = \phi_{lr} + \phi_{gl} + \phi_{cp}$$

$$\rho_g = \rho_p\rho_w(1+\beta/W_{pow})(1-\phi_{ch})/(\rho_w + \beta\rho_p)$$

$$A\delta_m^3 + B\delta_m^2 + C\delta_m + D = 0$$

$$A = \{n(1-\phi_{in}) + 3(1-\phi_{ou})\}/\{3(n+3)\}$$

$$B = \{n(1-\phi_{in}) + 2(1-\phi_{ou})\}r_0/(n+2)$$

$$C = \{n(1-\phi_{in}) + (1-\phi_{ou})\}r_0^2/(n+1)^2$$

$$D = -(\alpha r_0^3/3)[\phi_{in} + \beta\rho_p/\rho_w]$$

Before contact

$$\phi_{ou} = 1.0$$

After contact

$$\phi_{ou} = 1 - (X+Y)/Z \qquad \delta_{\max} = kr_o$$

$$X = -n(1-\phi_{in})\left[\frac{k^3}{3(n+3)} + \frac{k^2}{n+2} + \frac{k}{n+1}\right]$$

$$Y = \frac{\alpha}{3}\left[\phi_{in} + \beta\frac{\rho_p}{\rho_w}\right] \quad Z = \frac{k^3}{n+3} + \frac{2k^2}{n+2} + \frac{k}{n+1}$$

$$s_l = 510 f_{pc} + 1500 f_{sg} + 3100 f_{fa}$$

$$\zeta = 19.0 f_{pc} + 1.5 f_{sg} + 1.0 f_{fa}$$

$$SA_c = \frac{3\delta_m}{\zeta r_{eq}^3 (1-\phi_{in})}(A\delta_m^2 + B\delta_m + C)$$

$$SA_l = 2\phi_l \int_{r_m}^{\infty} B_l \exp(-B_l r)\, d\ln r \qquad SA_g = W_s \cdot sa_g$$

$$\phi(r) = \phi_{lr} + \phi_{gl}\cdot V_{gl} + \phi_{cp}\cdot V_{cp} \qquad V_i = 1 - \exp(-B_i r)$$

Thermodynamical equilibrium of mass

Moisture[7][8][9]

$$P_l = \frac{\rho_l RT}{M_l}\ln RH = -\frac{2\gamma}{r_c}$$

$$t_a = \frac{0.525\times 10^{-8}\cdot RH}{(1-RH/RH_m)(1-RH/RH_m + 15RH)}$$

$$RH_m = \exp\left[\frac{-\gamma M_l}{\rho_l RT(r-t_a)}\right]$$

$$S_{ads} = \int_{r_c}^{\infty}\left[1-\left(\frac{r-t_a}{r}\right)^2\right]dV$$

$$S_{cnd} = \int_0^{r_c} dV = 1 - \exp(-Br_c) = S_c$$

$$S_{lr} = RH$$

(Wetting stage)

$$S_{cnd} = S_c + \int_{r_c}^{\infty}\left(\frac{S_c}{V}\right)dV = S_c[1 - \ln S_c]$$

$$S_{lr} = RH^{0.05}$$

(Drying stage)

$$S_{cnd} = S_c + \int_{r_c}^{\infty}\left(\frac{S_{r_{\min}}}{V}\right)dV = S_c - S_{r_{\min}}\ln S_c$$

$$S_{lr} = S_{\max} RH^{0.05}$$

(Drying to wetting stage)

$$S_{cnd} = S_c + \int_{r_c}^{r_{\max}}\frac{S_c}{V}dV = S_c[1 + \ln S_{r_{\max}} - \ln S_c]$$

$$S_{lr} = 1 + (RH-1)\left(\frac{S_{\min}-1}{RH_{\min}-1}\right)$$

(Wetting to drying stage)

$$S_{cg} = S_{ads} + S_{cnd}$$

$$S_{tot} = \frac{\phi_{lr}S_{lr} + (\phi_{gl} + \phi_{cp})S_{cg}}{\phi_{tot}}$$

$$\omega_{free} = \frac{\phi_{cp}S_{cp}(1-V_g)}{W_{pow}}$$

Chloride ion [12]

$$\alpha_{fixed} = \begin{cases} 1 & C_{tot} \le 0.5 \\ 1 - 0.5(C_{tot} - 0.5)^{0.39} & 0.5 \le C_{tot} \le 4.5 \\ 0.141 & 4.5 \le C_{tot} \end{cases}$$

$$C_{tot} = C_{Cl} + C_{bound} \qquad \alpha_{fixed} = C_{bound}/C_{tot}$$

Fig.c State and compatibility law of mass, energy and solid (No.2)

Appendix Table. I List of input value

P_l	Initial pore pressure [Pa]	W_{pow}	Powder weight per unit volume [kg/m^3]
C_{cl}	Initial concentration of Cl ion[mol/l]	V_g	Aggregate volume per unit volume [kg/m^3]
T	Initial temperature [K]	f_{pc}	Weight fraction of portland cement
p_i	Mass ratio of chemical component of cement (C_2S, C_3S, C_4AF, C3A, gypsum)	f_{sg}	Weight fraction of blast furnace slag
		f_{fa}	Weight fraction of fly ash
ρ_p	Density of powder materials [kg/m^3]	p_{sp}	The amount of organic admixture
W/P	Water to powder ratio		

Table. II List of material constant

ρ_l	Density of the liquid[kg/m^3]	ρ_p	Density of powder materials [kg/m^3]
ρ_v	Density of saturated vapor[kg/m^3]	ρ_w	Density of chemical hydrated products [kg/m^3]
ρc	Heat capacity [kcal/K.m^3]	ϕ_{ch}	Specific porosity of gel products (=0.28)
D_0	Vapor diffusivity in free atmosphere[m^2/s]	ϕ_{in}	Specific porosity of inner products (=0.28)
D_{cl}	Chloride ion diffusivity in pore solution phase[m^2/s]	t_w	Thickness of interlayer porosity (=2.8×10^{-10}[m])
Ω	Parameter representing tortuosity of pore(=$(\pi/2)^2$)	s_l	Specific surface area of interlayer porosity [m^2/kg]
η_i	Viscosity under ideal conditions [Pa.s]	n	Parameter representing a generic pattern of deposition of products around particle
l_m	Mean free path length of a water molecule[m]	r_0	Radius of powder particle [m]
G_{max}	Maximum additional Gibbs energy for the activation of flow(=3500kcal/mol)	ζ	Ratio of volume to the external surface area of a typical hydrate products[nm]
τ	Reduction factor representing the interaction effects in the pore solution phase (0.3)	f_{pc}	Weight fraction of portland cement
K_H	Heat conductivity [kcal/K.m.sec]	f_{sg}	Weight fraction of blast furnace slag
p_i	Mass ratio of chemical component i of cement	f_{fa}	Weight fraction of fly ash
w_i	The amount of water consumption of chemical component i due to hydration [mol/mol]	r_{eq}	Equivalent spherical cell radius[m]
$\overline{Q}_{i,\infty}$	Maximum theoretical specific heat of component i [kcal]	sa_g	Specific surface area of the hydrates [m^2/kg]
W_{pow}	Powder weight per unit volume [kg/m^3]	r_m	Minimum radius of pores[m]
γ_i	Reduction factor representing the retardation effect to the hydration by flyash and organic admixture	M_l	Molecular mass of the liquid [kg/mol]
μ_i	Coefficient representing the effect of mineral composition (C_3S, C_2S) to hydration rate	γ	Surface tension of the liquid[N/m]
T_0	Referential temperature (=293 K)	V_g	Volume of aggregate per unit volume [m^3/m^3]
R	Gas constant [J/mol.K]	α_c	Coefficient of linear expansion of concrete [1/K]
		f_{pow}	Fractional volume of powder materials
		δ_{ij}	Kronecker delta
		ν	Poisson ratio

Table. III List of variables

t	Time [s]	η	Viscosity of fluid under non-ideal conditions[N·s/m^2]
P_l	Pore pressure [Pa]	η_e	Effective non-ideal viscosity of the pore fluid[N·s/m^2]
C_{cl}	Concentration of chloride ion [mol/l]	G_e	Additional energy for activation of flow
T	Temperature [K]	H_d	Fictitious humidity parameter
J_w	Flux of moisture [kg/m^2.s]	r	Pore radius [m]
J_{cl}	Flux of chloride ion [mol/m^2.s]	r_c	Pore radius in which the equilibrated interface of liquid and vapor is created [m]
J_H	Flux of heat[kcal/m^2.s]	V	Normalized pore volume [m^3/m^3]
D_w	Moisture conductivity [kg/Pa.m.s]	N_k	Knudsen number
K_l	Liquid conductivity [kg/Pa.m.s]	$\mathbf{u}_w$	Velocity vector of pore solution phase
K_v	Vapor conductivity [kg/Pa.m.s]		

J_l	Flux of liquid [kg/m^2.s]
Q_p	Sink term for moisture balance
Q_{pd}	The term representing bulk porosity change effects
Q_{hyd}	The term representing water consumption due to hydration
β	The amount of chemical combined water per unit weight of hydrated powder materials[kg/kg]
$\overline{Q}_i$	Accumulated heat generation of chemical component i [kcal]
Q_{cl}	The term representing the reduction of free chloride
C_{bound}	The amount of bound chloride [mol/l]
Q_H	The term of heat generation
$\overline{H}_i$	Heat generation rate of clinker component i [kcal/kg.s]
$H_{i,To}$	Referential heat rate of i-th component at temperature T_o[kcal/kg.s]
$E_i(\overline{Q}_i)$	Activation energy of component i [kcal.K/kg.s]
β_i	Reduction of probability of contact between unhydrated compounds and free pore water
ω_{free}	The amount of free water
η_i	Non-dimensional thickness of the cluster
V_s	Volume of hydrated products [m^3/m^3]
W_s	Weight of hydrated products [kg/m^3]
α	Average degree of hydration
ρ_g	Dry density of gel products [kg/m^3]
ϕ_{cp}	Capillary porosity
ϕ_{gl}	Gel porosity
ϕ_{lr}	Interlayer porosity
ϕ_{tot}	Total porosity
ϕ_{ou}	Porosity at the outermost boundary of the expanding cluster
δ_m	Cluster thickness
SA_c	Specific surface area of capillary porosity [m^2/m^3]
SA_g	Specific surface area of gel porosity [m^2/m^3]
B_i	Porosity distribution parameter
RH	Relative humidiry
t_a	Thickness of adsorbed layer [m]
RH_m	The humidity required to fully saturate the pore
S_{ads}	Degree of Saturation due to adsorbed water
S_{cnd}	Degree of Saturation due to condensed water
S_c	Degree of saturation due to condensed water in virgin wetting path
r_{max}	Pore radius of the largest pores that experienced a complete saturation in the wetting history [m]
S_{rmax}	Highest saturation experienced by the porous media in its wetting history
r_{min}	Pore radius of the smallest pores that experienced an emptying out in the drying history [m]
S_{rmin}	Lowest saturaion of the porous media in its wet-dry history
S_{max}	Highest saturation experienced by interlayer porosity in its wetting history
S_{min}	Lowest saturation experienced by interlayer porosity in its drying history
RH_{min}	Minimum RH experienced in its drying history
S_{cg}	Degree of saturation of gel and capillary pores
S_{lr}	Degree of saturation of interlayer porosity
S_{tot}	Degree of saturation of total pores
α_{fixed}	Equilibrium ratio of fixed chloride component to the total ion components
C_{tot}	Total amount of chloride ion [mol/l]
ε_T	Unrestrained strain due to temperature
ε_{sh}	Unrestrained shrinkage strain due to pore pressure
E_{sh}	Deformability against capillary stress [Pa]
f_c'	Compressive strength [Pa]
V_{pore}	Volume of capillary pore above 50nm
E	Elastic module [Pa]
f_t	Tensile strength [Pa]
X_i	External force in i-direction
u	Displacement in x-direction
v	Displacement in y-direction
w	Displacement in z-direction
t'	Loading period [s]
ε_{ij}	Strain tensor
σ_{ij}	Stress tensor
ε_V	Average volumetric stress
σ_V	Average volumetric strain
e_{ij}	Deviatoric stress
S_{ij}	Deviatoric strain
K	Bulk modulus
G	Shear modulus
φ	Creep coefficient
φ^T	Coefficient representing the effect of mixproportion and temperature
t_e	Coefficient representing the effect of temperature history
m	Coefficient representing the effect of compressive strength
χ	Coefficient representing the effect of W/C
η^T	Coefficient representing the effect of temperature history

9.References

1. Maekawa, K., Chaube, R.P. and Kishi T. (1995) Coupled mass transport, hydration and structure formation theory for durability design of concrete structures, *Proc. of the Int'nl workshop on rational design of concrete structures under severe conditions*, Hakodate, pp. 263-274
2. Maekawa, K., Irawan, P. and Okamura, H. (1997) Path-dependent three-dimensional constitutive laws of reinforced concrete - formation and experimental verifications, *Structural Engineering and Mechanics*, Vol.5, No.6, pp.743-754
3. K. Sakai (1995) Integrated design and environmental issues in concrete technology, *Proceedings of international workshop in Hakodate*, E&FN SPON
4. Okamura, H. and Maekawa, K. (1991) *Nonlinear Analysis and Constitutive Models of Reinforced Concrete*, Gihodo, Tokyo
5. Kishi, T. and Maekawa, K. (1996) Multi-component model for hydration heating of portland cement, *Concrete Library of JSCE*, No.28, pp. 97-115.
6. Kishi, T. and Maekawa, K. (1997) Multi-component model for hydration heating of blended cement with blast furnace slag and fly ash, *Concrete Library of JSCE*, No.30, pp. 125-139.
7. Chaube, R.P. and Maekawa, K. (1994) A study of the moisture transport process in concrete as a composite material, *Proc. of the JCI*, Vol. 16, No.1, pp.895-900.
8. Chaube, R.P. and Maekawa, K. (1996) A permeability model of concrete considering its microstructual characteristics, *Proc. of the JCI*, Vol. 18, No.1, pp. 927-932.
9. Ishida, T., Chaube, R.P. and Maekawa, K. (1996) Modeling of pore water content in concrete under generic drying wetting conditions, *Proc. of the JCI*, Vol. 18, No.1, pp. 717-722.
10. Ishida, T., Chaube, R.P., Kishi, T. and Maekawa, K. (1998) Micro-physical approach to coupled autogenous and drying shrinkage of concrete, *CONCRETE UNDER SEVERE CONDITIONS 2* edited by O.E.Gjørv, K.Sakai, and N.Banthia, Vol.3, pp.1904-1913, E&FN SPON, London.
11. Chaube, R.P., Kishi, T. and Maekawa, K. (1998) An integrated computational framework for predicting the long-term durability of cementitious materials, *CONCRETE UNDER SEVERE CONDITIONS 2* edited by O.E.Gjørv, K.Sakai, and N.Banthia, Vol.1, pp.104-113, E&FN SPON, London.
12. Maruya, T., Matsuoka, Y. and Msirikul, S. (1992) Simulation of chloride movement in hardened concrete, *Concrete Library of JSCE*, No.20, pp. 57-70.
13. Okamura, H. and Maeda, S (1987) *Reinforced Concrete Engineering*, Ichigaya Syuppan, Tokyo (in Japanese).
14. Okamura, H., Maekawa, K. and Ozawa, K. (1993) *High Performance Concrete*, Gihodo, Tokyo (in Japanese).
15. An, X., Maekawa, K. and Okamura, H. (1997) Numerical simulation of size effect in shear strength of RC beams, *Proceedings of JSCE*, No.564, V-35, pp.297-316.
16. Bazant, Z. P., Sener, S. and Kim, J. (1987) Effect of cracking on drying permeability and diffusivity of concrete, *ACI materials journal*, No.84-M35, pp.351-357.
17. Oshita, H. and Tanabe, T. (1997) Mathematical modeling for permeability behavior of non-homogeneous material and its applicability, *Concrete Library of JSCE*, No.29, pp. 69-92.
18. Bazant, Z. P. and Panula, L. (1978) A note on amelioration of the creep function for "Improved dischinger method", *Cem. & Con. Res.*, Vol.8, pp.381-386, 1978.
19. Mabrouk, R., Ishida, T. and Maekawa, K. (1998) Solidification model of hardening concrete composite for predicting creep and shrinkage of concrete, *Proc. of the JCI*, Vol.20, No.2, pp.691-696.

MORE RATIONAL DESIGN OF COMPLEX CONCRETE STRUCTURES
Rational concrete design

B. JAKOBSEN
BJ Konstruksjonsteknikk AS
Sandvika, Norway

Abstract
The present paper discusses modern design of complex concrete structures. It takes basis in the offshore concrete platforms which in many ways have led the development of the design of many other types of complex concrete structures. Reasons are discussed why reinforcement densities have increased significantly over the years. If we are not on the alert, we can end up with more expensive and less durable structures. Proposals to remedy this situation are given.
Keywords: Concrete structures, rational design, reinforcement densities

1 Introduction

The design and construction of the offshore concrete platforms built for the North Sea, Fig. 1, and more recently for the Grand Banks off Newfoundland and in Australian waters have in many respects contributed to and in fact led the recent development of modern concrete structures.

One important experience from offshore structures is that the reinforcement quantities have increased significantly since concrete was first used on a large scale offshore. For the first North Sea platforms the average reinforcement density was about 175 kg/m^3, while the corresponding number for the more recent platforms is more than 400 kg/m^3, see Fig. 2.

There are many reasons for this increase, among them increased concrete strength and more load situations to be considered. However, more important in the context of this paper is that the design process over the years has become highly automated and has left less room for engineering judgement. This experience should be kept in mind when deciding on design schemes for future complex structures of any kind.

Concrete Technology for a Sustainable Development in the 21st Century edited by O.E. Gjørv and K. Sakai.
Published in 2000 by E & FN Spon, 11 New Fetter Lane, London EC4P 4EE, UK. ISBN: 0 419 25060 3

2 The objective of structural design

The objective of the structural design process is to develop a structure which is sufficiently safe and fulfils the functional criteria over its intended life-time at a minimum cost.

My concern in this respect is that if we are not on the alert, we may increase the cost unintentionally and with the best of intentions end up with less durable and less safe structures.

3 The design process

The design process today is highly computerized and has been developed to a very sophisticated and advanced level. It involves a wide variety of steps. These include choice of structural system, definition of design philosophy, design methodology, safety levels, serviceability criteria, type and level of loading, idealization of the

Fig. 1. The Troll gravity base platform, from [1]

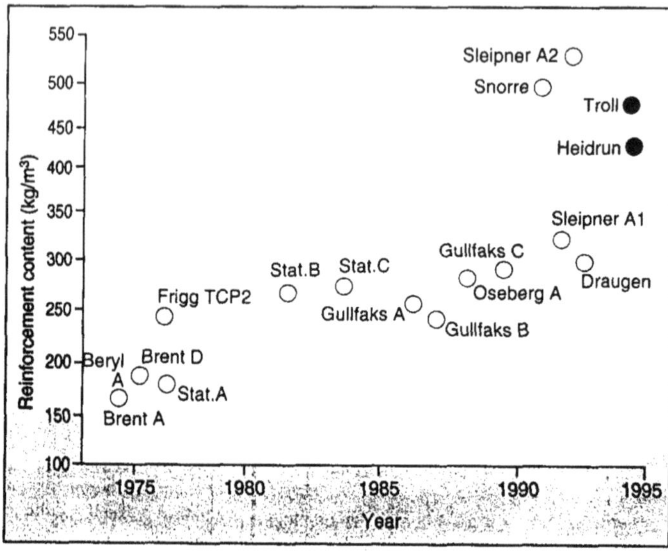

Fig. 2. Reinforcement density of some offshore platforms, from [1]

structure into an analysis model, choice of computer program, running of the analysis, post-processing including transferring of stresses into sectional forces, checking of concrete dimensions, selection of reinforcement lay-out and finally drawing production.

Ideally, all these steps should be in keeping with each other. It is thus important to determine the steps which are critical and have the largest impact on the structural safety and performance and to concentrate the effort on them. For instance, little is gained with respect to the obtained safety level by using sophisticated analytical models if the loading is encumbered with a high degree of uncertainty and is poorly represented. Nevertheless, there has been a tendency to do well what we are able to do well based on the thinking that a good engineer will do the best he can in all respects to arrive at the best possible result. The danger is that this very sincere attitude may lead to a false impression of high accuracy and reliability. And even worse, these good intentions may take the emphasis away from where it ought to be, and the result may be a design that does not comply with our objectives.

The modern design process may involve thousands of load combinations which have to be checked at thousands of "design points" throughout the structure. The size and complexity makes a highly automated design process necessary. It is thus a challenge to organize the process and the people involved in such a way, that an overview and a clear understanding is maintained.

4 The engineer

As elsewhere in the modern society, many young design engineers are educated to a very high level within a narrow field, and thus get increasingly specialized. This is a consequence of the vast amount of knowledge which is necessary to master in the modern world. The downside of this development is that the overview and the overall understanding may be at stake. The need to see the relative importance of the individual steps, to have the ability to look across the borderlines and to make sure that all steps fit into a totality becomes more and more important and turns into a real challenge as specialization goes on and as the structures and the design process get more complex. This is illustrated by the Sleipner-accident [2], which clearly demonstrates the need for experienced design engineers and a trained eye to superwise the design process. We must not forget or underestimate the importance of the experienced designer who over years has developed the feel for what is good and what is poor.

Based on many years of dealing with fine, young engineers, I have the impression that it is much more popular and associated with higher status nowadays to specialize within computer related matters such as structural analysis than for instance within structural detailing. Both areas are equally important and students should be encouraged to take an interest in all aspects of the design. Many structural failures are due to flaws in structural detailing and it is worthwhile to note the expression of a professional after the Sleipner accident: "The platform was overanalysed and under-designed". In a similar way, I have an unpleasant feeling that many newly educated engineers do not look critically enough on the results of their analyses. Often, they demonstrate lack of feeling for how a structure acts and responds to loading, and also lack of ability to check their results with simple long-hand calculations. This and similar points are reflected upon in a highly readable paper from 1993 called " How Engineers Lose Touch" [3]. Some of the reasons for the described situation lie in my opinion with the training at the universities, and I challenge the professors to examine the situation to see if they agree and to explore if they can do something to improve the education in this respect.

One obvious consequence of missing the overall understanding is that the basis for making the right engineering judgements is lacking. This manifests itself in many ways, for instance in "safe-guarding" and too strict and literal interpretations of given criteria. One example here is the interpretation of the crack-width criteria of the Norwegian design code NS 3473, which has turned out to govern the reinforcement quantities in large parts of the North Sea platforms. In the code it is clearly spelt out that these criteria are not to be understood as absolute and that engineering judgement should be used when applied. The "spirit" of this comment is frequently forgotten and in many cases the strict interpretation ot these criteria has caused excessive amounts of reinforcement and has in extreme cases also led to unnecessary and costly redesign.

5 The structural analysis

The global structural analysis, Fig. 3, constitutes a cornerstone in the modern design process. For offshore platforms and similarly complex structures the analysis for determining the global structural behaviour may require several hundred thousands degrees of freedom to represent the vital properties of the structure, and primarily the geometry satisfactorily. Similarly, some 50-150 different load cases may be used to represent the loading. Thus, both the geometry and the loading may be very thoroughly represented. For an analysis of this size the only feasible way today is to use the linear elastic method. The great benefit is then that the principle of superposition can be used. In this way the different loading and limit state conditions can be established by

Fig. 3. Finite element model of the Troll gravity base structure, from [4]

combination of the individual unit load cases. Furthermore, the important task of finally designing the quantities and lay-out of reinforcement can be done after the analysis has been performed.

The global analysis is in certain cases supplemented with local structural analyses to examine special effects. If deemed necessary, these analyses can also account for nonlinear effects.

Alternatively, a coarser global model supplemented by several local, refined analyses could be used instead of a large global model. The challenge with this alternative is the difficulty in representing the boundary conditions in the local models for general loading and also the increased need for book-keeping with superposition of results from different analyses.

6 The structural response

These linear elastic analyses do not account for nonlinear material behaviour such as the stress relieving effects of cracking and micro-cracking which again lead to evening out of stress peaks. Many load effects may also have the character of deformation loads where the internal forces are reduced through cracking of the concrete. It is very difficult to correct for these effects in the computerized design process presently used. The primary reason is that when a complex structure is analysed linear-elastically it is not easily done to ensure complete equilibrium during force redistribution and the corresponding sheding of forces when one force component, e.g. a bending moment, is reduced due to non-linear effects.

The analytical tools are now so powerful that they allow an efficient scan through all points of interest for the design. They thus "catch" most stress peaks in the "hot-spot" areas. In earlier days with coarser analysis models, missing some of these stress peaks in a way compensated for the lack of force redistribution in the linear elastic analysis. The consequence may therefore be that there are many areas, particularly in or near complex joints, with far too much reinforcement solely due to the design methodology. Unfortunately, these areas normally have high reinforcement densities also for other reasons. The placing of reinforcement bars and the concrete casting in such areas can thus become very difficult.

The modern post-processors check the concrete dimensions, the reinforcement quantities, the crack widths etc. in a large number of "design point" where a small element of the structure is subjected to forces determined in the linear elastic analysis, Fig. 4. In this way the small element is distorted independently of the surrounding parts of the structure. If significant cracking takes place, deformation compatibility may thus be significantly violated. One important consequence is that the calculated strains and crack widths may be vastly overestimated, thus leading to unnecessary reinforcement to meet the criteria.

Despite the lack of deformation compatibility, the applied methodology is firmly based on the lower bound theorem of plasticity theory. This theorem says that a state of stress which is in equilibrium with the loading and which does nowhere violate the sectional capacity represents a lower bound of the load carrying capacity of the structure provided that there is sufficient deformation capability inherent in the system. The load resisting capacity of the structure is therefore well represented by the methodology used. One of the prices to be paid may be, as discussed, unnecessary high amounts of reinforcement.

7 Consequences

The consequences of overly dense reinforcement can be summarized as follows:

- The cost increase of putting in extra reinforcement may be significant, particularly in already congested areas such as joints and other "disturbed" regions.
- The congestion may become so high that the designer has to undertake a tedious redesign, often in conjunction with the contractor, to find alternative reinforcement lay-outs to facilitate construction. The effect is that also the cost of the design process may increase significantly.

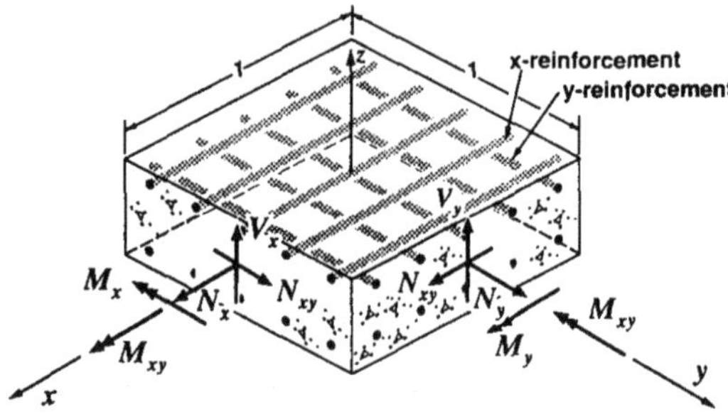

Fig. 4. Structural element subjected to forces from linear elastic analysis, from [5]

- Very dense reinforcement may cause difficulties during pouring of the concrete and the quality of the concrete cover may be severely reduced, thus reducing the protection against reinforcement corrosion.
- In extreme cases the density can become so high that the reinforcement practically forms a continuous layer of steel which to a large extent separates the concrete cover from the bulk concrete. The joint action of reinforcement and concrete through bond and thus the ability to resist extreme loading may be questioned in such cases. Furthermore, the real structural safety may be questioned in such situations, primarily in non-hyperstatic structures where force redistribution may not be facilitated.

8 The improved design process

Some needs for improvement in the design process have been suggested in the discussion above. I find it however particularly important that the items below be reflected upon in future design processes in order to improve the shortcomings mentioned.

- The importance of each step in the design process should be evaluated, and balanced focus should be put on each of them to ensure that they are in keeping. In particular, my experience indicates that there is easy to put too much emphasis on analysis issues, which in certain cases may happen at the sacrifice of important design issues
- The design criteria should be realistic and to the point. Again, take as an example the crack width criteria, which probably are those criteria which have been under most discussion in the era of the North Sea concrete platforms. The primary reason

for introducing these criteria is to prevent corrosion of the reinforcement. As discussed above, they may in many cases have contributed, also through too literal interpretation, to poorer protection of the reinforcement since dense reinforcement may have led to reduced quality of the concrete cover. It is now well accepted that crack widths are not so important for control of reinforcement corrosion as long as they are not too wide, say below 0,3-0,4 mm. It is much more important to ensure adequate thickness and quality of the concrete cover and to avoid single big cracks, for instance caused by unforeseen deformation loads. Crack control should primarily be ensured through sound criteria for minimum reinforcement. This is furthermore an example of where overfocus can take emphasis away from more important design issues.

- Engineers in key positions such as the verification and control function should have enough experience, a good overall understanding and a firm engineering basis to allow use of sound engineering judgement instead of having to safeguard themselves by too strict and too literal interpretation of the criteria.
- For complex structures requiring large structural analyses a relatively thorough pilot design should initially be performed by a small team of highly experienced engineers. They should select the load cases which are expected to be governing and perform a design based on those. If the subsequent automated design process, primarily executed as verification, gives significantly different reinforcement quantities than the pilot design, the differences should be understood, explained and accepted before any changes are done to the pilot design.
- For checking of crack widths and other corresponding serviceability limit state (SLS) criteria engineering judgement should be used to estimate and account for redistribution of forces due to cracking and the reduction of deformation type load effects also due to cracking. In SLS-checks, where load carrying capacities are not considered, there should be no reason to adhere strictly to the requirement that all load effects be in equilibrium. This relaxation seems to allow an easy way to estimate redistribution and thus reduce much of the congestion that SLS-criteria otherwise may cause.
- For checking of the load resisting capacity of the structure in the ultimate limit state (ULS) it is a fundamental requirement that equilibrium be fulfilled. Design often gives rise to large reinforcement quantities in joints when these are designed to resist forces such as bending moments from the linear elastic analysis. In complex shell-like structures these moments very often arise just to ensure compatibility between adjacent structural elements. They contribute very little to the total load resistance, which is primarily due to in-plane forces. When cracking occurs, these local bending moments are thus significantly reduced without any noticeable effect on the in-plane force flow or the total load resisting capacity of the structure. Although the corresponding sheding of forces in many cases is of minor significance, it should be accounted for in order not to violate the fundamental condition of equilibrium. The reduction of the bending moments and the corresponding force redistribution could be simulated in the linear elastic analysis by introducing local couple forces as load cases in the joints where the reduction is expected to be of importance. Through this means the increase in deformation locally , e.g. curvature, due to softening from cracking in or near the joint can be represented in an approximate manner.

- The bending moments discussed in the point above have the character of deformation loads. Also when in-plane forces are of this nature it could be beneficial to identify them and account for the force reduction due to reduced stiffness from cracking etc. Realising that this can be difficult it could be worthwhile in certain cases, for there may be potentials for significant savings of reinforcement quantities.

9 References

1. Jakobsen, B.(1994) The Evolution of The Offshore Concrete Platform, *From Finite Elements to the Troll Platform,* Ivar Holand 70th Anniversary, Tapir, Trondheim
2. Jakobsen, B.(1994) The Sleipner Accident and Its Causes. *Engineering Failure Analysis,* Vol 1, No.3 pp.193-199, Elsevier Science Ltd.
3. Ferguson, E.S.(1993) How Engineers Lose Touch, *Invention & Technology,* Winter 1993
4. Brekke, D.E., Åldstedt, E. and Grosch, H.(1994) Design of Offshore Concrete Structures Based on Postprocessing of Results from Finite Element Analysis, *Proc. of the Fourth Int. Offshore and Polar Engng. Conference*, Osaka, Japan, April 10-15.
5. Adebar, P. and Collins, M.P.(1994) Shear Design of Concrete Offshore Structures, *ACI Structural Journal*, Vol. 91, No.3, May-June

CONCRETE TECHNOLOGY FOR SUSTAINABLE DEVELOPMENT – AN OVERVIEW OF ESSENTIAL ELEMENTS

P. K. Mehta
Civil and Environmental Engineering
University of California, Berkeley, USA

Abstract

In less than two years a new century and a new millennium will begin. This is an appropriate point in time to consider the future societal needs and how they might affect the concrete industry. Current trends show that infrastructural needs of our rapidly industrializing world will continue to require large amounts of low cost and durable buildings materials such as portland cement concrete. This need will have to be balanced against another equally important human need, namely the preservation of the life-sustaining environment on the planet earth, which is being threatened by the uncontrolled use of natural resources and increasing amounts of environmental pollution. This answers the question why the issue of sustainable industrial development, in general, and of the concrete industry, in particular, has assumed great importance. Much more difficult questions to answer are: how and when sustainable development will happen. In this paper the author has made an attempt to identify the essential elements of the structure that must be in place before the edifice of sustainable development of the concrete industry can be built. The following issues are examined in detail: conservation of concrete-making materials, especially portland cement, enhancement of durability of new and existing concrete structures, and a paradigm shift from reductionistic to holistic education and research in concrete technology.

1 Introduction

A leaflet recently published by the British Institution of Civil Engineers reminds us that the civil engineers are responsible for building and maintenance of essential infrastructural elements of modern life such as:

- **the muscles and sinews which hold society together** (bridges, tunnels, roads, railways, airports, and docks).
- **the provision and maintenance of its hearts and lungs** (clean water, and disposal of sewage and industrial wastes).
- **the energy to make it all work** (offshore gas and oil, hydro, nuclear, tidal, and wind power).

With growing industrialization and urbanization, it is clear that there would be a corresponding growth in the world's demand for clean water, clean air, waste disposal, safe and rapid transport of people and goods, residential and industrial buildings, and sources of energy. Humans have been using many kinds of construction materials since antiquity. However, during the 20th century, portland cement concrete has undoubtedly emerged as the material of choice for modern infrastructural needs. It is not surprising, therefore, that the concrete industry today is the largest consumer of natural resources, such as water, sand, gravel, and crushed rock. According to one estimate, the concrete industry is currently consuming natural aggregates at the rate of approximately eight billion tons every year. The manufacture of portland cement, which is the commonly used binder for modern concrete mixtures, also requires large amounts of natural materials, as described next.

Concrete Technology for a Sustainable Development in the 21st Century edited by O.E. Gjørv and K. Sakai.
Published in 2000 by E & FN Spon, 11 New Fetter Lane, London EC4P 4EE, UK. ISBN: 0 419 25060 3

The world consumption of portland cement has risen from less than two million tons in 1880 to more than one billion tons in 1996. Besides other raw materials, the production of each ton of portland cement requires approximately 1.5 tons of limestone, and considerable amounts of both fossil fuel and electrical energy. Also, it is well known that the production of each ton of portland-cement clinker is accompanied by the release of approximately one ton of carbon dioxide, which is one of the gases primarily responsible for the greenhouse effect and **global warming**. The portland cement industry, considered as one of the energy-intensive and environment-polluting industries, is likely to come under increasing scrutiny in the industrially developed countries which have for the first time accepted legally-binding commitments to reduce the emission of gases contributing to global warming in accordance with the recently (December, 1997) concluded Kyoto Protocol to the UN Framework Convention on Climate Change.

Evidently, there is a rapidly growing public concern that we can no longer continue to ignore the environmental pollution problems on the one hand and the unrestricted depletion of natural resources on the other hand. A satisfactory resolution of this concern is essential because, if unresolved, it presents a clear threat to our standard of living and, more importantly, to the entire fabric of the life support systems on which the planet earth is dependent. Now that we are approaching another milestone in human history when a new century and, in fact, a new millennium is about to begin in about 550 days, we should pause to look back and reflect on lessons that can be learned from the past, construct a vision for the future, and lead the concrete industry into the new era of sustainable development. To move toward the goal of sustainable development, obviously a judicious balance will have to be struck between the two **equally** important needs of society, namely the infrastructure to support an acceptable standard of living for most of the people in the world, and the protection of the environment. Being an important player in the infrastructural development and a major consumer of natural resources of the earth, the concrete industry needs to be reoriented through the adoption of environment-friendly technologies.

In this paper certain key elements are identified that, in the author's opinion, would be crucial for the foundation upon which the structure of an environment-friendly concrete technology can be built. The concept is illustrated in Fig. 1. The three essential elements are:

- Conservation of concrete-making materials.
- Enhancement of durability of concrete structures.
- A paradigm shift from reductionistic to a holistic approach in concrete technology research and field practice.

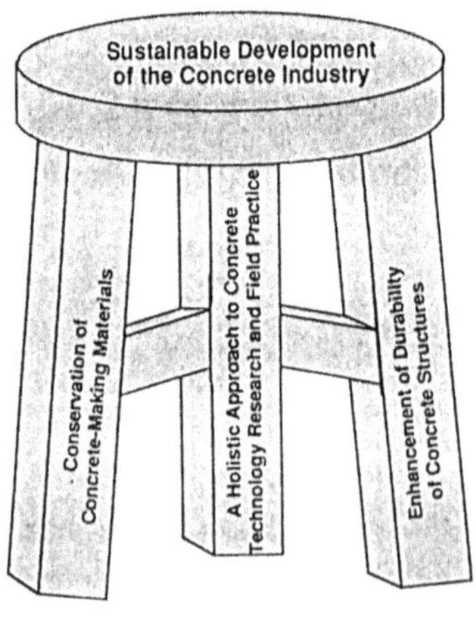

Fig. 1. Foundations of an environment-friendly technology for sustainable development of the concrete industry.

It should be noted here that only for the purposes of discussion and elaboration the three elements in Fig. 1 are shown as independent of each other. In reality they are interrelated and interdependent. For example, the enhancement of concrete durability is a resource-conserving step and, as will be discussed below, a holistic approach is necessary to address both the other issues.

2 Highlighting the problem

In an earlier paper (1) the author made the following comments which highlight the problem:

> "The problem is that, driven by demographic pressure and helped by technology, the earth's ecosystem has come under great stress. To support more and more people and to provide better living conditions for most, the engines of the industry for converting raw materials into consumer products must be run faster and faster. . . . Environmental pollution from industrial activity is not a new problem. However, the environmental crisis we now confront is quantitatively and qualitatively different from anything before, simply because so many people have been inflicting damage to the world's ecosystem during the present century that the system as a whole—not simply its various parts—may be in danger."

Another issue of considerable importance for the future is the large disparity in the standard of living in different parts of the world. In fact, it seems that we live in two worlds which exist side by side, and both are intent upon exploiting the earth's natural resources. The first world consists of the people in North America, western Europe, and Japan, which comprise 12% of the world's population but account for 60% of the total energy consumption and enjoy a high standard of living. Their economies are driven by consumerism and generate considerable waste and pollution on a per capita basis. The people in Asia (except Japan), Africa, and South America, numbering almost five billion, form the second world. For a variety of reasons, including high rates of population growth, these countries are less developed industrially and the people, in general, have a much lower standard of living. From the number and variety of ongoing or recently completed large infrastructural projects in western Europe, North America, and Japan, it is obvious that the industrially-advanced world is not slowing down in the use of the earth's natural resources. At the same time, in pursuit of a better life for the poor billions, many less developed countries have greatly accelerated the pace of industrialization. It is not difficult to imagine the end result of a process which encourages the continuation of high rates of consumption of natural resources of the earth on the one hand, and correspondingly high rates of environmental loading by a variety of polluting materials on the other hand. Clearly, a global environmental disaster would be unavoidable. It is in this context that both the industrially-rich and the industrially-poor worlds share an equal responsibility to search and adopt the technologies for sustainable development.

3 Defining sustainable development

The 1992 Earth Summit in Rio de Janeiro defined sustainable development as economic activity that is in harmony with the earth's ecosystem. In an ideal world the best way to insure sustainable development would be to practice a barter system with the earth in which we humans take as little as possible of the good things and return as little as possible of the bad things. We can never live up to this golden rule, but we can at least try to aim for it in all our economic and industrial undertakings.

Although glass, paper, plastics, and steel industries have been successfully recycling some of their used products back into the manufacturing process for making new products, this is not possible in the case of every industry. For instance, metallurgical furnaces produce large quantities of slag, and coal-firing thermal power plants produce huge volumes of coal ash. These by-products cannot be recycled back into the industries which produced them. However, we would be following the spirit of the golden rule if we were able to use the by-products of one industry as raw materials for some other industries, instead of disposing of them in landfills, ponds, and stream. In fact, to some extent this type of **"industrial ecology"** is already being practiced by the cement and concrete industries. However, as discussed below, much more needs to be done.

4 Concrete technology for sustainable development

As stated earlier (Fig. 1), the author has identified that a foundation comprising at least three elements is necessary to support the structure of an environment-friendly concrete technology for sustainable development. The three elements are discussed individually below.

4.1 Conservation of concrete making materials

Aggregates, cement, and water are the primary components of concrete. It is possible to conserve large amounts of each of these resources by the adoption of environment-friendly technologies described in recent publications. For instance, the published literature contains numerous case histories of successful substitution of natural aggregate with crushed concrete from demolition and construction sites. Water recycled from ready-mixed concrete plants has been satisfactorily used as a substitute for fresh mixing water for concrete. Detailed coverage of this topic is not intended here. However, on the matter of portland cement conservation, which is of vital importance, the author would like to reproduce below a few paragraphs from another recently published paper (2).

The goal of sustainable development of the cement and concrete industries can be reached if we make a serious effort for **complete utilization of the cementitious and pozzolanic by-products produced by thermal power plants and metallurgical industries.** In 1989, 560 million tons of coal ash were produced and only 25 million tons were used as a pozzolan by the cement and concrete industries, which is about 5% of the total available ash. The current annual production of coal ash is estimated to be about 650 million tons, of which at least 70% or 450 million tons is fly ash or fine ash that is generally suitable for use as a pozzolan. The worldwide yearly rate of consumption of fly ash today by the cement and concrete industries is reportedly about 30 million tons, which is dismally low. Another by-product which is useful for cement substitution is iron blast-furnace slag. Although the world production of this slag is about 100 million tons/year, the utilization rate as a cement substitute is still low because, in many countries, only a small portion of the slag is processed into the granulated or cementitious form.

The author was surprised to make an interesting discovery (2) that **relatively huge amounts of disposable coal ash and iron blast furnace slag happen to be available in those countries which would require large amounts of cement in the future.** For instance, China and India together produce about 200 million tons of coal ash every year. European countries, mainly Russia, Poland, the former Czechoslovakia, Romania, Germany, Spain, and the United Kingdom produce approximately 250 million tons of coal ash per year. Also, at least 50 million tons of the total yearly production of 100 million tons of blast-furnace slag comes from China, India, and Europe. At the same time, note that nearly 440 million tons of the total projected increase in the cement consumption by the year 2005 is expected from these countries (Table 1). **It should be immediately obvious that if we can find ways to use all or most of the available coal ash and iron blast-furnace slag, either in the form of blended portland cement or as mineral admixtures in concrete, we would be able to meet the projected cement demand in the year 2005 without any increase in the present capacity of portland-cement clinker production.** A sustainable development of the cement and concrete industries, as defined above, can thus be assured. Considering the additional ecological benefits described next, it is hard to imagine a better solution to the problem.

Nearly 90% of the coal ashes and metallurgical slags produced today end up either in low-value applications such as landfills and road bases, or simply disposed off by ponding and stockpiling. Disposal in this manner is not only wasteful but also harmful to human health because it contributes to land, air, and groundwater pollution. These by-product materials generally contain toxic metals. The concrete industry provides a preferred vehicle for their disposal because most of the harmful metals can be immobilized and safely incorporated into the hydration products of cement. In fact, owing to its large size, the concrete industry is probably the ideal home for safe and economical disposal of millions of tons of by-products. Based on a study by Schiessl and Hohberg (3), Fig. 2 shows the excellent environmental compatibility of a mortar made with a cement-fly ash mixture. In a realistic leaching test (tank test), the authors reported that only 0.09 mg/kg zinc and 0.15 mg/kg chromium were leached from a cement-fly ash mortar when the total amounts of the metals added to the mortar were 185 mg/kg and 53 mg/kg, respectively.

Table 1. Projection of regional cement consumption, million tons*.

Area	Year		
	1994	2000	2005
Europe, including countries of former Soviet Union	313	393	432
Asia	680	853	1000
Middle East	65	79	82
Africa	63	71	77
North America	90	92	92
South and Central America	92	118	142
Miscellaneous	7	9	10
Total	1310	1625	1835

*Adapted from: *World Cement*, V. 27, No. 5, May 1996.

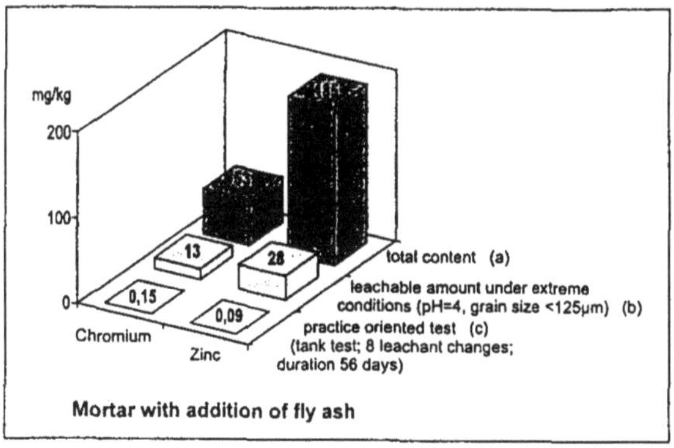

Fig. 2. Investigation of a specimen from a mortar with the Addition of fly ash, comparison of different leaching methods (3).

Given the overwhelming advantages that are possible from large-scale cement replacement by pozzolanic and cementitious by-products, is it really possible to accelerate their use in the cement and concrete industries, as advocated here? In the earlier paper (2) the major obstacles preventing the high rates of utilization of industrial by-products in concrete, suggestions for overcoming these obstacles, and recent cases of high-volume fly ash use are discussed. Only a few examples of high-volume fly ash concrete mixtures will be given here.

Studies by Malhotra (4) with superplasticized concrete mixtures have shown that, when the water/cementitious is limited to 0.3 or less, up to 60% cement can be replaced with a Class F or Class C fly ash (ASTM C 618) to obtain excellent strength and durability characteristics. For instance, a test mixture containing 150 kg/m^3 ASTM Type I cement, 200 kg/m^3 ASTM Class F fly ash, 102 kg/m^3 water, 1220 kg/m^3 coarse aggregate, 810 kg/m^3 fine aggregate, and 7 L/m^3 superplasticizer gave 8 MPa, 55 MPa, and 80 MPa compressive strengths at 1, 28, and 182 days, respectively. This innovative, high-volume application of fly ash unquestionably signals the best value-added use of the material in the construction industry.

If complete use of all coal ash is our goal, then we must also identify the segments of the construction industry which can profit from the use of nonstandard fly ash and pulverized bottom ash. A coal ash may not meet certain minimum chemical or physical requirements prescribed by a standard specification, but it may still be useful on account of its fine particle size. Examples of these applications are cited next.

Roller-Compacted Concrete Dams: Since the 1980's, roller-compacted concrete (RCC) has been accepted worldwide as the most rapid and economical method for construction of medium-height dams. According to Dunstan (5), until the end of 1992, 96 RCC dams had been built in seventeen different countries and 82 of the RCC mixtures contained a pozzolan. The high-paste-content RCC mixtures typically contain 250 kg/m^3 cementitious material of which 70-80% is a pozzolan. Fly ash has been used as a pozzolan in 90% of the 82 RCC mixtures. The Upper Stillwater Dam in the U.S. contains 1.24 million m^3 of concrete with 79 kg/m^3 portland cement and 173 kg/m^3 fly ash. In all, over 200,000 metric tons of low-calcium fly ashes from six different power plants were used. Imagine the volume of pozzolanic materials needed for the Zungeru Dam in Japan which contains 5 million m^3 RCC, and the 217-m high Longton Dam in China that will contain 7.5 million m^3 RCC. Further, according to Dunstan (5), even nonstandard fly ash is being successfully used as a component of RCC mixtures. For instance, the RCC mixture for the construction of 95-m high Platanovryssi Dam in Greece contains 35 kg/m^3 portland cement and 250 kg/m^3 of treated (pulverized and hydrated) fly ash which has an unusually high calcium content (42% total CaO). The fly ash is generated from thermal power stations using lignite as fuel.

Concrete Pavements for Highways: According to Golden (6), approximately 70% of the low-volume highways and local access roads in the U.S. require upgrading. Considering the cost savings resulting from the replacement of cement with high volumes of fly ash, EPRI (Electric Power Research Institute) funded several demonstration projects. In North Dakota, during the summers of 1988 and 1989, 20,000 m^2 of a 200-cm thick concrete pavement was constructed with "pozzocrete," which is a 0.43 w/c, air-entrained concrete mixture containing 100 kg/m^3 portland cement and 220 kg/m^3 high-calcium fly ash. Demonstration projects in Kansas have successfully used both low-calcium and high-calcium fly ashes in concrete pavement mixtures (10-20% fly ash by mass of concrete). An innovative feature of this project was the use of the crushed concrete from the old pavement as a source of coarse aggregate in the concrete mixture for the new pavement.

An additional but important approach toward the goal of resource conservation is through the enhancement of durability of concrete structures. This topic is discussed separately below.

4.2 Enhancement of durability of concrete structures

As stated above, the earth's natural resources are conserved when the service life of a manufactured product is prolonged. Recently, numerous materials and methods have been developed for enhancement of durability of concrete structures. Due to the high cost of materials and complex technologies involved, they have found limited applications. Ordinary concrete is the material of choice for construction because it is a relatively inexpensive product of simple technology. **The challenge, therefore, lies in making the ordinary concrete a highly durable, high-performance building material for the future structures.**

It is generally known that the major causes of deterioration of reinforced concrete structures are the corrosion of reinforcing steel, exposure to cycles of freezing and thawing, alkali-silica reaction, and sulfate attack. From a review of case histories of concrete degradation, the author developed a holistic approach encompassing the major causes of concrete deterioration (1). This approach is based on field experience that, with every one of these four causes of concrete deterioration, a high degree of water saturation is a prerequisite to the mechanisms responsible for expansion and cracking of concrete. Therefore, the water-tightness of concrete, which is its first line of defense against a hostile environment, must somehow become breached before the material is seriously damaged. This shows that, compared to other properties, the soundness of concrete, i.e. the freedom from cracking, is closely related to concrete durability. It seems that modern concrete construction practice does not pay adequate attention to the two primary causes of early cracking in concrete, namely, thermal contraction and drying shrinkage.

Driven by the high speed of construction, concrete mixtures today tend to contain a high content of normal or high-early strength portland cement. The extensibility or crack resistance of such concretes would be low due to increase in the drying shrinkage, thermal shrinkage, and elastic modulus on the one hand, and a reduction in the creep coefficient on the other hand. This is the reason that high-early-strength concrete mixtures are more vulnerable to cracking than moderate or low-strength concrete mixtures. Traditionally, structural cracking is controlled by the use of sufficient steel reinforcement but, as explained below, substitution of a few wide cracks with numerous invisible and unmeasurable microcracks is not a good solution for concrete durability problems.

The preceding theoretical considerations are confirmed by field experience. In 1995 the U.S. National Highway Cooperative Research Program conducted a survey of recently built concrete bridge decks. Noting that more than 100,000 bridge decks showed transverse cracks even before the structure was one month old, Rogalla et al. (7) drew the following conclusions:

1. A combination of thermal shrinkage and drying shrinkage caused most of the cracks, not traffic loads or vibration during the hardening of the concrete.

2. Generally, decks are made of high-strength concrete. These concretes have a high elastic modulus at an early age. Therefore, they develop high stresses for a given temperature change or amount of drying shrinkage, and most important, the concrete creeps little to relieve these stresses.

3. High-strength concretes typically contain more cement. Therefore, they shrink more and produce higher temperatures during early hydration. Modern cements are apt to cause cracking because they are finer and contain higher sulfate and alkali contents.

In conclusion, according to the holistic approach to concrete deterioration, a well constituted and properly consolidated and cured concrete will remain essentially watertight as long as the pores and cracks present in the interior do not form an interconnected network of pathways leading to the surface. Structural loads as well as weathering effects such as exposure to cycles of heating-cooling and wetting-drying facilitate the propagation of microcracks that normally pre-exist in the transition zone between the cement mortar and coarse aggregate in concrete. This happens during the first stage of the structure-environment interaction. Once the watertightness of concrete has been lost, it can become saturated and harmful ions also can move into the interior. This marks the beginning of the second stage of the structure-environment interaction during which the deterioration of concrete takes place through successive cycles of expansion, cracking, loss of mass, and increase in permeability.

Based on the holistic approach to concrete deterioration, a two-stage model of the damage process (Fig. 3) was proposed by the author (1). Such models cannot be used for making exacting predictions of the service life of concrete structures. Nevertheless, they can be helpful in decisions regarding cost-effective strategies for prolonging the service life of concrete exposed to aggressive environments. When thermal cracking and durability are of primary concern, experience shows that the most cost-effective solution is the replacement of a part of the portland cement in a concrete mixture with fly ash or slag, while meeting the setting and hardening requirements of the job under given ambient conditions. Concrete mixtures containing fly ash or slag tend to have stronger transition zone (the interfacial zone between the cement paste and particles of coarse aggregate) are less prone to microcracking and, therefore, influence durability of the structure through prolonged watertightness, namely through their effect on Stage 1 of the damage model in Fig. 3. To enhance the service life of a reinforced concrete structure, the use of corrosion-inhibiting admixtures, epoxy-coated reinforcing steel, external coatings for concrete, and cathodic protection of the structure are not only much more expensive but also offer a limited extension of service life through their effect on Stage 2 of the damage model. The results of an unpublished study by the author comparing the relative impact of each of the recent technological advancements on the concrete industry as a whole are summarized in Table 2.

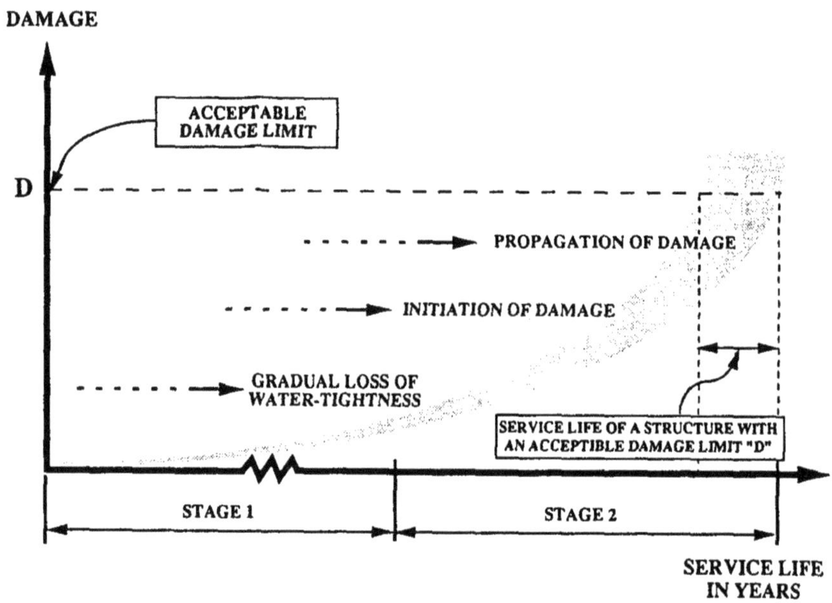

Fig. 3. A two-stage model of concrete deterioration (1).

The foregoing comments on enhancement of durability of concrete structures are applicable to new construction. In the interest of cost economy and conservation of resources, enhancement of service life of existing concrete structures by repair and rehabilitation should not be ignored. Due to the tremendous growth of the concrete repair industry in recent years, it is not possible to cover the subject in this paper. Nevertheless, it may be noted that superplasticized shotcrete containing fly ash and silica fume, external coatings for concrete and cathodic protection of structures are being increasingly considered for **the enhancement of service life of existing structures exposed to severe environmental conditions.**

Table 2. Suggested ratings for recent advancements in concrete technology.

	Complexity of the technology	Initial cost of materials and construction	Life-cycle cost	Environmental friendliness of the product	Future impact on the concrete industry
Macro-Defect-Free Cement Pastes and Mortars	High	High	High	Poor	Negligible
Chemically-bonded Ceramics	High	High	Unknown	Poor	Negligible
Reactive Powder Mortars	High	High	Unknown	Poor	Negligible
Superplasticized, Concrete with or without Silica fume	Moderate	Moderate	Low	Moderate	Moderate
Self-Compacting Concrete	Moderate	Moderate	Unknown	Moderate	Moderate
Superplasticized, High-Volume Fly Ash Concrete	Low	Low	Low	**Excellent**	High
Superplasticized, High-Volume Slag Concrete	Low	Low	Low	**Excellent**	High
Corrosion-Inhibitors	Moderate	High	Unknown	Poor	Unknown
Epoxy-Coated Reinforcement	High	High	High	Poor	Unknown
Surface Coatings for Concrete	High	High	High	Poor	Unknown
Cathodic Protection of the Structure*	High	High	High	Poor	Unknown

*This technology has proven to be effective for extending the service-life of chloride-contaminated reinforced-concrete structures in moist environment. If the concrete can be completely dried, and kept dry during the remaining service life, it would clearly be a simpler and less expensive alternative.

4.3 A holistic approach to concrete technology research and field practice

Among the experienced researchers in the area of concrete durability, there is now an increasing appreciation of the value of a holistic approach to concrete technology research and field practice. As discussed below, the prevailing reductionistic approach is, in fact, responsible for many wasteful practices in concrete technology today. According to this approach, all aspects of a complex system can be fully understood and controlled by reducing it to parts and by considering only one part at a time. As a result, test methods for concrete durability, materials specifications, and codes of recommended practice have failed to consider that durability is a holistic (pertaining to the whole structure) performance criterion which must take into consideration many factors including weather conditions, structural design, and concrete processing.

On the basis of his wide experience with concrete deterioration due to alkali-silica reaction (ASR), in his recently published book Idorn (8) has put together a very interesting account of the rise and fall of six decades of concrete technology research which, in his opinion, has become gradually infected with reductionistic philosophy and laboratory empiricism. According to him, approximately forty test methods have been developed during the course of research on ASR, but none of them including ASTM Methods C 289 and C 227, can establish whether a reactive aggregate will cause harmful or harmless reaction if used in field concrete. In spite of the fact that a harmless reaction has predominated in field practice, the adoption of **a no-risk ASR policy** in the U.S. has led to rejection of high-alkali cements and many deposits of aggregates which were found to be reactive in laboratory testing. On the other hand, smaller countries like Denmark and Iceland, where low-alkali cements were not available and reactive aggregates were plentiful, have successfully pursued **a low-risk ASR policy** according to which a pozzolanic admixture (viz., calcined clay or silica fume) was incorporated into concrete containing a high-alkali cement and a reactive aggregate. **The low-risk ASR policy is clearly holistic because it discourages materials waste and promotes the use of industrial by-products, such as silica fume and fly ash, to enhance concrete durability.**

Similarly, in a lucid review of alkali-aggregate attack in concrete, Swamy (9) stated, "to attack and cause damage, **all three members of the triad must be present, namely: sufficient alkali in the concrete, critical amount of the reactive aggregate, and sufficient moisture.**" He also observed that, in a moist environment, the swelling pressure of ASR gel is unlikely to create structural distress in a well-designed reinforced concrete member. The economic and ecological implications of Swamy's conclusions are apparent. For example, it is not necessary to reject high-alkali raw materials for cement making, or reactive aggregates for making concrete mixtures, provided the concrete structure remains dry during its service lives, and it is properly reinforced. This is a good illustration of a holistic approach to durability which considers climatic conditions and structural design in addition concrete-making materials, mix proportions, and processing methods.

A highly controversial topic in concrete technology today is the damage caused by the DEF phenomenon (delayed ettringite formation). Using the holistic approach, Collepardi (10) has concluded that **a high risk of damage due to the DEF would occur only if all three of the following conditions are present together: late sulfate release, microcracks in concrete, and exposure to water.** Again, the economic and ecological implications of this conclusion are profound. For example, for the production of portland clinker, it is not necessary to stop the use of secondary fuels, such as old automobile tires and petroleum coke, which usually contain a high sulfur content. From economic and ecological standpoints, a preferable solution is to reduce the chances of excessive cracking and microcracking, and of subsequent water penetration during the service life of a concrete made with a portland cement containing higher than normal sulfate content.

It may be noted here that a holistic approach should not be confused with a systems approach, which is commonly practiced in the resolution of complex problems. An integrated approach considering both structural design and structural durability aspects is a good example of a systems approach. However, it is not comprehensive enough to be called holistic. **The holistic approach has its roots in the idea that the whole exists before the parts.** For instance, the holistic approach would consider society as a whole, and the concrete industry as a part of the whole. Therefore, in addition to providing a low cost building material, the concrete industry must assume responsibility for other society needs—for example, conservation of the earth's natural resources and safe disposal of polluting wastes produced by other industries, as discussed above. **In short, if sustainable development is a wheel, and conservation of concrete-making materials and durability are spokes, then the holistic approach in concrete technology is the kingpin or the interlinking hub of the wheel.**

A difficult question to answer is: how can we accomplish a paradigm shift to holistic from the currently prevailing reductionistic practices in the industry? To develop a holistic field practice for the concrete industry, first the concrete technology research must become holistic, and the concrete technology research will not become holistic without a major transformation in the mindset which guides today's engineering education in general, and concrete science education in particular. Clearly, to develop a holistic concrete technology, the reform process will have to begin from the universities. In a personal communication, Gunnar Idorn seems to concur. He says:

> "Partly due to financial starvation cement and concrete research in the universities today is predominantly reductionistic which is low cost, of limited value in regard to technology innovations, unattractive for the best students, and makes journal and conference papers that are unexciting and full of fragmentary information."

It is hard to say when and wherefrom the financial and human resources will come for development of holistic education and research in concrete technology. One thing seems certain. **As long as the concrete community continues to pursue reductionistic solutions to major technological and societal problems confronting us, it will neither attract highly motivated researchers nor the financial support needed for sustainable development. Change will eventually come as a consequence of visionary leadership from within the concrete community, and public pressure from outside.**

5 Concluding remarks

In a paper presented at the Mohan Malhotra Symposium on **Past, Present, and Future of Concrete Technology** in San Francisco in 1994, the author (1) briefly discussed the powerful forces shaping tomorrow's world. The forces discussed were population growth and urbanization, pollution and waste management, technology innovations and globalization, and a paradigm shift in scientific research. The theme of this paper, namely the foundations needed for sustainable development of the concrete industry in the 21st century, has provided an opportunity to elaborate on the ideas expressed earlier. The goal of sustainable development will not be easy to reach, but it has to be pursued in response to the challenges created by the combination of above-mentioned forces. The author hopes that the ideas presented in this paper will stimulate rigorous debate on some of the issues raised here, and will prove helpful in ushering the concrete industry into an era of sustainable development during the 21st century.

6 References

1. Mehta, P.K., "Concrete Technology at the Crossroads — Problems and Opportunities," *American Concrete Institute, SP-144*, pp. 1-30, 1994.

2. Mehta, P.K., "Bringing the Concrete Industry into a New Era of Sustainable Development," *Proceedings of the Mario Collepardi Symposium on Advances in Concrete Technology*, pp. 49-68, 1997.

3. Schiessl, P., and Hohberg, I., "Environmental Compatibility of Cement-Based Building Materials," ibid., pp. 27-48, 1997.

4. Malhotra, V.M., "CANMET Investigations Dealing with High-Volume Fly Ash Concrete," *Advances in Concrete Technology, Natural Resources, Ottawa, Canada*, pp. 445-482, 1994.

5. Dunstan, M.R.H., "Future Trends in Roller-Compacted Concrete Dam Construction," *American Concrete Institute, SP-144*, pp. 307-324, 1994.

6. Golden, D.M., "U.S. Power Industry's Activities to Expand Coal Ash Utilization," *Proceedings of the International Workshop on Utilization of Fly Ash*, Seoul, Korea, 1997.

7. Rogalla, E.A., Kraus, P.D., and McDonald, D.B., "Reducing Transverse Cracking in New Concrete Bridge Decks," *Concrete Construction*, Vol. 40, pp. 735-738, 1995.

8. Idorn, G.M., *Concrete Progress from Antiquity to the Third Millennium*, Thomas Telford Publishing, London, U.K., 395 pages, 1997.

9. Swamy, R.N., "Alkali-Aggregate Reaction — The Bogeyman of Concrete," *American Concrete Institute, SP-144*, pp. 105-140, 1994.

10. Collepardi, M., "A Holistic Approach to Concrete Damage Induced by Delayed Ettringite Formation," *Proceedings of the Mario Collepardi Symposium on Advances in Concrete Technology*, pp. 373-396, 1997.

THE ROLE OF VOLUNTARY INTERNATIONAL ASSOCIATIONS IN PROMOTING A SUSTAINABLE DEVELOPMENT IN CONCRETE

Role of voluntary international associations

J. MOKSNES
Jan Moksnes Consulting, Stavanger, Norway

Abstract
A number of international associations are active in the field of promoting research and development and in the dissemination of knowledge related to concrete and concrete structures.

To a large extent these associations are run on small budgets and rely on the voluntary service of enthusiastic individuals to execute their work and promote their interests. Their activities include the publication of Codes of Practise, Recommendations, Technical Reports and Journals as well as the organisation or sponsorship of Congresses or Symposia where engineers and scientists get together to share their knowledge and experience. Most of us belong to one or several of these associations and have served on their committees.

The paper describes how some of the international associations are organised and interact with their members and their users and discusses how they can continue to play important roles and sustain future developments in concrete in a changing world. By and large the international concrete associations are not well known outside their own circle of supporters and in view of their strong technical basis they could do well to re examine their Modus Operandi to ensure a more efficient interaction with society and their ultimate clients.
Keywords: Voluntary organisations, concrete networks, interactive loops, team work.

1 Introduction

Most of the papers at this workshop are of a technical nature and are presented by specialists in the field of technology. This is what one would expect, particularly in view of the agenda which emphasises the service life of concrete structures and the growing impact of environmental issues.

My contribution on this occasion is of a different nature and looks at the way we organise our work in the professional organisations that we belong to and use for improving our skills and disseminating our knowledge. In our day to day activities we focus on technical and scientific developments in the design offices, the laboratories and in the field and we are continuously developing our ability to design and construct more durable and serviceable structures. Most people will readily agree that in general good people can achieve more collectively, working as a team, than they can by the sum of their individual efforts. Hence, in the concrete industry, like most other industries, people with common interests seek together in groups or associations where they meet to produce papers, learn from each other and collectively address current issues. This can take the form of local or national groups like the national

Concrete Technology for a Sustainable Development in the 21st Century edited by O.E. Gjørv and K. Sakai.
Published in 2000 by E & FN Spon, 11 New Fetter Lane, London EC4P 4EE, UK. ISBN: 0 419 25060 3

concrete societies or professional associations, or international bodies with a much wider membership. Most of these associations are run by a small full time staff and are supported by individuals who work on a voluntary basis and put in massive efforts in terms of committee work and the production of technical papers. This impressive voluntary effort demonstrates a keen devotion to the concrete profession and constitutes an invaluable resource which we must continue to nourish and to utilise.

The output from the most prestigious associations often constitute the 'state of the art' and form the basis for National and International Codes and Recommendations. In my view, however, the role of the voluntary associations and their potential impact on current technology is not always fully utilised.

Most of us work for or have worked for some of these associations and contributed to their progress. My own background includes committee membership end executive positions in National and International bodies and participation in their Symposia and Congresses. My general experience is this:

- The concrete industry has a number of dedicated and enthusiastic people who will contribute their time and knowledge to voluntary professional work.
- These people and their sponsors get together or publish their work in journals or proceedings that are read by and adopted by the concrete profession. Seldom do we talk to or reach the decision-makers (politicians, authorities, planners, owners), our competitors or our critics.
- The work of the international concrete associations tends to overlap with each other and to focus on technical issues rather than 'people' or 'society' issues.
- Important events frequently do not get the attention or participation they warrant due to over bookings of events in the calendar.
- There appears to be an increasing problem attracting younger engineers and scientists into active participation in the work of the voluntary associations.
- Engineers and scientists are almost exclusively preoccupied with the technical and scientific aspects of their work and seldom engage in business and marketing activities. We need to realise that we belong to an industry selling materials, concepts and structures on a competitive basis.

Concrete is the most widely used construction material in the world and capable of great achievements. Never the less, concrete is subject to controversy in many respects, including aesthetics, durability, serviceability and environment aspects, which are indeed the issues which will be addressed at this workshop. I believe that the voluntary international associations can play a bigger role in solving some of these global problems and improving technologies than they do at present. I also think that they can provide a unique platform for young engineers and scientists for promoting their views, building networks and participate in knowledge sharing and teamwork.

2 The organisation of the international voluntary associations

2.1 General

The organisational models adopted for the different associations are quite similar and consist essentially of a small permanent staff supporting a governing and an executive body headed by an elected President. The all important technical work is done by Committees, Commissions or Work Groups with a given mandate or scope of work and are headed by a Chairman appointed by the executive body. The membership of the committees tends to be a mixture of nominated, appointed or co-opted members, all with the appropriate experience and serving for a specified term. Through membership of a Committee, Commission or Work Group, the National Groups or individuals can have a profound influence on the direction of the work and the associated out put in papers and reports, some of which become Codes and Recommendations adopted by the industry. The sponsors or fee paying membership of the associations themselves are normally a mixture of National Member Groups, Industrial or Corporate Members and Individuals.

Most of the international associations are now accessible on the Internet and detailed information on their activities and publications can be found there. Some of this information is presented below:

2.2 *fib (Federation Internationale du Beton)*

fib was created this year (1998) through the merger of FIP* and CEB** realising that the two associations to a large extent drew on the same limited resources and that together they could better serve the needs of the industry and of their sponsors.

**FIP (Federation Internationale de la Precontrainte) was inaugurated in England in 1952 for the development of new techniques and knowledge concerning prestressed concrete and subsequently widened to embrace the development of structural concrete. The principal members of FIP were the National Member Groups existing in about 40 countries and the work of the Commissions focused particularly on practical applications of technology. All the work of FIP was done on a voluntary basis and supported by a small part time secretariat in London.*

***CEB (Committee Euro-International du Beton) was set up in 1953 to co ordinate and initiate research for problems related to technology and innovation in structural concrete in post war Europe. A need was soon identified for the harmonisation of the principles and rules for the conception, calculation, construction and maintenance of concrete structures, resulting in the development of International Recommendations and later Model Codes which have been widely adopted. CEB had National Member Groups in about 30 countries plus a number of Corporate and Individual members. The work of CEB was also done purely on a voluntary basis, supported by a small full time secretariat in Lausanne.*

The fib Mission Statement reads: "To synthesise research findings, technical information and practical experience and to translate the results for application through dissemination in various ways, thus contributing to the improvement of the technical performance, aesthetics and economy of concrete structures and the promotion of the role of the civil and structural engineers who are responsible for them. The means of dissemination include the preparation of appropriate guidance documents, approved through specific procedures, and the organisation of International Congresses, Workshops and Symposia."

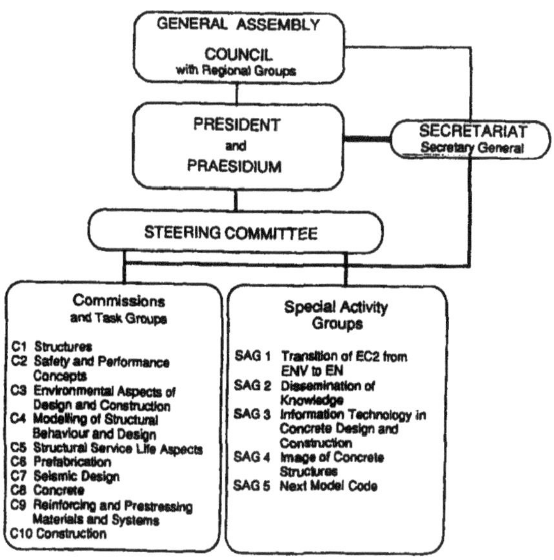

Fig. 1. *fib* Organisation

The fib organisational model, which in many respects is similar to that of some of the other associations, is shown in Fig. 1 in order to illustrate some general points.

The General Assembly consists of members from the National Groups and is the governing body which elects the Presidium and the Steering Committee and approves formal documents. The Council which consists of Heads of National Delegations, approved budgets and accounts, debates policy issues and co ordinates the national contributions. The Presidium serves as an executive committee and is responsible for the organisation of congresses, marketing of publications, the response to technical or political issues arising from time to time and the co ordination of activities with other bodies. The Steering Committee monitors the work of the Commissions and Special Activity Groups, proposes their Chairmen for approval by the Council, suggests new activities, guarantees the quality of the fib published documents and assists with editorial work.

The Commissions have a responsibility to monitor and collect ongoing work and developments in its specific area of interest, to synthesise available and developing information into suitable documents and to react on actual problems in the industry recommending appropriate actions. The current Commissions (10) and Special Activity groups (5) are shown in Fig.1 and it will be evident that they all have fairly broad mandates. Smaller, dedicated Task Groups are therefore commonly appointed within Commission to deal with specific topics. It is through the Commissions or Task Groups that the specialist researcher or engineer can contribute his knowledge, be stimulated through a network of people with common interests, and contribute to the advancement of his subject for the benefit of the profession.

2.3 RILEM (The International Union of Testing and Research Laboratories for Materials and Structures)

RILEM is a non profit making non governmental association founded in Paris in 1947 and providing an international network of communication throughout 70 countries, linking up some 1200 specialists from research and industry with a view to promote progress in the design, testing, manufacture and use of building materials. RILEM works through a number of Technical Committees which are groups of experts and of which some deal directly with cement and concrete, others deal with timber, masonry, composites and others. With respect to the agenda for this workshop it is interesting to note that the current RILEM technical programme includes: - Recycling of building and waste materials - Performance, maintenance and repair of building materials and structures during the service life – and Impact of building materials on the environment and on the health of human beings.

2.4 IABSE (The International Association for Bridge and Structural Engineering)

IABSE is a non profit making scientific society established in 1929 to promote the international exchange of knowledge in the field of structural engineering. It has more than 4000 members in over 90 countries and the secretariat is located in Zurich. IABSE deals with all aspects of structural engineering and all types of structures composed of any structural material. It achieves its objectives through organising or sponsoring conferences, publication of a quarterly Journal, and the production of publications and the work is performed on a voluntary basis through a number of Standing Committees, Commissions and Working Groups.

2.4 ACI (American Concrete Institiute)

ACI is the veteran of the international associations and was established in 1904. It has some 20 000 members worldwide and is organised in 61 Chapters of which 43 are in the US. It is a technical and educational society dedicated to improving the design, construction, manufacture and maintenance of concrete structures. ACI works through a large number of Standing Committees and has set up about 100 Technical Committees which operate under 5 main headings: Research and administration – Materials and properties of concrete- Design and construction practices – Structural analysis – Special products and services. ACI has published over 400 technical standards, reports and manuals and organises and sponsors conventions and conferences on most aspects on concrete.

2.5 Others

There are several other international associations which take an interest in concrete and concrete structures. There is also a Liaison Committee of International Associations of Civil Engineering which was formed in 1961 to co ordinate activities, to arrange joint activities and to harmonise activities and publications for greater efficiency and awareness of the roles of the different members. It is not the intent of this paper to deal with all international associations but rather to discuss with reference to selected examples their role and their inter reaction with their members and with society.

3 The concrete networks

The relationship between the individual engineers and scientists, their employers, national associations and, finally, the international umbrella associations is illustrated in Fig. 2.

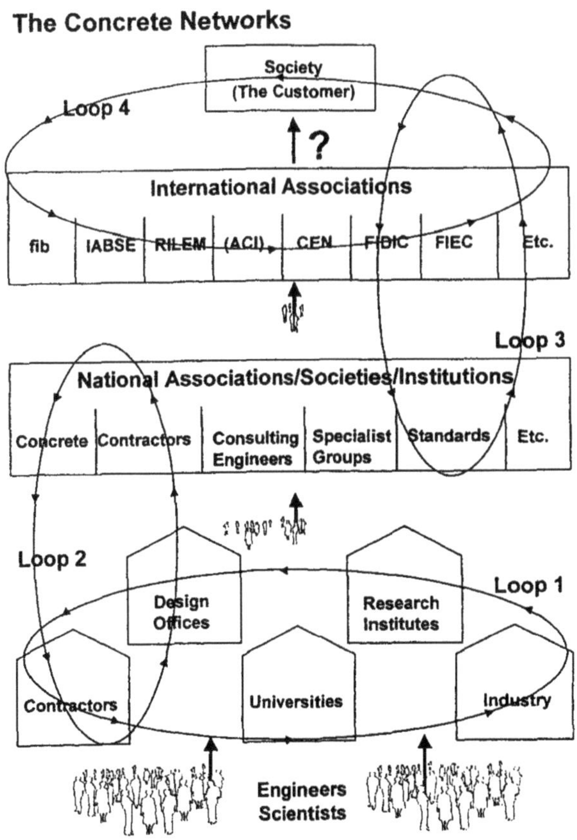

Fig. 2. The Concrete Networks.

Four Network loops are shown to illustrate the main connections between people, institutions and associations and the interaction which takes place on a day to day and case to case basis. The 4 loops have the following characteristics:

Loop 1: The vast majority of engineers operate throughout their career in this loop which embraces their place of work and their relations with colleagues in other companies and other disciplines. Typically, the work is related to projects which require different skills and learning and development goes on within and between the different entities. This loop is driven by schedules and budgets and success is measured on results and profits and by promotion of the individuals.

Loop 2: This loop embraces the National Associations which companies, institutions and individuals join and support through activities on committees and commissions and through membership fees and grants. The National Associations join together people with similar interests, be it technical or commercial, and sometimes membership constitutes a qualification which indicates a professional level of proficiency. National Codes of Practise and guidelines or recommendations for commercial and technical issues are developed within this loop. Some political or commercial lobbying may take place but most of the activities are of a technical or administrative nature.

Loop 3: This loop brings in the voluntary international associations and their inter relations with their Member Groups. The Member Groups, which are normally a relevant National Association, nominate representatives on to Councils and Committees and individuals are elected, nominated or drafted on to Commissions and Task Groups where, together with other experts in the same field, they deal with current technical issues and produce papers and reports. The integrity of these documents is protected by the integrity of the groups of international experts involved and by the procedures adopted for their review and approval. These documents sometime constitute the basis for subsequent national Codes or guidelines. The uniqueness of this loop is the way it involves a very broad membership of voluntary experienced people in a close network which is driven by technical rather than profit motives. Commercial interests in the form of promoting concrete and concrete structures through product developments and improved efficiency is on the other hand a driver also in this loop. The loop offers great opportunities for engineers to contribute to the work and to benefit from the dissemination of the collected knowledge. The main challenge is to perform the work on a shoe string budget and at the same time do it efficiently so that also younger people (volunteers) with less funds behind them may participate.

Loop 4: Loop 4 is in my opinion the problem and the main challenge when examining the role of the voluntary international associations. The problem is on two levels,

- the interaction and efficient co operation between the different associations, and
- the interaction with society (politicians, planners, authorities, customers, media).

The first problem of cooperation is a management issue and progress has been made, through efforts like the Liaison Committee and more frequent co sponsorship of events. Closer co operation and perhaps mergers between some of the most closely associated bodies will help, but is not easy to achieve. The recent merger between CEB and FIP took six years to negotiate and resolve.

The second problem is more difficult and more pressing. It is a fact that when concrete people get together in committees, for meetings or for Congresses and Symposia they talk to themselves and very seldom are 'society' or 'people' issues addressed. When we have successes to report we do not utilise the power of the media.

Some examples may serve to demonstrate how the Commissions of the international associations can be utilised when the needs arise. In 1972 FIP set up a Commission on Concrete Sea Structures which already in 1973 published the first edition of the "FIP Recommendations for the design and construction of concrete sea structures". This document was subject to revisions in 1974, 1977 and 1985 and helped to convince a very steel minded oil and gas industry that concrete structures could be used safely for offshore oil and gas production, thus giving rise to a whole new industry. When alarm bells ring , like in the case of the 1992 UK ban on post tensioned grouted bridges and the devastation following the 1995 Kobe earth quakes, the teams of experts already assembled in the relevant FIP Commissions could have been readily available task forces. Similarly, in relation to the agenda at this workshop there are many issues concerning concrete which are more than just technical and which will need more dialogue between society and the technologists. The international associations should pay attention and take initiatives in this direction.

Loop 4 is the least efficient and the most under utilised in the concrete networks. As a profession we may be competent and in many respects successful, but society's perception of us, our material and our structures is not very clear or very satisfactory. We should do something about it.

The 4 loops and their current rating, in the view of this speaker, are illustrated in Fig. 3.

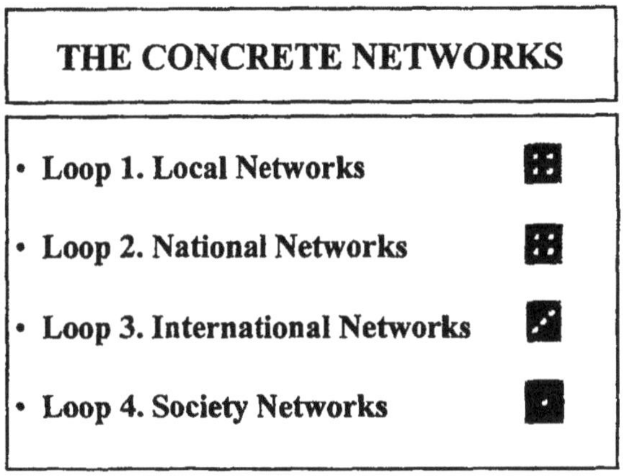

Fig. 3. The Network Loops
(Rated on a scale from 1 to 6)

4 Conclusions

The voluntary international associations are important elements in the concrete network. They play a most important part in bringing together experts who collect and disseminate state of the art technical knowledge and produce recommendations and guidelines for the industry in the spirit of knowledge sharing. Through these documents we ensure that progress is utilised and implemented.

The voluntary international associations provide platforms for individuals within the profession for their own advancement through active and continuous work in the Commissions and Task Groups. They also offer unique opportunities for individuals in the profession to participate in successful teams and to benefit from the strength of team work.

Other initiatives, like the present Workshop, clearly have their own value but must be recognised for what they are: in depth internal discussions between colleagues on important specific subjects.

The real challenge to us all is to go from here with all the good ideas we have assembled and make sure that we bring this knowledge into a wider context and to a wider audience. We must encourage our best talents to participate not only in the customary and probably more comfortable Local and National networks (Loops 1 and 2), but also include the more demanding International network (Loop 3) which we can all access through the voluntary international associations. Most of all we must unify our efforts and as a profession pay more attention to Society (Loop 4), or the outside world of customers, politicians, planners, decision makers and so on, and pay attention to their concerns and demands and their perception of our industry and our concrete structures.

IS HIGH STRENGTH CONCRETE DURABLE?

S. P. Shah, K. Wang , and W. J. Weiss
NSF Center for Advanced Cement-Based Materials
Department of Civil Engineering
Northwestern University
Evanston, Illinois, USA

Abstract

A long-term investigation is currently being conducted at the NSF Center for Advanced Cement-Based Materials (ACBM) to examine the long-term durability of concrete. This paper provides an overview of recent results to illustrate the influence of mixture proportioning on permeability, mechanical properties, and early-age shrinkage cracking. The permeability study investigated a large array of concrete mixture proportions with various aggregate volume fractions, gradings, and maximum sizes for several different water-to-cement ratios (w/c). The effects of dilution, tortuosity, interfacial transition zone, and percolation on the permeability of concrete are discussed. In addition, a relationship between crack width and water permeability of concrete was investigated. Restrained shrinkage testing has been performed to understand the influence of mechanical properties on early-age cracking, especially for mixes with chemical and mineral additives. Results are provided to illustrate the influence of the water-to-cement ratio, silica fume replacement, latex, and a shrinkage reducing admixture; on strength, permeability, and restrained shrinkage cracking. These results illustrate that a specification based solely on strength may not ensure a durable concrete. It is proposed that three criteria: strength, permeability, and cracking resistance; are necessary to design a durable concrete.

Keywords: aggregate volume fraction, cracking, high strength concrete, latex, mix design criteria, normal strength concrete, permeability, shrinkage, silica fume, water-to-cement ratio

Concrete Technology for a Sustainable Development in the 21st Century edited by O.E. Gjørv and K. Sakai.
Published in 2000 by E & FN Spon, 11 New Fetter Lane, London EC4P 4EE, UK. ISBN: 0 419 25060 3

1. Introduction

Originally, high strength concrete (HSC) was specified for use in high-rise buildings and long-span bridges because it had reduced weight and increased load carrying capability. Recently however, use of HSC has spread to underground mass transit facilities, airport facilities, waste water treatment systems, offshore oil platforms, and marine structures; fueled primarily by durability concerns. Research has identified that the majority of concrete durability problems are related to the resistance of concrete to permeation of water and chemical ions, examples of which include: corrosion of reinforcement, freezing-thawing damage and alkali-silica reaction. The mechanisms responsible for increasing the strength of HSC typically also decrease permeability. As a result, it becoming increasing commonly to find HSC specified when durability is the primary concern. However, does the use of HSC always result in more durable structures? The use of HSC has been accompanied with new problems associated with increased cracking sensitivity in structures with a large surface area. This forces us to question whether a criterion based on either strength or permeability is enough for the specification of durable concrete.

This paper presents an overview of results from several ongoing ACBM research projects dealing with concrete permeability, mechanical properties, and restrained shrinkage cracking. The paper relates the permeability of concrete mesostructure and crack characteristics. It also illustrates the shrinkage cracking potential for mixes composed with different water-to-cement ratios, silica fume, latex, and a shrinkage reducing admixture (SRA) on concrete properties. The research suggests that atleast three criteria: strength, permeability, and cracking resistance; should be considered simultaneously in concrete mix design to achieve durable concrete.

2. Concrete permeability

2.1 Permeability and mixture proportions

The permeability of concrete is a key factor which influences the durability of concrete. Concrete permeability is dependent on the permeability of its each constituent material and their geometric arrangement. The permeability of cement paste is influenced primarily by pore structure (porosity, size, and connectivity), which is a function of the water-to-cement ratio (w/c) and the degree of hydration. The aggregates which are commonly used in concrete typically have a much lower permeability than the cement paste. However, the aggregates influence the permeability of the composite concrete in four ways: dilution, tortuosity, interfacial transition zone (ITZ), and percolation (Figure 1) [1]. A description of each effect is provided in the following section.

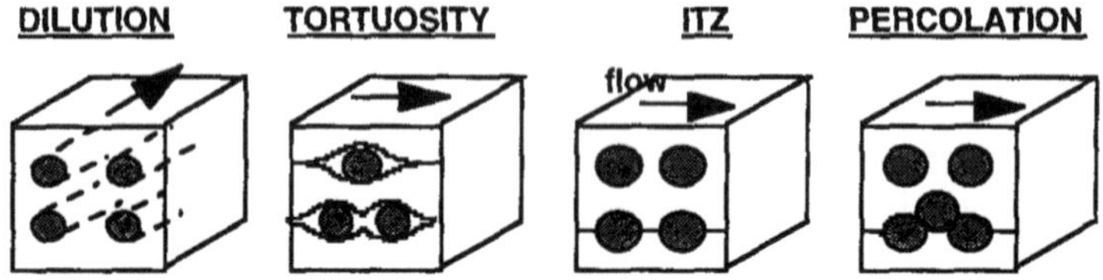

Fig.1. Effects of Concrete Mesostructure on Permeability

The dilution effect occurs because the aggregates are less permeable than the paste. As a result, the aggregate particles block the flow paths, effectively reducing the permeable area in a cross-section of concrete. The degree of dilution depends on the actual aggregate permeability and volume fraction (V_F). If we assume the aggregate is impermeable and a linear parallel ion flow exists, the conductivity of concrete (σ), which is directly associated with permeability, can be represented using the equation: $\sigma = \sigma_p(1-V_F)$, where σ_p is conductivity of cement paste, and V_F is aggregate volume fraction [2].

The tortuosity effect occurs as a result of the impermeability of the aggregates which forces flow around the aggregate particles; thus increasing the length of flow paths and reducing the flow rate. The degree of tortuosity is dependent on aggregate size and spacing. Combining the tortuosity effect with the dilution effect, the conductivity of concrete can be expressed by the Bruggeman equation [3]: $\sigma = \sigma_p(1-V_F)^{3/2}$.

The interfacial transition zone (ITZ) which develops between the aggregate and concrete also influences permeability. The ITZ properties are primarily influenced by the aggregate and w/c of the cement paste. The volume of ITZ is related to the aggregate volume fraction while the shape, size, and orientation of the aggregate particles influence the thickness and strength of the ITZ. To account for the influence of the ITZ, Garboczi et al. [2] developed an analytical expression (***m***). The ***m*** term is a function of aggregate size (α), cement paste conductivity (σ_p), ITZ thickness (h_{ITZ}), and ITZ conductivity (σ_{ITZ}). After incorporating the ITZ factor, the concrete conductivity can be expressed as $\sigma = \sigma_p(1 \pm m \bullet VF)$.

The term percolation describes the flow path which connects the ITZ's. The degree of percolation depends primarily on the aggregate volume, size, and spacing. Winslow et al. [4] reported that percolation occurs in mortars when the sand volume fraction reaches 0.5. Mclachlan et al. [5] modified the Bruggeman theory in a model for the conductivity of a composite: $\sigma = \sigma_p(1-VF/VF_c)^{m \bullet VF_c}$, where VF_c is the volume fraction at percolation and ***m*** is the same as above.

The dilution and tortuosity effects reduce concrete permeability while the ITZ and percolation effects increase permeability. However, if these factors are combined it becomes unclear as to which factor is the most dominant in concrete. To further explore the relationships between concrete mix proportion, mesostructure, and permeability; ACBM researchers have investigated a large array of mix proportions which cover a large range of aggregate volume fractions ($0\% < V_F < 78\%$) without having considerable segregation [1]. Four different gradings, two different maximum sizes of aggregates, and three different w/c (0.38, 0.45, and 0.52) were used. The rapid chloride permeability test was used as a measure of permeability. Figures 2-4 [1] describe the relationship between w/c, aggregate volume fraction (V_F), compressive strength, and rapid chloride permeability. Currently, the actual structure of these concretes is being investigated using image analysis techniques in addition to a three-dimensional reconstruction model. Using this type of an approach will make it possible to assess the influence of coarse aggregate volume fraction, total aggregate surface area, coarse and fine aggregate ratio, location of aggregate particles, and distance between aggregates. This approach is being combined with statistics to develop a model to predict the permeability of concrete.

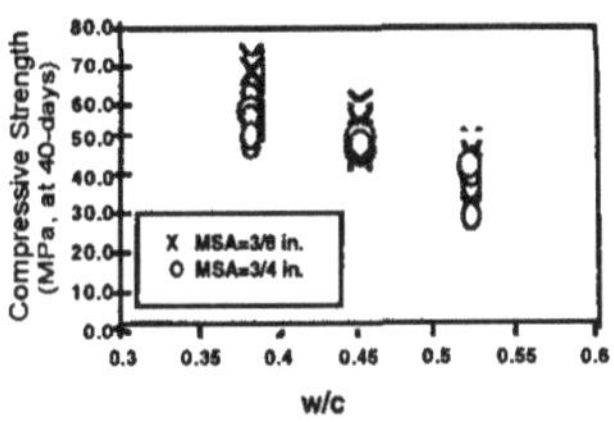

Fig. 2. Relationship Between the W/C and Concrete Strength

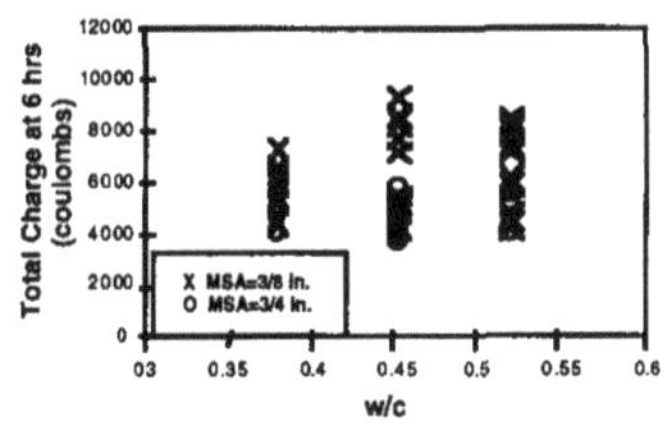

Fig. 3. Relationship Between the W/C and Concrete Permeability

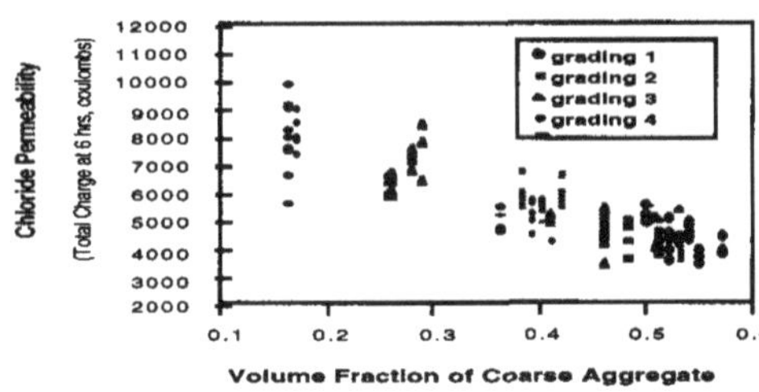

Fig. 4. Effect of Volume Fraction of Coarse Aggregate on Permeability

From these graphs, the following observations can be made:

(a) concrete compressive strength exhibits a linear relationship with w/c (Fig. 2),
(b) no direct relationship exists between rapid chloride permeability and w/c (Fig. 3),
(c) a linear relationship exists between coarse aggregate volume fraction and rapid concrete permeability (Fig. 4).

These results imply that the aggregate volume fraction is more directly related to permeability than the w/c for NSC. This suggests that using the w/c as the sole mix design criterion may provide the desired strength but not necessarily the desirable level of permeability. Using linear regression, the rapid chloride permeability of concrete can be approximated using the following equation [1]:

$$\log(Q) \cong 9.0 + 0.435(w/c) - 1.55(Vf)$$

in which **Q** is the total charge passed in coulombs, ***w/c*** is the water-to-cement ratio and (***Vf***) is aggregate volume fraction. This equation illustrates that aggregate volume fraction has a more dominant influence on concrete permeability than the w/c. The maximum size of aggregate and location of concrete specimen exhibit a little effect on the permeability values, while aggregate grading appears to be insensitive [1].

Figure 4 implies that neither the w/c or ITZ had a strong effect, however, the dilution effect appears to strongly influence the rapid chloride permeability measurements. A similar trend was observed for mortar specimens in which the normalized conductivity decreased linearly with the sand volume fraction, indicating that the dilution effect dominates permeability [6] (Figure 5). The normalized mortar conductivity is defined as the ratio of the mortar conductivity and the cement paste conductivity, $\sigma_{mortar}/\sigma_{paste}$. To investigate the effect of ITZ, a computer model simulation was conducted by using a factor K_{appa}. Here, $K_{appa} = [\sigma_{mortar}/\sigma_{paste}]/[(1-V_{f,\ sand})^{3/2}]$ and $K_{appa} = 1$ indicates no appearance of ITZ effect. Figure 6 shows that ITZ affects mortar conductivity in two different ways. One effect comes from the high porosity of the ITZ, which causes K_{app} to increase with sand volume fraction. The other effect results from the change of w/c in bulk cement paste. Because of the ITZ, where w/c is higher than that designed for mortar, the w/c in bulk cement paste becomes lower than designed for mortar. Thus, $K_{app} = \sigma_{bulk}/\sigma_{paste}$ decreased with sand volume fraction. Preliminary results suggest that the two effects are similar in order of magnitude, yet opposite in sign. Therefore, they offset and no ITZ effect appears in the final conductivity result (Figure 6).

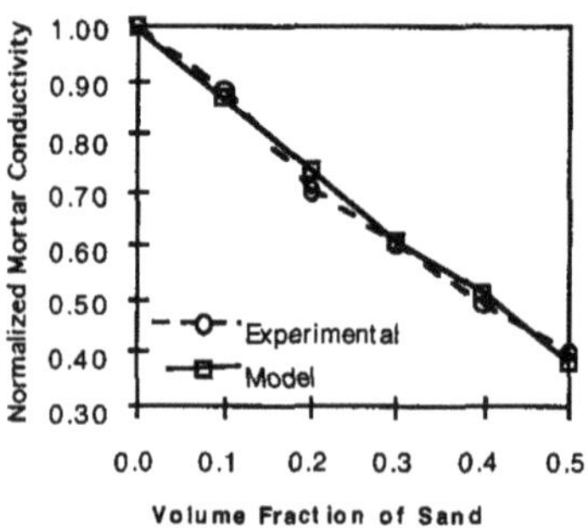

Fig. 5. Effect of Volume Fraction of Sand on Conductivity of Mortar (w/c=0.4) [6]

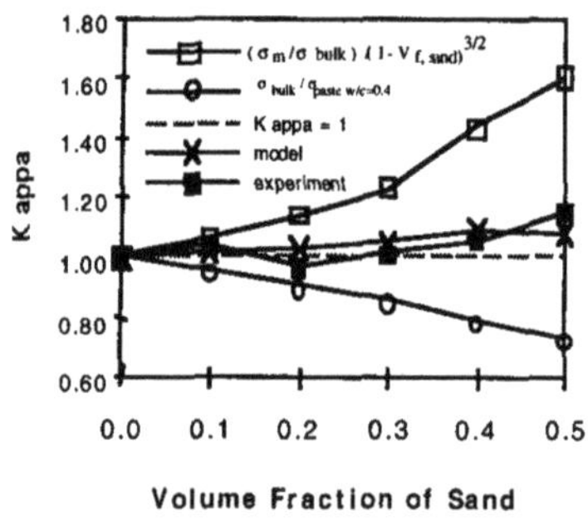

Fig. 6. Effects of Tortuosity and ITZ on Conductivity of Mortar [6]

It is likely that the ITZ effect may be reduced in HSC as a result of improved aggregate interface properties. In addition, the effects of dilution, tortuosity, and percolation will be reduced as the permeability of cement paste approaches that of aggregate. As a result, the permeability of concrete is probably best controlled by controlling the permeability of the cement paste. The work of Powers and Brownyard [7] demonstrated that a decrease in the w/c was accompanied by lower porosity. As the porosity decreases, the pore size decreases and the pores disconnect which results in a decrease in permeability. The addition of mineral admixtures, especially silica fume, can improve both the pore structure and ITZ; resulting in a drastic permeability reduction. For example, consider three concrete mixes: w/c = 0.5, w/c = 0.29, and a 15% silica fume replacement with a w/b = 0.29 (water-to-binder ratio in which binder is cement + dry silica fume)[16]. As the w/c is decreased from 0.5 to 0.29, the rapid chloride permeability decreases from 11,322 to 2,967 coulombs. Replacing 15% of the cement with silica fume results in a further reduction of the rapid chloride permeability from 2,967 to 254 coulombs (more details are provided in section 4).

2.2 Permeability and crack width

Using the considerations mentioned above, a well proportioned concrete mix can be designed to have a low permeability. However, concrete is one of the few materials which is typically designed to crack in normal loading and service conditions. Additional cracking often develops due to improper curing, finishing, and placement techniques. Cracks in concrete provide a path, which can accelerate the ingress of corrosive agents which accelerate the deterioration of the structure. However, difficulties associated with generating the desired cracking patterns and determining an appropriate method to measure permeability have limited research in the permeability of cracked concrete. Frequently, the maximum allowable crack width is based on past experience and general 'rules-of-thumb' for different applications [8].

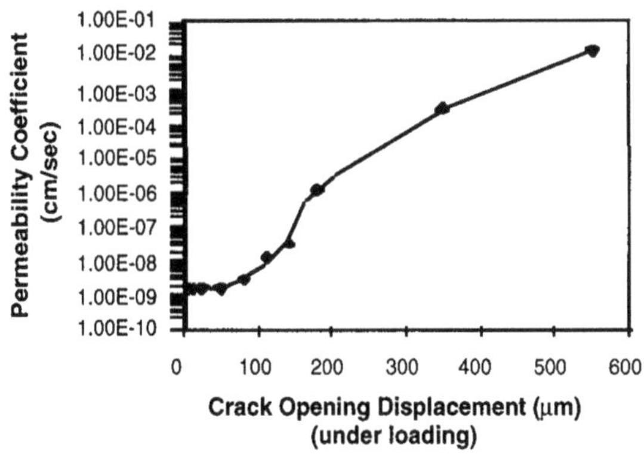

Fig. 7. Effect of Crack Width on Concrete Permeability [9]

Researchers at ACBM [9] have used closed-loop testing to control the size of a crack which develops in a tensile splitting test. By controlling crack width the specimens can be unloaded and the water permeability of the cracked concrete sections can be measured. The relationship between water permeability and crack opening is plotted in Figure 7. It can be seen that if the crack opening displacement is less than 50 microns the crack had little effect on the water permeability coefficient. However, as the crack opening displacement increased from 50 microns to approximately 200 microns, the water permeability increased rapidly. Cracks greater than 200 microns, appear to have a steadily increasing water permeability rate. More recently a study was conducted comparing NSC and HSC [10]. After cracking, the difference in permeability values for NSC and HSC becomes less significant.

3. Early age cracking in concrete

Cracking frequently develops in concrete structures as a result of restrained shrinkage. Shrinkage describes a volume contraction which occurs as a result of moisture loss or temperature reduction. If this movement is prevented by restraint, tensile stresses develop. Cracking occurs when these stresses exceed the material's resistance. Research at ACBM has focused primarily on shrinkage which occurs as a result of moisture loss. Structures with a large surface area to volume ratio are particularly sensitive to drying shrinkage since they lose water the most rapidly. Examples of such structures include: highway pavements, industrial floors, parking garages, and bridge decks.

The free shrinkage strain (without restraint) of concrete typically ranges between 400 and 1000 x 10^{-6}, depending on the mix composition. If we compute the stress that develops as a result of restraint by using the free shrinkage strain, elastic modulus, and a strength of materials approach ($\sigma = \varepsilon E$); the stresses developed typically exceed the tensile strength of concrete; however, cracking may not occur. This difference can be explained by creep relaxation which reduces the stresses over time. The actual stress level in concrete is the difference between the tensile stress induced by drying shrinkage and the stress relief due to creep as shown in a simulation for NSC and HSC (Fig. 8).

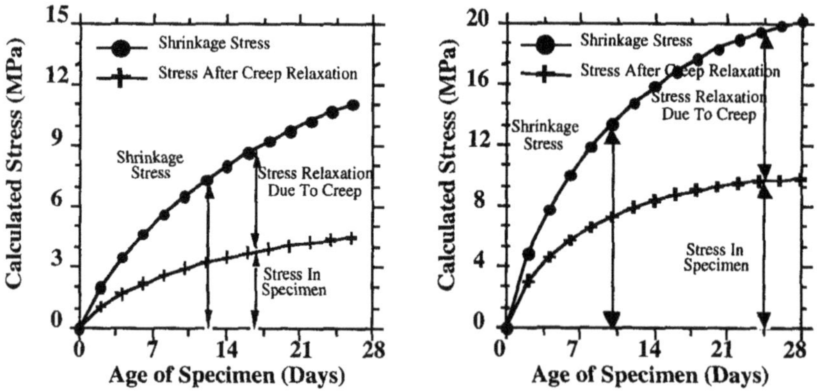

Fig. 8. Simulation of Stress Development for NSC and HSC Considering Creep Relaxation [12]

Research at ACBM is focused on developing both experimental procedures and theoretical models to assess the potential for shrinkage cracking. Restrained ring and slab specimen geometries have been used in the experimental program. The ring test provides a high degree of restraint while eliminating the effects of end conditions, thereby simulating an infinite strip of concrete [11]. The slab specimens can be used to simulate a slab of finite length with or without subgrade friction [12]. Theoretical modeling consists of two main components: stress development and failure analysis. A coupled shrinkage and creep formulation has been developed to calculate the internal stresses which develop over time [13]. The failure analysis is based on a fracture mechanics failure criterion which relates the materials resistance (R-curve) to the energy supplied by shrinkage to drive the crack (G-curve). Failure is assumed to occur when two conditions are satisfied: an energy balance (G = R) and an equivalent energy rate for the same crack length ($\partial G/\partial(\Delta a) = \partial R/\partial(\Delta a)$)[14]. Results of this model compare favorably with experimental observations [14,15,18].

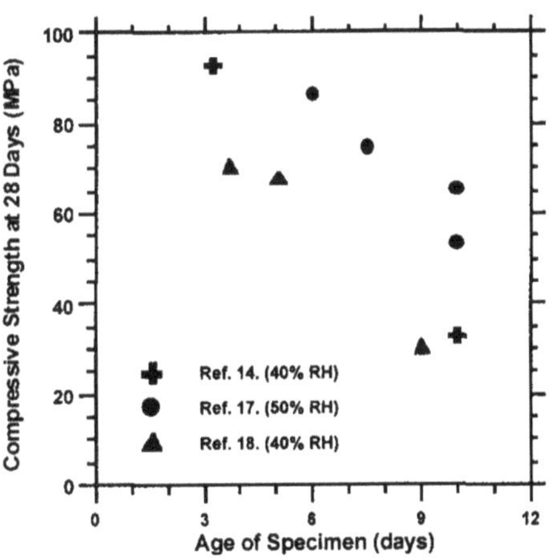

Fig. 9. The Relationship Between 28 Day Comp. Strength and the Age of Cracking

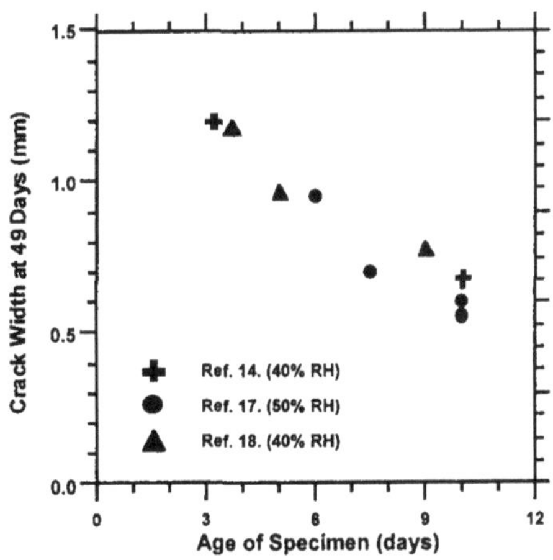

Fig. 10. Relationship Between Age of Cracking and Crack Width

Figure 9 has been developed to illustrate the influence of increased compressive strength on the age of first cracking using the results of several recent studies [14,17,18] (note: latex and SRA mixes do not appear). Specific details on the mix constituents, mix proportions, and curing conditions are available in literature. A trend exists to imply that an increase in the strength of concrete results in an increased potential for early age cracking. Possible explanations for the increased shrinkage cracking potential may include increased brittleness, increased microcracking, decreased creep relaxation, and higher autogenous shrinkage. As illustrated in section 2, the width of the crack which develops plays a key role in determining the permeability. An optical microscope was used to investigate the width of the shrinkage crack. Figure 10 illustrates that wider cracks result when the specimen cracks at an earlier age.

4. Influence of mix design on strength, permeability, and shrinkage cracking

The compressive strength and water-to-cement ratio (w/c) of concrete have been widely utilized in international specifications for structural design, mix design, and quality control during construction. This often misleads engineers into thinking that strength and w/c are sufficent parameters to specify the performance of concrete. Actually, the majority of concrete structures failure as a result of durability related problems rather than insufficient strength. Recently, there has been a shift in concrete mix design from a strength approach to an integrated approach which involves both strength and durability. As stated earlier, the durability of concrete is closely related to its resistance to permeation of water and chemical ions; therefore, permeability of concrete is often used as an index for durability of concrete. Since HSC generally demonstrates an increased strength and a decreased permeability, it is thought to be a durable material. However, as discussed in section 3, HSC may not be durable due to a high risk to early age shrinkage cracking.

A recent study demonstrated that admixtures can significantly alter the performance of concrete without changing the w/c or strength. Six mixture proportions (Table 1) were selected to illustrate the effect of w/c (0.5 and 0.29), a styrene-butadiene latex, a shrinkage reducing admixture (SRA), and silica fume replacement. Each mix used identical aggregate type, grading, and volume fraction. The experimental measurements included mechanical properties, rapid chloride permeability, and early age shrinkage cracking. Further information on the details of this study is available in literature [16].

Table 1. Concrete Mixture Proportions and Material Properties [16]

	w/c=0.5	w/c=0.5 2% SRA	w/c=0.29	w/c=0.29 2% SRA	w/c=0.29 latex	w/b=0.29 silica fume
Cement	1.00	1.00	1.00	1.00	1.00	0.85
Coarse aggregate	2.00	2.00	1.52	1.52	1.74	1.57
Fine aggregate	2.00	2.00	1.52	1.52	1.74	1.57
Water	0.50	0.48	0.28	0.26	0.28	0.28
Superplasticizer (solid)	-	-	0.018	0.018	-	0.026
SRA	-	0.02	-	0.02	-	-
Silica Fume (solid)	-	-	-	-	-	0.15
Latex (solid)	-	-	-	-	0.10	-
Water/binder ratio	0.50	0.50	0.29	0.29	0.29	0.29
Aggregate volume (%)	65%	65%	65%	65%	65%	65%
Compressive strength at 28 days, dry (Mpa)	30.4	30.5	68.1	59.1	47.2	70.8
Compressive modulus at 28 days, dry (Gpa)	22.6	21.8	34.5	32.3	26.4	32.3
Rapid chloride perm. (Total Charge, Coul.)	11,322	8,754	2,967	2,261	2,509	254

As anticipated, mixes with a lower w/c exhibited an increase in compressive strength. As previously discussed (section 2), a decrease in the w/c also corresponded to a reduction in rapid chloride permeability while the mix which contained silica fume showed the lowest permeability. Mixes with SRA and latex exhibited lower permeability than the reference mixes despite having similar or lower compressive strengths. Figure 11 shows the free shrinkage of the six mixes, while Figure 12 shows the age of first cracking for restrained shrinkage ring specimens. It can be observed that the shrinkage behavior of the mixes fell along two distinct trends. The mixes with a high initial rate of free shrinkage (w/c = 0.5, w/c = 0.29, w/b = 0.29) cracked the earliest, while the mixes with a lower initial rate of free shrinkage (SRA and latex) showed a substantial reduction in shrinkage cracking potential. The following section summarizes the influence of each mix constituent on the material behavior.

Water-to-Cement Ratio: When the w/c ratio was reduced from 0.50 to 0.29, the compressive strength of concrete approximately doubled, the elastic modulus increased by approximately 50%, and chloride permeability value decreased to nearly 1/3. Despite substantial differences in long term shrinkage, both w/c's had a similar initial rate of free shrinkage. The w/c = 0.29 cracked earlier than the mix with a w/c = 0.50.

Silica Fume Replacement: The mix which contained silica fume replacement (w/b = 0.29) demonstrated a lower initial strength and elastic modulus despite having a similar strength at 28 days. Despite having a similar initial rate of free shrinkage, the silica fume concrete had a lower ultimate shrinkage and cracked earlier than the mix without silica fume (w/c = 0.29). The replacement of silica fume showed a significant reduction in rapid chloride permeability (1/10).

Latex Modification: The addition of 10% latex resulted in a nearly 40% reduction in strength, a 20% reduction in elastic modulus, a 10% reduction in permeability, and a significant reduction in free shrinkage. The latex-modified concrete showed high resistance to shrinkage cracking.

Shrinkage Reducing Admixture: The use of a SRA significantly reduced free shrinkage for both water-to-cement ratios. The w/c = 0.5 mix showed similar properties with and without SRA, however, the addition of SRA reduced concrete strength and elastic modulus by approximately 15%. Both SRA mixes demonstrated a reduction in rapid chloride permeability and reduced potential for shrinkage cracking.

These results clearly illustrate that no absolute relationship exists between concrete strength, permeability, and cracking resistance. A HSC may not have desirable overall durability, despite having very high strength and low permeability. Excellent cracking resistance results from the combined effects of a low shrinkage rate, high creep relaxation, and high toughness.

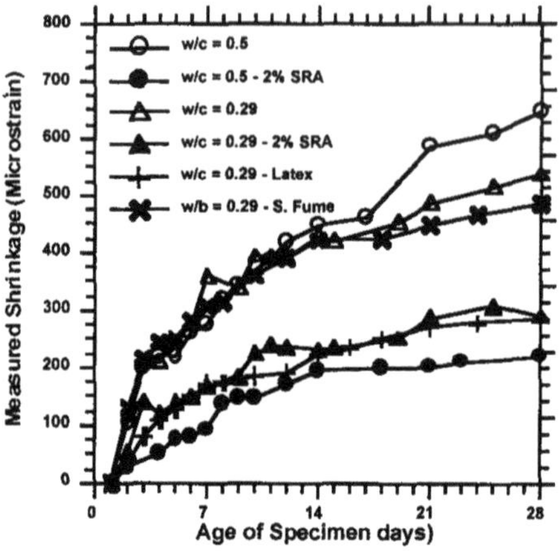

Fig. 11. Measured Free Shrinkage for Mixes from Table 1 [16]

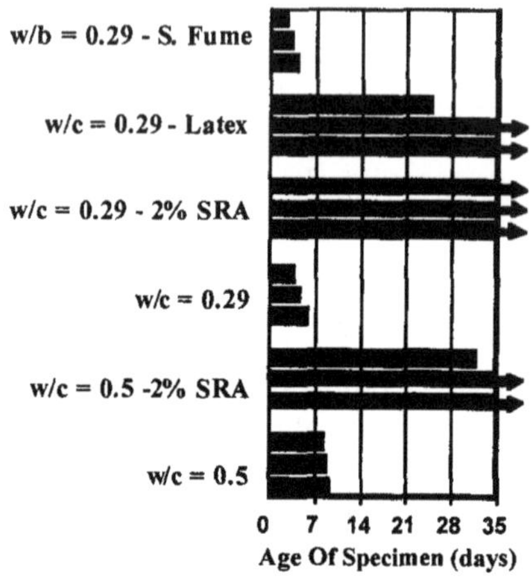

Fig. 12. Age of Cracking for Mixes from Table 1 [16]

Shah and Wang [17] proposed that to achieve a durable concrete, three criteria may need to be considered in concrete mix design. The three criteria are strength, permeability, and crack resistance. A strength criterion ensures that concrete is able to resist designed stress without failure. A permeability criterion ensures that concrete has a limited flow penetration rate, minimizing vulnerability to water/ion-related attacks during a designed period of service life. And a crack resistance criterion ensures that concrete has a minimum capability to resist the cracking due to environmental conditions, such as thermal and moisture shrinkage.

A great deal of work has been performed regarding strength and permeability criteria, however, limited work has been done regarding quantitatively describing the cracking resistance of concrete. With an increasing number of concrete structures exposed to aggressive environmental conditions and the desire for increase performance it is urgent that reliable criteria are established for the specification of concrete mixes with improved overall durability.

5. Concluding remarks

The present paper provides an overview of results from ongoing research projects being conducted at ACBM. The features of HSC and their influence on concrete performance is presented. The relationships between concrete mesostructure, permeability, cracking, and durability are discussed. The research results indicate that HSC, with a low w/c ratio and high cement content may be more susceptible to early-age cracks than traditional NSC. This implies that despite having higher strength and lower permeability, HSC may not have desirable overall durability. In addition, it is suggested that strength is not the optimal specification parameter when durable concrete is desired, especially when specialty mineral and chemical additives are used. It is proposed that three criteria should be considered to produce durable concrete, these factors include: strength, permeability, and cracking resistance.

Acknowledgment

The authors gratefully acknowledge the support of the National Institute of Statistical Sciences (NISS), ARCO Chemical Corporation, Grace Construction Products, a division of W. R. Grace Co. Conn., and the National Science Foundation Center for Science and Technology of Advanced Cement Based Materials. The work summarized here has been conducted by several ACBM researchers.

References

[1] Wang, K., Igusa, T., and Shah, S. P., "Permeability of Concrete -- Relationships to Its Mix Proportion, Microstructure, and Microcracks" (to be published in Sidney Diamond Symposium, Honolulu, Hawaii, August 30-September 3, 1998

[2] Garboczi, E., Bentz, D., and Landgren, E. S. "Multiscale Microstructural Modeling of Concrete Diffusivity: Identification of Significant Variables, " ASTM Cement, Concrete and Aggregate (in press)

[3] Bruggeman, D. A., Annals of Physics, 24, 636, 1935

[4] Winslow, D., "Percolation and Pore Structure in Mortars and Concrete," Cement and Concrete Research, 24, 25-27, 1994

[5] Mclachlan, D. S. "Electrical Resistivity of Composites," Journal of American Ceramic Society, 73, 8, pp. 2178-2203, 1990

[6] Shan, J and Mason, T. (private communication, Northwestern University, USA 1998)

[7] Powers, T. C., and Brownyard, T. L., "Studies on the Physical Properties of Hardened Portland Cement Paste", PCA Research Bulletin 22, March 1948

[8] ACI Committee 318, "Building Code Requirement for Reinforced Concrete (ACI 318-89)", Am. Concr. Inst. , Detroit, pp 106-108 (1989)

[9] Wang, K., Jansen, D.C., Shah, S.P., and Karr, A.F., "Permeability Study of Cracked Concrete", Cement and Concrete Research , Vol. 27, No. 3, pp. 381-393, 1997

[10] C. Aldea, S. P. Shah and A. Karr, "Permeability of Cracked Concrete" (submitted to Materials and Structures")

[11]Grysbowski, M, and Shah, S. P., "Shrinkage Cracking of Fiber Reinforced Concrete", ACI Materials Journal, Vol. 87, No. 2, pp. 138-148, 1990

[12]Weiss, W. J., Yang, W., and Shah, S. P., "Shrinkage Cracking of Restrained Concrete Slabs", ASCE Journal of Engineering Mechanics, to appear July 1998

[13]Yang, W., Wang, K., and Shah, S.P., "Prediction of Concrete Cracking Under Coupled Shrinkage and Creep Conditions", Proc. of the 4th Material Engineering Conference, ASCE, pp. 564-573, Nov. 10-14, 1996, Washington D.C.

[14]Shah, S. P., Swartz, S. E., and Ouyang, C., "Fracture Mechanics of Concrete: Applications of Fracture Mechanics to Concrete, Rock, and other Quasi-Brittle Materials", John Wiley and Sons Inc., 1995

[15]Wiegrink, K., Marinkunte, S., and Shah, S. P., "Shrinkage Cracking of High Strength Concrete", ACI Materials Journal, Vol. 93, No. 5, pp. 147-157, 1996

[16] Weiss, W. J., Schießl, A., Yang, W., and Shah S. P.,: "Shrinkage Cracking Potential, Permeability, and Strength of HPC: Influence of W/C, Silica Fume, Latex, and Shrinkage Reducing Admixtures" Int. Symposium on High-Performance and Reactive Powder Concrete, Sherbrooke, Canada, accepted for publication, August 16-20, 1998

[17] Shah, S. P., and Wang, K., "Microstructure, Microcracking, Permeability, and Mix Design Criteria of Concrete" The Fifth International Conference on Structural Failure, Durability and Retrofitting, pp. 260-272, Singapore, November 27-28, 1997

INTERNAL STRESSES AND LONG TERM PERFORMANCE OF CEMENTITIOUS MATERIALS

ARNON BENTUR
National Building Research Institute-Faculty of Civil Engineering
Technion, Israel Institute of Technology, Haifa, Israel

Abstract
The long term performance of cementitious materials is usually addressed from the point of view of physical and chemical processes that may occur over time and may lead to deleterious effects. Such influences can be considered as a material property and they can be evaluated by studying small specimens in the laboratory or in exposure sites. As a result, when new materials are evaluated for long term performance the tendency is to identify whether such physical and chemical processes may occur. However, attention should be given also to another type of effect which is associated with internal stresses that may develop in the material. Such stresses can be the result of autogenous shrinkage or environmentally induced shrinkage, which may induce relatively high stresses when the material is used to produce components which are restrained during their service. The present paper highlights such influences in two advanced cementitious materials: high quality concretes and fiber reinforced thin sheet cementitious composites. The magnitude of the stresses that might be developed and the need to deal with this problem combining the material and overall design points of view are highlighted.
Key words: internal stresses, autogenous shrinkage, cracking, restrain, high strength concrete, fiber reinforced cement.

1 Introduction

Long term performance of cementitious materials is usually addressed from a chemical and physical point of view, i.e. identifying chemical and physical processes that may occur over a long time span and lead to deterioration of the material. Examples of physical processes are freeze-thaw and abrasion whereas sulfate attack and alkali-aggregate reactions can be cited as examples of chemical processes. A common feature of these processes is that they might be considered as a material characteristic and thus can be readily evaluated by relatively simple tests in which a specimen is exposed to the deteriorating environment which accelerates such processes.

There is however another type of long term performance problem which is encountered more frequently in recent years and it tends to show up in some of the

Concrete Technology for a Sustainable Development in the 21st Century edited by O.E. Gjørv and K. Sakai.
Published in 2000 by E & FN Spon, 11 New Fetter Lane, London EC4P 4EE, UK. ISBN: 0 419 25060 3

more advanced cementitious materials. It is associated with the development of relatively large internal stresses due to excessive shrinkage. This is due mainly to autogenous shrinkage in high performance concretes (characterized by low water/binder matrix), or environmental induced shrinkage which is quite significant in various types of fiber reinforced cementitious thin sheet composites which are characterized by a high cement content matrix. The direct damage due to such influences is seen as macro- and micro-cracks, which by themselves may be unacceptable to the performance; they may also facilitate ingress of deleterious chemicals which may induce detrimental chemical processes.

There is inherent difficulty in dealing with this type of long performance problem, since it is not only a function of the materials property; the magnitude of the internal stresses depends not only on the extent of shrinkage (which is a materials property) but also on the overall design of the component (shape, size, bonding and connections to surrounding elements or structure). The overall design features and shape of the component may lead to restrain which is an essential characteristic needed for stresses to develop as an outcome of shrinkage. Thus, laboratory testing of such influences may be misleading, and there is a need to consider the materials properties as well as the overall design to prevent cracking due to such influences. This is frequently not taken into account due to a "gap" which exists between the materials engineer and the design engineer, and also lack of awareness to this issue on both sides.

The object of this presentation is to review some of the systems where such effects may occur, present data of the magnitude of stresses that may develop, and provide some insight as to the measures that might be taken to mitigate such influences.

2 High Performance Concretes

Modern high performance concretes (or high strength concretes as they are frequently called) are characterized by a low water/binder ratio, usually less than 0.35. Low water/binder ratio and sufficient workability are achieved by combining into the mix high range water reducers and various fillers, in particular microsilica. Self desiccation and resulting autogenous shrinkage and cracking in such systems is a phenomena that is well documented (e.g. references 1-4). The significance of this influence on potential early age cracking is appreciated by many practitioners. However, quantification of this effect to realize its potential hazard was limited mainly to studies of the shrinkage strains that may develop. This is a materials parameter, which can be readily evaluated in the laboratory. However, this characteristic can not provide sufficient quantitative insight as to the risk of cracking, and as a result it can not be developed into a parameter which the engineer can consider in the design of a component to prevent cracking due to self desiccation. Field experience has sometimes indicated erratic early age cracking in high strength concrete which was attributed to self desiccation (since it occurred even if proper curing was applied), but it was not possible to correlate it directly with this effect, because of the erratic occurrence.

With these limitations in mind, a number of groups (e.g. references 1,5,6,7) have started with fundamental investigations, to resolve the magnitude of stresses induced under restraining conditions, and develop quantitative tools and parameters to analyze the conditions for occurrence of cracks. These studies are based on restrained shrinkage tests of longitudinal specimens, as shown schematically in Fig. 1.

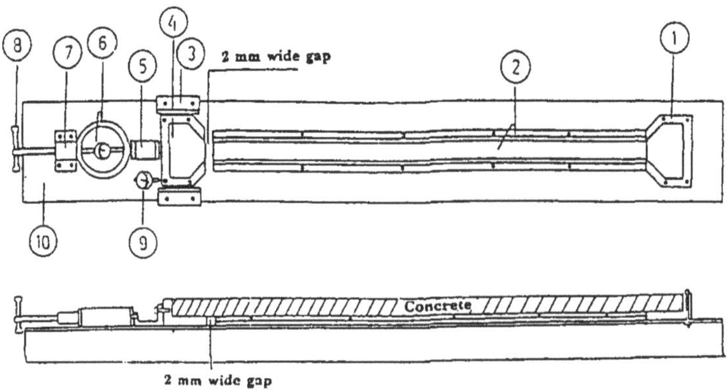

Figure 1. Schematic description of uniaxial restraining testing apparatus (after Bloom and Bentur (1)).

Instrumentation of this kind can enable to detect the stresses developed under particular restraining and environmental conditions. However, for in depth analysis, one must take into consideration not only the stresses which are induced by autogenous shrinkage, but also the relaxation (creep) which can occur at the same time. The creep relaxation reduces the stress that can be predicted to occur if one takes into account only the shrinkage and the modulus of elasticity of the material. For that purpose, improved uniaxial restrain testing methods were developed using instrumented and computer controlled systems (8). An overview of such systems is beyond the scope of this presentation and the reader is referred to references (1,3,5,8).

Early age cracking due to internal stresses has so far been given attention with regard to thermal effects induced by the exothermic hydration reaction. This is well documented and appreciated by designers, and various quantitative engineering design tools are available to deal with this issue. The risk for similar effects due to autogenous shrinkage is a relatively new phenomena characteristic to the high quality concrete, and therefore a question may be first paused as to its significance relative to thermal cracking. Springenschmid (5) compared between the two phenomena and concluded that in isothermal conditions the main reason for high restraining stresses in high performance concretes with silica fume is not heat of hydration but rather autogenous shrinkage.

Typical curves showing the autogenous shrinkage under free conditions and the development of stresses under restrained conditions of sealed high strength concretes are shown in Fig. 1. Similar range of stresses were reported in other studies (Table 1). The stresses which develop within the first week are in the range of 1 to 3 MPa, which can amount to more than 50% of the strength of the concretes during this age. The level of the autogenous shrinkage and the restrained stresses develop in conventional concretes of water/binder ratios greater than 0.5 is much smaller or non at all (1,10).

This high level of stress may obviously cause instability in the concretes and may lead to internal damage (microcracking) or even failure (macro-cracks), even if they are well cured. The extent of damage and the time of its occurrence (in the setting stage, i.e. plastic shrinkage cracking, or in the hardening stage, i.e. early age shrinkage cracking) will depend on a "competition" between several processes: the level of autogenous shrinkage (which increases with increase in silica fume and reduction in water/binder ratio - Fig. 1), the level and nature of restrain and relaxation effect (i.e. tensile creep, which may be significant in early age concrete). An analysis of the combination of these effects was presented by Kovler (8).

Table 1. Compilation of restrained shrinkage stresses reported in the literature.

Composition		Restraining Stress, MPa	Reference
water/binder ratio	silica fume content, % wt.		
0.30	0	2*	(5)
0.30	4	~2.2*	
0.30	8	~3*	
0.50	0	1.1*	(1)
0.33	0	2.0*	
0.33	15	2.3*	
0.51	0	0.84**	(6)
0.56	0	1.0**	

* autogenous shrinkage
** autogenous + drying shrinkage

The practical evaluation of this effect is usually based on tests to observe whether cracking has occurred (e.g. reference 4). This is of limited significance, as the observation is with regards to macro-cracking only, and not to damage that may be caused by invisible micro-cracks. Also the testing conditions, in particular the type of restrain, may not be representative of the actual conditions on site and in the structure. For example, field experience with high strength concretes sometimes indicate the occurrence of macro-cracks which seems erratic; i.e. for the same mix and conditions it sometimes occurs and sometimes prevented. The interpretation of this apparent inconsistent behavior may be readily explained on the basis of the magnitude of stresses developed (e.g. Fig.2, Table 1). The stresses at certain conditions may be as high as 50% of the strength or more, but not sufficiently high to cause cracking by themselves. If on site, an additional incidental stress is superimposed on the already high shrinkage induced stress (e.g. thermal stress, stress induced by deformation of the formwork), this addition may tilt the delicate balance and induce a crack. This implies that in order to deal with this issue systematically there is a need to consider in the design the stresses that may develop at early age, taking into account basic quantitative parameters of the material properties as well as the overall characteristics of the structure and the nature of restrain. At present, data of this kind and routine testing methods to evaluate it are not sufficiently available. Additional work such as in references 1,5,6,7,8 to develop testing methods and relevant data is highly recommended. In particular, attention should also be given to characterization of early age visco-elastic properties (e.g. reference 7). This should be complemented by the development of design methods which take into account these characteristics. This by itself is a challenge, since the methods will have to address the structure with input of material characteristics which are "dynamic" i.e., changing quite drastically within the period of the first week after casting.

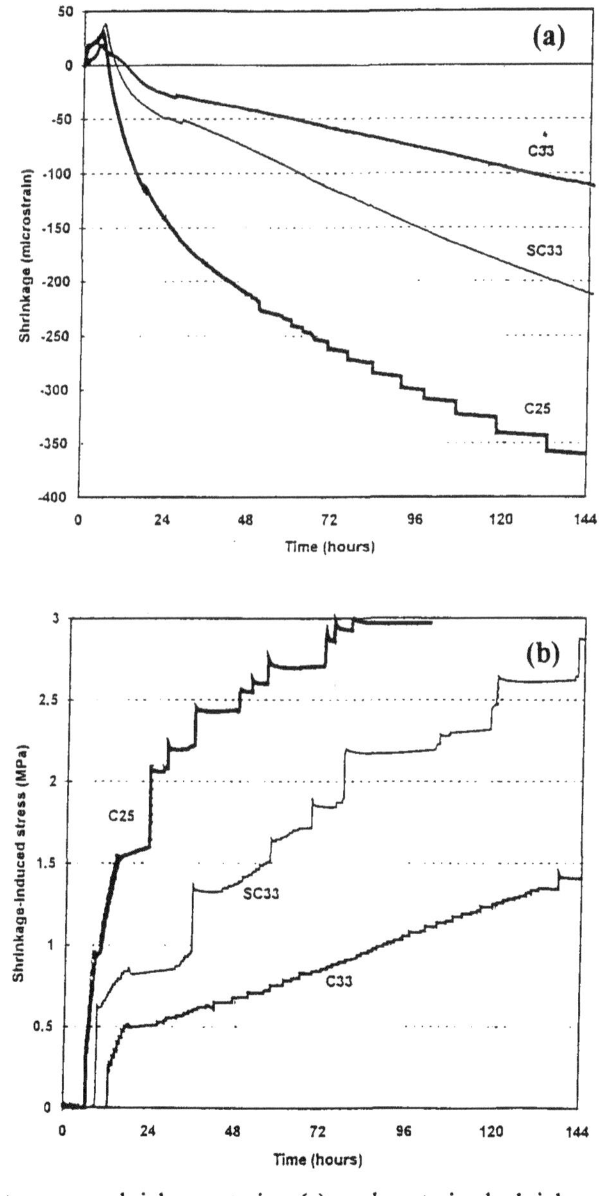

Figure 2. Autogenous shrinkage strains (a) and restrained shrinkage stresses (b) developed in high quality concretes (after Igarashi et al (9)).
C- concrete without silica fume; SC- concrete with 10% silica fume;
25- w/c=0.25; 33- w/c=0.33

In the context of this paper, the significance of these stresses and the impact of the damage associated with them on the long term performance of high strength concretes should be addressed. The presence of macro-cracks which may be generated by such stresses is clearly unacceptable. It may affect the stability of the structure and provide easy access to deleterious substances into the concrete and through the concrete. The influence of internal damage in the form of micro-cracks which may be induced by such stresses is not sufficiently resolved, neither the nature of the damage nor its influence. Relatively few studies have been advanced to characterized the long term performance of cracked concretes (e.g. references 11,12,13). Most of the investigations of permeability and diffusivity which provide essential input for calculation and prediction of service life deal with uncracked concrete. The issue of the influence of cracking and crack width on the corrosion of steel in concrete is still in debate. Yet, there is sufficient evidence to indicate that internal damage and microcracking certainly has an influence on the rates of deleterious processes. For example, the data of Sanjuan et al (13) indicates that restrained shrinkage induced cracks have an influence on the rates of corrosion of steel; their reduction by incorporation of fibers reduces these rates (Fig.3).

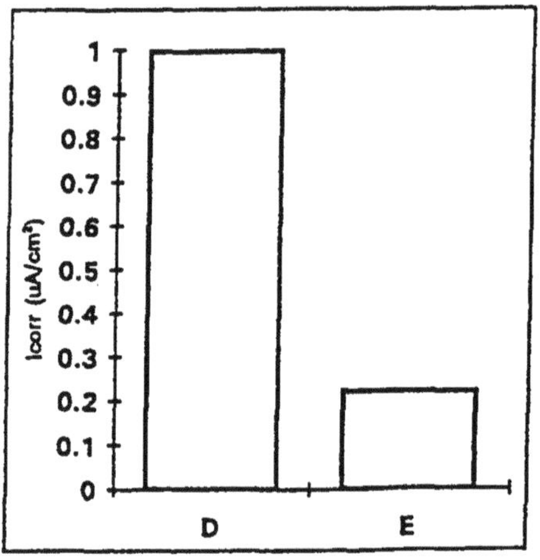

Figure 3. Steel corrosion rates of shrinkage restrained cracked specimens without polypropylene fibers (D) and with polypropylene fibers (E), after 25 days in chloride environment; cracking was smaller in the specimen with polypropylene fibers (after Sanjuan et al (13)).

3 Fiber Reinforced Composites

Shrinkage induced stresses and cracking are usually addressed in the context of fiber reinforced cements from two different points of view: (i) the introduction of fibers into concretes to reduce cracking, and (ii) performance of thin sheet cementitious composites which are used for wall and roof cladding. The first issue is well documented in numerous publications dealing with the phenomena itself and with the developments of various types of restrained shrinkage tests to evaluate the influence of the fibers (references 14-17). Here, the fibers can be viewed as additives incorporated in the matrix to mitigate cracking. The second issue, which is the one to be dealt with here is of a different type; the fibers in the thin sheet material are an essential part of the composite which possesses unique characteristics, one of them being high shrinkage/swelling. This is a separate class of advanced cementitious materials whose high shrinkage/swelling tendency deserves special attention with regards to long term performance associated with induced internal stresses.

The thin sheet cementitious composites are made usually with a low water/binder ratio matrix, and the content of the cementitious material (~ 1000kg/m^3) is much higher than in high performance concretes (~ 450 kg/m^3). These compositional characteristics as well as the high surface area of the component, make them quite vulnerable to autogenous as well as externally induced shrinkage. In practice, many of the long term performance problems of these composites have been cracking due to shrinkage induced stresses. Although this statement is clear to many practitioners, it is not obvious to many researchers, since most of the work published in the literature on the long term performance of thin sheet composites dealt with the chemical and physical stability of high performance man made fibers (e.g. glass, carbon) in the highly alkaline cementitious matrix (18,19). Yet, It should be borne in mind that assuring chemical/physical durability in the newly developed fiber and fiber-matrix combinations is a separate issue that will not necessarily mitigate the composite sensitivity to long term performance due to shrinkage induced stresses (e.g. reference 21). On the contrary, it sometime may even aggravate the problem, as is the case when cellulose pulp is used.

The basic material parameter to generate shrinkage stresses is the magnitude and rate of shrinkage. This can be evaluated by testing small coupons in the laboratory which is done routinely in various types of standard tests. A compilation of typical shrinkage/swelling data for thin sheet composites and its comparison with conventional concrete is provided in Table 2. The much greater values in the thin sheet materials are illustrative of the special nature of this effect in this type of composites, and the need to give it a special attention.

Table 2. Shrinkage/swelling stresses of different types of fiber reinforced cement thin sheets (after Baum and Bentur (20)).

Type of composite	shrinkage/swelling, %
concrete	0.05
asbestos-cement	0.09
glass-cement (GRC)	0.13
cellulose-cement	0.30
wood particle-cement	1.8

However, stresses and the resulting cracking are induced only when the material is restrained; therefore free shrinkage tests is not sufficient by itself to evaluate or predict the generation of stresses and cracking, although shrinkage is the driving force for it.

Restrain is a function of the size, the shape and type of connections of the thin sheet element to the structure. Significant differences exist between small sized slates and shingles and between big facade sheets. The strategy to deal with restraining effects is to provide flexible connections, which enable some freedom of movement in the plane of the facade. Examples are provided in Fig.4. In small sized slates and shingles, the recommendation of the producers should be followed, using their standard connection devices. These are usually based on "good practice" acquired from full scale laboratory tests, and experience gained in practice and in evaluation of performance in exposure sites. It should be emphasized that recommended practice for one type of material may not be adequate to another type of material due to differences in properties, in particular the shrinkage/swelling potential (Table 2). This remark is highlighted here, since there is a tendency by many inexperienced practitioners to refer to all thin sheet materials similarly.

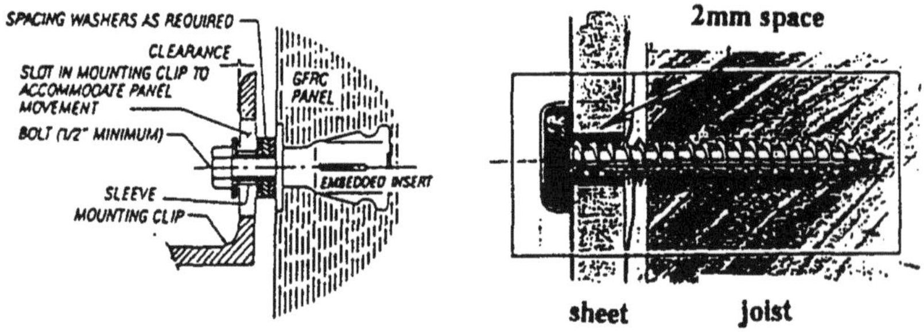

(a) (b)

Figure 4. Examples of connections allowing local flexibility recommended for use in (a) glass fiber reinforced panels and, (b) cellulose fiber reinforced panels.

Special attention should be given to large facades, where the restraining stresses are usually higher, due to size effects. These facades are in many instances custom made and are different one from another. As a result, there is a need here for the structural engineer to get involved in the design and construction. Recommendations and design procedures and parameters are available by the producers of the thin sheets, as well as in the literature (22-24). In these facades, special attention should be given to restrain which may be induced within a facade, regardless of its connections to the structure. This may be due to influences associated with the tri-dimensional geometry (i.e. special restraining effects may occur in an architectural panel which is not flat). Also, in the design it is essential to take into consideration variability in moisture between the internal and external parts of the facade, which may cause bowing and warping.

The strategy of producers of thin sheet materials to deal with these effects is first of all based on reducing the shrinkage. This can be achieved by controlling the composition of the matrix as well as application of various sealing methods to reduce water movements. Some thin sheet materials, like would chip reinforced cements, can not be applied externally unless they are combined with external protective layer of rendering, or even rendering system with some insulating boards. Additional strategy is to produce composites which exhibit pseudo-plastic behavior in the post cracking zone. This ductility can compensate for the high shrinkage by mitigating the restraining stresses with a material that can absorb considerable strain without exhibiting macro-cracks. This is an important characteristic of the material which should be considered in relation to the detrimental potential of high shrinkage. Although commercial thin sheet materials can be quite different with regards to this characteristic (Fig.5), this is frequently overlooked as the tendency in many specifications is to address only strength and shrinkage parameters (25). This is not only a limitation of specifications, but also the result of limited tools to evaluate the trade-off between shrinkage and ductility.

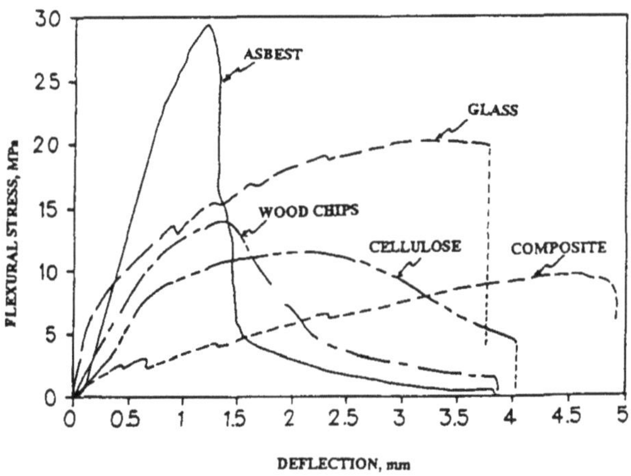

Figure 5. Load-displacement curves of different types of fiber reinforced thin sheet cements (after Baum and Bentur (20)).

The standards and specifications for thin sheet materials sometimes reflect these types of sensitivities, and put forward requirements for a full scale testing of a component which is connected to a structural frame. The component is exposed externally to cycles of wetting-drying-heating which induce moisture movement and generate restraining stresses (26). This is a full-scale accelerated test in which all of the above parameters and influences are combined. The evaluation of performance is qualitative or semi-quantitative; it is based on observation of damage (cracking, bowing, delamination, etc.).

4 Conclusions

1. Long term performance of cementitious materials may be detrimentally affected by internal stresses that are induced by shrinkage effects. Such effects can be particularly significant in advanced cementitious materials such as high quality (high strength) concretes and thin sheet fiber reinforced cements.

2. In high performance concretes considerable autogenous shrinkage develops, resulting in an induced stress of 1 to 3 MPa in fully restrained conditions. Such stresses

can lead to internal damage and cracking which may affect the long term performance of this material. This phenomena must be addressed on the material level as well as on the design level, since the magnitude of the stress will depend also on the size and the restraining conditions in the structure.

3. Considerable shrinkage can be induced in thin sheet fiber reinforced composites due to the low water/binder ratio of the matrix and the high content of the cementing material. Frequent long term performance problems in such components are the result of cracks induced in the structure due to restrained shrinkage which results in build up of tensile stresses over time. Such influences are of practical significance which matches or exceeds problems caused by chemical deterioration of fibers. It should be taken into consideration in the design of the shape of facades and their connections to the structure, to eliminate significant restrain.

5 References

1. Bloom, R. and Bentur, A., (1995) Free and restrained shrinkage of high strength concrete with and without silica fume, *Amer. Concr. Inst. Materials. J.*, Vol. 92, No. 2, pp. 211-217.

2. de Larrard, F. (1990) Creep and shrinkage of high strength field concrete, in *Utilization of High Strength Concrete*, 2nd International Symp., ACI SP-121, American Concrete Institute, Detroit, pp. 577-598.

3. Paillere, A.M., Buil, M. and Serrano, J.J. (1989) Effect of fiber addition on the autogenous shrinkage of silica fume concrete, *Amer. Concr. Inst. Materials J.*, Vol. 86, No. 2, pp. 139-144.

4. Folliard, K.J. and Berke, N.S. (1997) Properties of high performance concrete containing shrinkage-reducing admixtures, *Cem. Conc. Res.* Vol. 27, No. 9, pp.1357-1364.

5. Springenschmid, R. (1994) The influence of cement, pozzolans and silica fume on the cracking tendency of high strength concrete, in *Durability of High Performance Concrete*, (ed. H.Sommer), RILEM.

6. Altoubat, S.A. and Lange, D.A. (1997) Early age shrinkage and creep of fiber reinforced concrete for airfield pavement, in *Aircraft Pavement Technology*, (ed. F.Hermann), ASCE, 1997.

7. Bissonnette, B. and Piegeon, M, (1995) Tensile creep at early ages of ordinary, silica fume and fiber reinforced concretes, *Cem. Conc. Res.* Vol. 25, No. 5, pp.1075-1085.

8. Kovler, K. (1994) Testing system for determining the mechanical behaviour of early age concrete under restrained and free uniaxial shrinkage, *Materials and Structure*, Vol. 27, pp.324-330.

9. Igarashi, S., Bentur,A. and Kovler,K. unpublished results

10. Koenders, E.A.B. and van Breugel, K. (1997) Numerical modeling of autogenous shrinkage of hardening cement paste *Cem. Conc. Res.* Vol. 27, No. 10, pp.1489-1500.

11. Reinhardt, H.W., Sosoro, M. and Zhu, X. (1998) Cracked and repaired concrete subject to fluid penetration *Materials and Structure,* Vol. 31, No. 296, pp. 74-83.

12. Wang, K., Jansen, D.C., Shah, S.P. and Karr, A.F. Permeability study of cracked concrete, *Cem. Conc. Res.* Vol. 27, No. 3, pp. 381-393.

13. Sanjuan, M.A., Andrade, A. and Bentur, A. (1997), Effect of crack control in mortars containing polypropylene fibers on the corrosion of steel in cementitious matrix, *Amer. Concr. Inst. Materials Journal,* Vol. 94, No. 2, pp. 134- 141.

14. Shah, S.P, Karaguler, M.E. and Sarigaphuti, M. (1992) Effects of shrinkage reducing admixtures on restrained shrinkage cracking of concrete, *Amer. Concr. Inst. Materials. J.,* Vol. 89, No. 3, pp. 289-295.

15. Kovler, K., Sikuler, J. and Bentur, A. (1993) Restrained shrinkage tests of fibre-reinforced concrete ring specimens: effect of core thermal expansion, *Materials and Structures,* Vol. 26, pp.231-237.

16. Banthia, N., Azzabi, M. and Pigeon, M. (1993) Restrained shrinkage cracking in fibre-reinforced cementitious composites" *Materials and Structures,* Vol. 26, No. , pp.405-413.

17. Banthia, N., Yan, C. and Mindess, S. (1996) Restrained shrinkage in fiber reinforced concrete: a novel test technique, *Cem. Conc. Res.*, Vol. 26, No. 1, pp. 9-14.

18. Bentur, A. and Mindess, S. (1990) Fiber Reinforced Cementitious Composites, Elsevier Applied Science.

19. Beaudoin, J.J. (1990) *Handbook of Fiber Reinforced Concrete,* Noyse Publications.

20. Baum, H. and Bentur, A. (1994) Fiber reinforced cementitious materials for lightweight construction: development of criteria for evaluation of their long term performance", Research Report, National Building Research Institute, Technion, Haifa, Israel.

21. Williamson, G.R., (1985) Evaluation of glass fiber reinforced concrete panels for use in military construction, in *Durability of Glass Fiber Reinforced Concrete Symposium,* (ed. S.Diamond), Prestressed Concrete Institute, Chicago, 1985, pp.54-63.

22. Fordyce, M.W. and Wodehouse,R.G. (1983) *GRC and Buildings,* Butterworths, UK.

23. Oesterle, R.G., Schultz, D.M. and Glikin, J.D. (1990) Design considerations for GFRC facades, in *Thin Section Fiber Reinforced Concrete and Ferrocement,* (eds. J.Daniels and S.P.Shah), ACI SP-124, American Concrete Institute, Detroit, pp.157-182.

24. Hansen, N.W., Roller, J.J. Daniel, J.I. and Weinmann, T.L. (1990) Manufacture and installation of GFRC facades, pp. 183-213 in ref. 23.

25. Bentur, A. (1995) Fiber reinforced cement thin sheet composites, in *Fiber Reinforced Concretes: Modern Developments,* (eds. N. Banthia and S. Mindess), Proc. 2nd University-Industry Workshop on Fiber Reinforced Concrete, The University of British Columbia, pp. 187-199.

26. UEAtc Directive, (1989) Directive for the assessment of the durability of thin fibre reinforced cement products", European Union of Agrement, Centre Scientifique et Technique du Batiment, CSTB, Paris.

CRACKING TENDENCY OF CONCRETE: CORROSION OF REINFORCEMENT

R.J. DETWILER
Construction Technology Laboratories, Inc., Skokie, Illinois, USA

Abstract

High-performance concrete has a reputation among field engineers as being more prone to cracking than conventional concrete. This paper reviews recent research in order to examine whether this is true and how significant cracking is to the durability of reinforced concrete structures. While the results do indicate an increased tendency to cracking in some cases, good concreting practice is effective in mitigating cracking. Silica fume concrete in particular may crack more readily at early ages if not adequately cured. Other results show that cracking need not be deleterious to the service life of a structure provided the concrete is otherwise of high quality. The high electrical resistivity and low permeability and diffusivity of silica fume concrete have beneficial effects on corrosion resistance. In concrete that is continuously submerged, the rate of corrosion is controlled by the availability of oxygen to the cathodic areas of the steel. Thus the thickness and quality of the cover concrete are far more important than the crack width.
Keywords: chloride ions, cracking, corrosion, durability, high-performance concrete, silica fume

1 Introduction

Numerous advances in concrete technology in the past few years have led to increases in strength and durability as measured by chloride ion diffusion coefficients in the laboratory. However, there is some question as to whether these improvements can be seen in the field. The measures taken to reduce the rate of deterioration of concrete by reducing its permeability and diffusivity are believed to increase the tendency of the concrete to crack. This paper reviews recent work focusing on three questions: (1) Does high-performance concrete have a greater tendency to crack? (2) If so, how detrimental is cracking to the performance, particularly the durability? and (3) What needs to be done to ensure high performance in the field, not just the laboratory?

2 Cracking tendency

The tendency of concrete to crack is a function of its strength and the stresses imposed at a given time. Stresses may arise from applied loads, from the settlement of supports, or from the restraint of volume changes such as thermally-induced strains or drying shrinkage. If at any time in the life of the concrete the total principal tensile stress exceeds the tensile strength at that time, the concrete will crack. Field engineers often associate high-performance concrete with a greater tendency to crack.

2.1 Effect of silica fume

In order to determine whether this was so, Whiting and Detwiler [1] performed a full-factorial examination of variables on a series of concretes designed to simulate

Concrete Technology for a Sustainable Development in the 21st Century edited by O.E. Gjørv and K. Sakai.
Published in 2000 by E & FN Spon, 11 New Fetter Lane, London EC4P 4EE, UK. ISBN: 0 419 25060 3

those typically used in full-depth bridge decks, with water/cementitious materials ratios ranging from 0.35 to 0.45 and silica fume contents ranging from 0 to 12% replacement of cement by mass. They cast these concretes in ring molds which provided a high degree of restraint against shrinkage such as would be encountered in concrete placed in a full-depth bridge deck. They moist cured these specimens for either 1 or 7 days, monitoring cracking by means of strain gages attached inside the inner ring. For the specimens cured only 1 day, the presence of silica fume was associated with cracking at earlier ages. However, for specimens cured 7 days, the silica fume content did not affect the tendency to crack. They found no statistically significant effect of water/cementitious materials ratio on the tendency to crack. Their results confirm the generally held belief that silica fume concretes are more sensitive to the effects of curing than concretes without silica fume.

The same authors also investigated the effects of silica fume content and water/cementitious materials ratio on the unrestrained drying shrinkage of concrete. Volume changes due to temperature change and drying shrinkage do not cause stresses or cracking if allowed to take place without restraint. However, in structures there is generally at least partial restraint due to the conditions of support or even the thickness of the member. Stresses due to restraint of volume changes from any source can contribute to cracking, particularly at early ages when the concrete has little strength. Whiting and Detwiler performed multivariate regression analyses of their data, finding that shrinkage increased with increasing water/cementitious materials ratio, and the effect was greater as the silica fume content increased. However, for a fixed water/cementitious materials ratio the effect of silica fume content was significant only for the lowest water/cementitious materials ratios. The authors concluded that silica fume has little effect on the long-term shrinkage of concrete, but at early ages silica fume concretes may shrink more than comparable concretes without silica fume.

2.2 Effect of water/cement ratio

Justnes et al. [2] investigated the effect of water/cement ratio on the chemical shrinkage of cement pastes, that is, the shrinkage due to the difference in volume between reactants and products in the cement hydration reactions. They found that the shrinkage per gram of cement was identical regardless of water/cement ratio, so that on a volume basis the shrinkage of a cement paste with a low water/cement ratio is relatively high because of its higher cement content. The rate of shrinkage is perhaps more significant, however, since the shrinkage of pastes with a lower water/cement ratio tends to level off at an earlier age. This corresponds to the earlier, gradual formation of a rigid skeleton, but with insufficient strength to resist the stresses due to chemical shrinkage. Pastes with higher water/cement ratios have a more abrupt but later transition from liquid to solid. As these pastes continue to shrink they can collapse with little effect on the ultimate internal structure; pastes with lower water/cement ratio will develop internal stresses that may lead to cracking. These effects may be expected in concrete as well, somewhat diluted by the presence of aggregate.

2.3 Shrinkage stresses

Tazawa and Miyazawa [3] examined the generation of internal stresses due to restrained shrinkage in fresh concrete. They found considerable stress was generated within the first few days after casting even when no evaporation was permitted. For concrete with a water/cementitious materials ratio of 0.20 and a silica fume content of 10%, there was little difference in stress between the dry-cured and sealed concretes. Comparison with an identical concrete cured wet showed that most of the stress was caused by autogenous shrinkage, not drying. The main difference in stress generation took place in the first 40 hours after casting; after that the rate of stress generation was about the same regardless of curing condition. The authors conclude that in low water/cementitious material ratio concrete, most of the strain of plastic shrinkage is due to autogenous shrinkage, not evaporation of water from the concrete surface. They also state that the stresses due to restraint of autogenous shrinkage alone could be sufficient to cause cracking in heavily reinforced concrete with low water/cementitious materials ratio.

Miyazawa et al. [4] found that stress due to autogenous shrinkage is initiated within 24 hours after casting in concretes with water/cementitious materials ratios less than 0.30. For a water/cementitious materials ratio of 0.17 and a steel ratio of 4.97%, the stress due to autogenous shrinkage reached 3.25 MPa, resulting in cracking of the concrete 28 days after casting. As the water/cementitious materials ratio decreases, the proportion of stress due to restraint of shrinkage attributed to autogenous shrinkage increases from 30% for a water/cementitious materials ratio of 0.40, to 50% for a water/cementitious materials ratio of 0.30.

3 Effects of cracking on performance

Wang et al. [5] studied the water permeability of concrete specimens that had been previously cracked in feedback-controlled splitting tension tests to produce different crack widths. They found that cracks with crack opening displacements smaller than 50 μm prior to unloading had little effect on concrete permeability. For cracks greater than 50 μm, increasing crack widths resulted in increasing permeability. The authors noted that cracks induced by this technique have more or less parallel walls, unlike the flexural cracks that are more common in durability testing. The cracks tended to close somewhat on unloading.

In structures, the relation between cracking and durability is not necessarily so straightforward. Fidjestøl and Nilsen [6] studied reinforced concrete beams cracked in flexure and submerged in seawater. They were particularly concerned that the small area of a reinforcing bar near a crack would corrode rapidly because of the large cathodic area that could be expected to form on the rest of the bar. However, they found the long-term rate of corrosion was much less than would be predicted due to the restricted availability of oxygen at the steel surface. The low diffusivity of oxygen in saturated concrete, the presence of oxygen-consuming microorganisms on the concrete surface, the deposition of calcium and magnesium salts in the cracks, and the continuing hydration of the cement all acted to limit the availability of oxygen at the steel surface. Over time, the net effect of cracks on the corrosion of reinforcement in submerged concrete is negligible.

Once the steel near the cracks is depassivated by chloride ions, it acts as the anode, while the steel between the cracks acts as the cathode. Thus the rate of corrosion is strongly influenced by the condition of the concrete between the cracks, where the limited availability of oxygen limits the rate of the cathodic reaction.

$$O_2 + 2H_2O + 4e^- \rightarrow 4OH^-$$

Schießl and Raupach [7] experimented with reinforcing bars of different diameters to obtain different crack widths. They measured both corrosion currents and loss of steel area to determine the effect of crack width on the rate of corrosion. They found that the thickness and quality of the concrete cover were more significant to the rate of corrosion than was the crack width. They also found that the effect of crack width became less significant over time; after a period of two years there was no clear relationship between crack width and loss of steel area in the cracked region.

Gérard et al. [8] reviewed a large number of references on the effects of cracks on transport properties. They found that the effect depends on the transport mechanism. The rate of diffusion through a crack is proportional to the crack width, while the rate of permeation is proportional to the cube of the crack width. Thus the effect of cracks on diffusion is much less than on permeation.

Fidjestøl and Tuutti [9] stated that in cracked concrete in a marine environment, chloride ions reached the surface of the steel reinforcement in a very short time. However, failure due to rapid corrosion did not necessarily follow. They found that cracked reinforced concrete specimens submerged in seawater for 20 years showed no visible sign of corrosion even though the chloride concentrations at the steel surface were higher than in the seawater. Within a short time of submersion, magnesium and calcium salts had deposited in the cracks, hindering further ingress of oxygen and chlorides.

Sandberg et al. [10] reported that in long-term field investigations, exposed concrete reinforcement had a coating of iron oxide up to 200 μm thick, but no visible rusting or pitting, despite a chloride ion concentration of 1% by mass of cement at the steel surface. They pointed out that the greatest risk of corrosion damage was associated with cycles of wetting and drying because this concentrates the chlorides in solution and made oxygen more readily available at the steel surface.

Pettersson et al. [11] studied reinforced concrete specimens of different grades of concrete partially submerged in seawater. Some of these specimens were cracked. They found that the use of silica fume reduced the rate of corrosion because of its beneficial effects on electrical resistivity and transport properties. Increasing the thickness of the cover over the reinforcing bars was also beneficial; the authors attributed this result to a narrower crack near the steel surface and more rapid healing in the wet environment. They did not find an effect of crack width on the rate of corrosion. They were able to find a relationship between the rate of corrosion and the electrical resistivity of the concrete, although it was less distinct than expected. They also found the rate of corrosion was lower in the submerged portion of the specimens than in the portion above the surface of the water; this result reflects the importance of the availability of oxygen. Sandberg [12] pointed out that in submerged concrete, corrosion is controlled by the cathodic reaction, since the rate of diffusion of oxygen through saturated concrete is very limited.

Eriksen et al. [13] examined specimens taken from a number of structures exposed to seawater or deicing salts to determine the effect of cracks on chloride ion ingress. They found that only cracks connected to the surface have any effect on chloride ion ingress. This effect is limited to the immediate vicinity of the crack. They discussed a hypothesis based on percolation theory, that in a well-hydrated cement paste of sufficiently low water/cement ratio, chloride ions can readily migrate in from the surface to the point where there is no continuous channel (due to either capillary pores or cracks) leading from the surface. At this point the dominant transport mechanism is diffusion, which is several orders of magnitude slower than capillary transport. Over time, the region which is connected to the surface via capillary pores will experience an increase in chloride ion concentration as the chlorides are bound by the cement hydration products. This hypothesis is supported by their data from different specimens obtained from the same structure. Plots of chloride ion concentration versus distance from the surface of specimens having different degrees of cracking were very similar except for the few mm nearest the surface, where the presence of numerous cracks increased the chloride ion concentration locally.

They also found that the chloride binding capacity of hydrated cement paste is quite high, as both calcium silicate hydrate and calcium aluminate hydrate bind chlorides. Since only free chlorides affect the corrosion of reinforcing steel, the chloride binding capacity is important. Carbonation radically reduces the chloride binding capacity of hydrated cement paste. Both carbonation and the availability of oxygen are much greater in the splash zone than in parts of the structure that are continuously submerged.

François et al. [14] loaded 3 m long reinforced concrete beams in flexure and mapped the resulting cracks, which ranged in width from 0.05 mm to 0.50 mm. They then exposed the loaded beams to a salt fog environment for a total of 12 years, examining the beams periodically. At the end of the exposure period they extracted samples from the portions of the beams subjected to tension, compression, and zero stress. They found that although the chloride concentration was greatest where the tensile stress was greatest, the corrosion of the reinforcement was not influenced by the width of the cracks, or by the presence of cracks. The tensile stresses affected chloride ion ingress not by inducing flexural cracking, but by causing bond cracks along the interfaces between paste and aggregate, and between steel and concrete.

4 Factors affecting field performance

Structural design procedures are based on the implicit assumption that the concrete will crack. Such standard practices as using a large number of small reinforcing bars rather than a few large bars do not reduce the amount of cracking, but favor the formation of many small cracks rather than a few large ones. Other structural design details intended to minimize corrosion include specifying a thick cover for the main reinforcing steel and providing control joints. Unfortunately, structural engineers generally do not analytically determine the required spacing of control joints based on the tendency of the concrete to shrink.

A number of changes in materials and construction practices in recent years have led to the perception that concrete is cracking more than in the past. At least some of this perception is based in reality. For example, the use of stronger steel means that less steel is necessary to provide the same structural strength. However, since the modulus of elasticity of metals is a function of interatomic forces and strength is a function of microstructure, increased strength does not correspond with increased modulus of elasticity. In concrete, increased strength does correlate with increased modulus of elasticity, but the two are not proportional. Depending on the exact configuration of the cross section, the structure may be significantly less stiff, especially if the choice of higher strength materials led to a reduction in the dimensions of the cross section. This change in stiffness may result in more cracking, as less stiff sections lead to increased deformations and the concrete cracks when its maximum strain is exceeded.

Fast-track construction schedules require the concrete to gain strength quickly, so that forms are removed and service loads applied at earlier ages. In order to meet the compressed schedule, concrete with low water/cement ratios and/or silica fume may be necessary. Justnes et al. [2] have shown that low water/cement ratio (that is, high cement content) pastes will have a greater tendency to crack due to chemical shrinkage; a similar effect, diluted by the presence of aggregate, may be expected in concrete. Compressed schedules may also militate against adequate curing; the work of Whiting and Detwiler [1] has demonstrated that silica fume concrete is more sensitive to poor curing, confirming a widely held view. In general, more sophisticated materials may be less forgiving of poor construction practices.

Sanjuán et al. [15] studied the effectiveness of polypropylene fibers in controlling cracking and corrosion in reinforced concrete exposed to chloride solutions. They found that increasing the sand content of the concrete was more effective in reducing early-age drying shrinkage than was adding fibers. However, after 8 months' exposure to corrosive conditions, the fiber-reinforced specimens experienced the least damage. Apparently cracks in the fiber-reinforced specimens closed, while those in the other specimens propagated. The authors found no clear correlation between cracking in the vicinity of a reinforcing bar and its corrosion when taken on an individual basis. However, in general, for a given concrete mix proportion, the presence of polypropylene fibers correlated with less cracking and lower corrosion currents.

Fidjestøl and Tuutti [9] stated that structures must be designed in such a way that the combination of concrete quality and cover thickness minimize the variation of moisture at the steel surface in order to prevent the concentration of chloride ions from increasing above tolerable levels. They also pointed out that in concrete of high quality with a correspondingly high electrical resistivity, large cathodic areas are not created on the steel, with the result that corrosion rates are relatively low even near cracks.

The work of Schießl and Raupach [7] also emphasizes the importance of the quality and thickness of the cover concrete. They found that in time the effect of crack width in submerged concrete was not significant; the rate of corrosion was controlled by the availability of oxygen to the steel beneath the intact cover.

Sandberg et al. [10] pointed out another advantage of high quality for concrete exposed in the splash zone, where cycles of wetting and drying provide chlorides, moisture, and oxygen in abundance. Concrete that is highly permeable will experience a high degree of capillary action, which will draw seawater in to a significant depth. The portion of the same concrete above the water surface will experience drying to a significant depth, so that the salt from the seawater is concentrated in this region. However, in concrete of low permeability, both the capillarity and the evaporation are reduced to the point where diffusion becomes the main mode of chloride ion transport.

Jacobsen et al. [16] reviewed a large number of studies on the effect of cracking on durability. They concluded that for conventional concrete the long-term effect of cracks less than 0.2 to 0.4 mm wide on the corrosion of reinforcement is minimal. Such factors as exposure conditions and the quality and thickness of the cover are much more important. In high performance concrete, the initiation period is shortened by cracks. However, the high electrical resistivity and low oxygen permeability of the concrete reduce the rate of corrosion. Thus the prediction of service life for these concretes should be based on the propagation rate rather than the short initiation time. They also found that deicing salts represent a more severe environment than sea salts, even for cracks narrower than 0.3 mm. Cracks that are parallel to the reinforcement are more detrimental than transverse cracks.

Pettersson [17] supported the view that the propagation rate is the main factor in predicting the service life of high performance concrete structures, even though the presence of cracks drastically reduces the length of the initiation period. She found that corrosion initiates almost instantaneously in high performance concrete with 0.2 to 0.4 mm wide cracks. However, the rate of corrosion is limited by the availability of oxygen to the cathodic areas of the steel. Thus active corrosion may proceed for a long time before the structure suffers any significant damage.

5 Conclusions and recommendations

It is clear that concrete of high quality, even if cracked, can still provide good protection for reinforcing steel against corrosion. The high electrical resistivity and low permeability and diffusivity of silica fume concrete all have beneficial effects on corrosion resistance. However, concrete with low water/cement ratio and concrete containing silica fume shrink more at early ages than normal concrete, and silica fume concrete in particular may crack more readily at early ages if not adequately cured.

Over time, the net effect of cracks on the corrosion of reinforcement in submerged concrete is negligible. The low diffusivity of oxygen in saturated concrete, the presence of oxygen-consuming microorganisms on the concrete surface, the deposition of calcium and magnesium salts in the cracks, and continuing hydration of the cement all limit the availability of oxygen at the steel surface. In cracks subjected to cycles of opening and closing and cracks not continuously submerged, the rate of corrosion is faster.

For concrete subjected to cycles of wetting and drying, it seems prudent to take steps to limit the size of cracks by means of such common sense structural practices as providing numerous small reinforcing bars with adequate cover thickness, including control joints at appropriate spacing, and minimizing thermal stresses by appropriate proportioning of the concrete mix. In particularly hostile environments, reducing the applied stresses by increasing the cross section may be a cost effective means of minimizing cracking.

The use of fiber reinforcement may also be a practical option. However, the first line of defense is simply to proportion the concrete with a low water/cementitious materials ratio; use supplementary cementing materials as appropriate; and use good practice in mixing, placing, finishing, and curing in order to obtain high-quality cover concrete. High-performance concrete, being less forgiving of poor practice than is normal concrete, may require more stringent quality control measures and better on-site inspection in order to realize its potential.

References

1. Whiting, D. and Detwiler, R. (1997) *Silica Fume Concrete for Bridge Decks*, NCHRP Project 18-3 Final Report, National Cooperative Highway Research Program, Transportation Research Board, National Research Council, Washington, DC.
2. Justnes, H.; Van Gemert, A.; Verboven, F.; and Sellevold, E.J. (1996) "Total and External Chemical Shrinkage of Low W/c Ratio Cement Pastes," *Advances in Cement Research*, vol. 8, no. 31, pp. 121-126.
3. Tazawa, E. and Miyazawa, S. (1996) "Influence of Autogenous Shrinkage on Cracking in High-Strength Concrete," *4th International Symposium of High-strength/High-performance concrete, Paris, 1996*, pp. 321-330.
4. Miyazawa, S.; Tazawa, E.; Sato, T.; and Sato, K. (1993) "Autogenous Shrinkage Stress of Ultra-High-Strength Concrete Caused by Restraint of Reinforcement," *Transactions of the Japan Concrete Institute*, vol. 15, pp. 115-122.
5. Wang, K.; Jansen, D.; Shah, S.; and Karr, A. (1997) "Permeability Study of Cracked Concrete," *Cement and Concrete Research*, vol. 27, no. 3, pp. 381-393.
6. Fidjestøl, P. and Nilsen, N. (1980) *Performance of Concrete in Marine Environment*, American Concrete Institute, SP-65, pp. 205-221.
7. Schieβl, P. and Raupach, M. (1997) "Laboratory Studies and Calculations on the Influence of Crack Width on Chloride-Induced Corrosion of Steel in Concrete," *ACI Materials Journal*, vol. 94, no. 1, pp. 56-62.
8. Gérard, B., Jacobsen, S., and Marchand, J. (1998) "Concrete Cracks II; Observation and Permeability—A Review," *Concrete under Severe Conditions 2, Environment and Loading, Vol. 1*, Gjørv, E & FN Spon. O.E., Sakai, K., and Banthia, N., Eds., pp. 183-197.
9. Fidjestøl, P. and Tuutti, K. (1995) "The Importance of Chloride Diffusion," *Chloride Penetration into Concrete*, Proceedings of the RILEM International Workshop, 15-18 October 1995, L.O. Nilsson and J.P. Ollivier, Eds., pp. 470-480.
10. Sandberg, P.; Pettersson, K.; Sørensen, H.E.; and Arup, H. (1995) "The Importance of Chloride Diffusion," *Chloride Penetration into Concrete*, Proceedings of the RILEM International Workshop, 15-18 October 1995, L.O. Nilsson and J.P. Ollivier, Eds., pp. 453-459.
11. Pettersson, K.; Jørgensen, O.; and Fidjestøl, P. (1996) "The Effect of Cracks on Reinforcement Corrosion in High-Performance Concrete in a Marine Environment," *Concrete in Marine Environment*, American Concrete Institute, SP-163, V.M. Malhotra, Ed., pp. 185-200.
12. Sandberg, P. (1996) "Research needs regarding the influence of macro cracks on salt penetration and reinforcement corrosion in concrete," presented at the Nordic Miniseminar on Service Life Prediction Utilising the Corrosion Propagation Stage, CBI, Stockholm, 10 December 1996.

13. Eriksen, K.; Thaulow, N.; and Grelk, B. (1996) *Mikrodefekters inflydelse på chloridindtrængning i beton*, Rapport nr. 75, [in Danish]. Vejdirektoratet, Copenhagen, Denmark, 48 pp.
14. François, R., Arliguie, G., and Castel, A. (1998) "Influence of Service Cracking on Service Life of Reinforced Concrete," *Concrete under Severe Conditions 2, Environment and Loading, Vol. 1*, E & FN Spon. Gjørv, O.E., Sakai, K., and Banthia, N., Eds., pp.143-152.
15. Sanjuán, M.; Andrade, C.; and Bentur, A. (1997) "Effect of Crack Control in Mortars Containing Polypropylene Fibers on the Corrosion of Steel in a Cementitious Matrix," *ACI Materials Journal*, vol. 94, no. 2, pp. 134-141.
16. Jacobsen, S., Marchand, J., and Gérard, B. (1998) "Concrete Cracks I: Durability and Self Healing—A Review," *Concrete under Severe Conditions 2, Environment and Loading, Vol. 1*, E & FN Spon. Gjørv, O.E., Sakai, K., and Banthia, N., Eds., pp.217-231.
17. Pettersson, K. (1998) "Service Life of Concrete Structure Including the Propagation Time," *Concrete under Severe Conditions 2, Environment and Loading, Vol. 1*, E & FN Spon. Gjørv, O.E., Sakai, K., and Banthia, N., Eds., pp. 489-498.

CONCRETE DAMAGE INDUCED BY DELAYED ETTRINGITE FORMATION

M. COLLEPARDI
Enco, Engineering Concrete, Spresiano (TV), Italy

Abstract
By adopting the holistic approach, a new model is proposed to explain the concrete damage induced by delayed ettringite formation (DEF). This model is based on three essential elements: late sulfate release, microcracking, and exposure to water. Late sulfate release from a cement with high sulfate content (especially that with high content of clinker sulfate in less available form) can cause the delayed deposition of ettringite in pre-existing microcracks after sulfate ions diffuse through the pore solution in concrete, either intermittently, or continuously exposed to environmental water. Microcracking may be promoted by alkali-silica reaction, steam curing at high temperatures, localized high stress in prestressed structures or other causes.

Theoretically the DEF-induced damage occurrence can be reduced or prevented by controlling at least one of the above three parameters. In practice, the best way of reducing the DEF-induced damage risk is either to avoid cements with high clinker sulfate which are responsible for the late sulfate release, or to adopt lower and more homogeneous stress distribution derived from the prestressing process in precast elements, such as concrete ties.
Keywords: Durability, delayed ettringite formation, microcracking, prestressed concrete elements.

1 Introduction

Brittleness, combined with poor tensile strength, is responsible for microcracking in concrete structures subjected to thermal and drying shrinkage, as well as to static and

Concrete Technology for a Sustainable Development in the 21st Century edited by O.E. Gjørv and K. Sakai.
Published in 2000 by E & FN Spon, 11 New Fetter Lane, London EC4P 4EE, UK. ISBN: 0 419 25060 3

cycling dynamic loading in service. According to Mehta (1) microcracks, promoted by weathering effects and loading in service, represent preferential paths for the penetration of aggressive environmental agents such as air, humidity, sulfate and chloride ions. Therefore, pre-existing discontinuous microcracks act as precursors in the corrosion of the reinforcement and in the deterioration of the cement matrix itself as well as that of the reactive aggregates, if any. Once any of these processes is initiated (all having an expansive-disruptive nature), microcracks grow to become macrocracks. After an initial period of a few years or more, the degradation process - in the form of cracking, spalling and loss of mass - increases very rapidly.

A holistic approach has been used by Mehta (1) to re-examine the four principal causes of concrete deterioration: sulfate attack, alkali-silica reaction, corrosion of reinforcing steel, and freezing-thawing cycles. In the present paper, based on a previous work of the author (2), the holistic model is used to address another problematic cause of concrete deterioration, namely delayed ettringite formation (DEF) - a phenomenon which has attracted considerable attention all over the world during the last decade. In the present paper the term DEF will be used to indicate the ettringite-related deterioration of concrete in a **sulfate-free environment regardless of whether or not the concrete member had been subjected to steam curing.**

2 Reductionist approach for DEF-promoted concrete damage

Different researchers have adopted a reductionist approach, instead of a holistic one, to explain concrete distress induced by DEF. Each has reached a different conclusion by relating the concrete distress with a specific predominating mechanism. Among these, the following hypothesis can be mentioned as the most important:

(*i*) **DEF promoted by high temperature steam curing:** In normal portland cement pastes ettringite, which forms during the plastic stage of fresh concrete, is destroyed by steam curing at 65-100°C; then ettringite forms again at later ages in concrete structures stored in water either intermittently or permanently and causes disruptive expansion of the hardened concrete in service. According to Heinz *et al.* (3,4), DEF expansion in concrete is due to the transformation of metastable monosulfate into ettringite when steam-cured concrete is exposed to normal temperature moist-curing at later ages. According to Lawrence (5), the correlation between expansion and sulfate content of the cement points out the importance of ettringite in the expansion mechanism of steam-cured mortar prisms; however, the expansive hydration of MgO may increase the sensitivity of cements to heat cure. On the other hand, Fu *et al.* (6-8) argued that, although steam curing of concrete at high temperatures is a key element, the DEF mechanism process is different from that based on thermal decomposition of ettringite: at temperatures above 65°C, calcium silicate hydrate (*C-S-H*) would adsorb very quickly sulfate from gypsum, so that this would not be available to react with the aluminate phase and produce normal ettringite; later sulfate ions, slowly released from the *C-S-H* phase, would diffuse through pore solution and feed the nucleation of ettringite crystals in the tip-zone of the pre-existing microcracks. In the case of DEF associated with steam curing the specific mechanism is highly controversial. On the one hand, Scrivener and Taylor

(9) and others (10, 11) think that the DEF effect in steam-cured concretes is related to an uniform and homogeneous paste expansion which occurs in the post steam-curing exposure. Accordingly to Johansen *et al.* (10) the uniform paste expansion results in rim cracks around aggregates; subsequently ettringite deposition would fill the rim cracks, but this is considered to be benign since ettringite deposition is considered to play no role in the expansion and cracking of concrete. On the other hand, Diamond (12) and others (6-8) think that the expansion and cracking induced by DEF are related to the crystal pressure exerted by the growing ettringite crystal on their surroundings. In particular, on the basis of free-energy considerations, it was shown (7) that ettringite crystal nuclei should first form in the tip-zone of the pre-existing microcracks; after nucleation, the growth of these ettringite crystals would be responsible for the opening of the microcracks.

(*ii*) **DEF promoted by alkali-silica reaction (ASR) or other microcrack-causing mechanisms:** *ASR* is the primary cause of deterioration in the form of cracks and microcracks, whereas ettringite, which is found in the cracks in moist-curing conditions at later ages, is considered to have formed as a consequence of the pre-existing cracks rather than being the cause of it (13,14). Thaulow *et al.* (15) found mixtures of alkali-silica gel and ettringite in steam-cured prestressed concrete railroad ties. However, Diamond and Ong (16) found that, in the presence of siliceous reactive aggregates, initial expansion and cracking were caused by the ASR, but after about one month no more *ASR* gel formed and the subsequent further expansion was caused by deposition of ettringite. Steam curing at high temperatures is suspected to aggravate reactivity of silica and silicates with alkali (17,18). Therefore, according to Taylor (17), limestone aggregates would be preferable to siliceous aggregate in manufacturing DEF-free steam-cured concrete. Additional conditions which are considered essential to the DEF-damage process include: microcracking promoted by thermal stresses in steam-cured precast concrete (18,19), or freezing-thawing in service (3,4), or dynamic loading and fatigue stress (13). The use of ASTM Type III portland cement (20) or relatively high contents of SO_3 (> 3.6%), MgO (> 1.6%) and equivalent Na_2O (> 0.8%) in portland cement (21) are factors that would aggravate expansion of steam-cured concrete.

(*iii*) **DEF promoted by large amounts of sulfate in clinker and/or high cement sulfate content:** Portland cement contains a mixture of sulfur compounds from the cement clinker and from added calcium sulfate (e.g. gypsum, $CaSO_4 \cdot 2H_2O$) which acts as set regulator. The sulfates from different sulfur sources have different water-solubility kinetics: sulfate from the gypsum dissolves sufficiently fast to participate in the regulation of the setting of the cement, whereas sulfate from the clinker phase is usually unavailable for the setting regulation. Therefore, cements with the same *total* sulfate content, but with different proportions of sulfate from the clinker phase and that from the gypsum source, may perform in different ways.

In many present-day clinker kilns there are poly-functional burning systems which are capable of using either gaseous or liquid hydrocarbons, as well as solid small-particle coals, depending on the cheapest source of available fuels. The sulfur content of these different fuels can change and cause unwitting variations from one day to another of sulfate incorporated in the clinker phase. Moreover, the clinker sulfate content may increase with the use of high-sulfur organic residues, such as tires that are burned in cement kilns to destroy environmentally harmful products in a safe and cost-effective way.

According to Hime (22) present-day cements, produced in kilns that burn sulfur-rich fuels or waste materials, can incorporate large amounts of sulfates, up to 3% by clinker mass. When high sulfate levels are not balanced by a high alkali content, the excessive SO_3 may occur as $CaSO_4$ or react with calcium aluminates or even occur as interstitial impurity in the alite and belite phases. According to Hime all these forms of sulfate in the clinker phase are slowly soluble in the mixing water and, therefore, can lead to **late sulfate release** which is essential for the delayed ettringite formation (DEF)-related damage. According to this recent perspective, based on field occurrence, DEF is not restricted to overheated steam cured concretes, since precast Friday ties* , cured at the factory temperature, present the same distress incidence as the steam-cured products (23). Moreover, there is experiential knowledge that cast-in-place concrete structures (12,22,23) may evidence the same DEF-induced damage occurrence as that of steam-cured precast concrete products. According to this hypothesis, DEF by itself causes microcracks radiating from localized areas of ettringite development, expansion of the cement matrix relative to coarse aggregates, and macrocracks due to the internal expansion relative to the exterior surface region (23). On the other hand, Miller and Tang (24) found that North American and European present-day clinkers may contain sulfate levels from a few hundredths of a percent to about 2.5%. However, they did not find $CaSO_4$ in the clinker phase and concluded that, under ambient curing conditions, the sulfur-containing phases of the present-day commercial clinkers are unlikely to cause expansive stress and cracking induced by internal sulfate attack.

Each of the above hypothesis - from (*i*) to (*iii*) - can be partly correct; however, they cannot solely explain all the available experiential data on the DEF-induced damage of concrete structures. For instance, it may be true that many DEF-induced concretes were subjected to steam curing at high temperatures, but there are also apparently proven cases of DEF-related deterioration in the absence of steam curing. On the other hand, it may be true that *ASR* acts as the precursor of some DEF-related damages, but there are also DEF-distress cases in the absence of *ASR*. Finally, it may be true that present-day cements incorporate large amounts of slowly soluble sulfate in the clinker, but not all present-day concrete structures are subjected to DEF-induced damage. It is opinion of the author that, due to the reductionist approach, each researcher or research-team over-emphasized their own results and under estimated or rejected others. Moreover, each of the above hypotheses cannot give a satisfactory answer to the following questions:

---

* *Friday ties are manufactured on Friday and cured at room temperature because by Monday they would attain adequate strength to allow for cutting the prestressing strands.*

1. Why are some specific concrete products (e.g., prestressed concrete **railway ties) more vulnerable to DEF-induced distress** than other precast or cast-in-place concrete structures?
2. Why is **DEF-induced damage so erratic** in the sense that, everything apparently being the same, it occurs only under some specific circumstances and not others?

3 Field Experience of DEF-Related Damage

An attempt will be made in the present paper to integrate all the above mentioned experiential data, as well as results from the field experience of this author, with the available knowledge on concrete science and technology.

The results from field experience directly available to the author include two different types of concrete structures: **a**) prestressed precast concrete ties and **b**) cast-in-place concrete structures.

For each of the above concrete structures, DEF-related damage cases were found with ettringite appearing, as a gel-like mass under the optical microscope, in the cracks of the damaged structures. By use of a scanning electron microscope and an X-ray elemental analyzer the gel was identified as fibrous consisting of calcium, sulfur, and aluminum. It was definitely confirmed by X-ray diffraction analysis indicating that the product was primarily ettringite.

a) Concrete ties. For the DEF-damaged concrete ties almost all the results published by Mielenz *et al.* (23) were confirmed, and in particular field observations revealed that:

- The DEF-induced damage incidence was identical whether concrete ties were steam cured or not.
- Unused stock-piled ties underwent the same distress-incidence as the ties in service subjected to vibrational distress caused by the passage of high-speed trains.
- Ties not exposed to rain (such as those in railroad tunnels or un-used ties below and in the middle of outdoor storage stacks) did not deteriorate.

In addition to these results, field experience showed the important role played by the
microcracks produced during the prestressing process in determining the DEF-induced damage. Microcracks (i.e. cracks invisible to the naked eye) were detected by using an optical microscope, specially adapted for field experience. The appearence of microcracks (10-100 μm in thickness) on the surface of concrete ties was monitored either immediately after cutting the prestressing strands or later on in stock-piled products and those in service. In all the concrete ties of a plant, where there were complains for the DEF-induced damage, **microcracks were detected immediately after cutting the prestressing strands**. They initially appeared as microcracks parallel to the prestressing wires and later as irregular cracking.

Field observations of microcracks of un-used concrete ties revealed that their presence was related with the specific type of manufacturing process, and in particular with the local stress level induced in the concrete by the sudden cutting of the prestressing strands. Two typical processes are adopted for manufacturing prestressed concrete ties. In one of these - named *long-line method* and widely used all over the world - the same set of prestressing strands is used for 15 to 20 lined up concrete ties. Each concrete batch is in general placed in a row of 6 or 8 parallel ties. When the concrete attains to adequate compressive strength, the prestressing strands are cut. At this time **excessive local stress can occur in the concrete area close to one end of the tie**, especially in the eight ties located at the two extreme parallel rows.

In the other process - named *anchored steel-plate method* - each concrete tie is individually prestressed using reinforcing wires anchored to steel plates by special cold-worked heads. The anchored steel plates, embedded in the concrete mixture at the ends of each tie, transfer the prestressed force from the wires to the concrete when this reaches adequate compressive strength. This process consumes more time and also workmanship, and therefore prestressed concrete ties processed according to this method are more expensive. However, as a matter of fact, in concrete ties manufactured with this process microcracks do not appear or occur to less extent at the time of prestressing, and the frequency occurrence of the DEF-induced damage seems to be lower than in concrete ties manufactured with the *long-line method.*

Not all the microcracked concrete ties subsequently evidenced macrocracks and deposition of ettringite into the cracks. The frequency of occurrence of the DEF-related distress was found to be high only in the presence of other two events accompanying the initial microcracking. These events are:

- relatively high cement sulfate level that sometimes surpassed the limit of 4%, in terms of SO_3 content, according to the European Norm EN 197/1;
- intermittent exposure to rain water, aggravated by alternate drying caused by sunny conditions.

The role played by the micro-climate (Fig. 1) was studied by examining unused, two-year old, stock-piled concrete ties, all microcracked as evidenced by the field optical microscope. Those exposed to the alternate actions of rain and sun (on the sides and specially at the tops of outdoor storage stacks) were severely macrocracked and DEF-damaged; those exposed to rain but in a permanent shadow condition evidenced a less severe distress; concrete ties permanently protected from both rain and sun exposure (below and in the middle of stacks) remained only microcracked without any further crack growth and apparent DEF-related damage.

b) Cast-in-place concrete structures. Many pedestals for electric power-line installations deteriorated very severely over a period of 2-3 years after being cast in place in 1993. Since they were located in an area (near Ancona, Italy) where siliceous reactive aggregate are frequently found, an ASR-damage was initially diagnosized. However, due to the unusual and severe macrocracking not always accompanied by the typical gel appearance of the ASR product, some cored samples were analyzed and massive local deposits of ettringite in the cracks were detected by XRD. No significant sulfate content (< 0.01%) was detected in the environmental ground surrounding the pedestals as well as in the coarse aggregate extracted from the concrete. The latter, on the other hand, was found to be a slowly reactive siliceous aggregate when tested according to the ASTM C 289 procedure.

Fig. 1 - Un-used stock piled concrete ties exposed to different microenvironment (sun and shadow, dry and rain).

On the basis of the SO_3 content in the concrete, the known nominal portland cement content, and the concrete specific gravity, the SO_3 level for the cement used was assumed to be as high as 4.4%. The unusually high SO_3 level of the cement was related with the clinker sulfate as high as 2% according to the available informations for the cement used at that time in the area. Most of the clinker sulfate is not available for early cement hydration reactions because of its lower (25) and /or slower (22) water solubility, especially when the alkali content in the clinker phase is relatively low (26,27). Therefore, the amount of total SO_3 in the cement (from the clinker source and the gypsum used for setting regulation) increases by increasing the clinker sulfate and sometimes surpasses the EN limit (4%). The required amount of gypsum *available* for the setting regulation, besides the *unavailable* sulfate from the clinker phase, can become relatively high in high-strength portland cement with high fineness and high C_3A content.

4 New Proposed Mechanism for DEF-Related Damage

The following factors are considered to be essential for the DEF-related damage: microcracks promoted by the manufacturing process itself (e.g. prestressing in railway ties) or ASR or other causes including those related to steam curing;

- exposure to wetting-drying cycles;
- late sulfate release from the cement clinker or other sources;
- migration of reactant ions ($SO_4^{=}$, Al^{+3}, Ca^{+2}) through the pore aqueous solution of concrete exposed to water or saturated air;
- deposition of ettringite inside the existing microcracks, and subsequent crack opening by ettringite swelling or crystal growth.

The holistic approach for the DEF-induced damage is based on the occurrence of three essential events. The three events are:

- **Microcracking**
- **Exposure to water or saturated air**
- **Late sulfate release**

Each event, in turn, can be determined by one or more possible causes which will be considered in details further on. First, a synthetic representation of the holistic approach for the DEF-related damage will be examined, through the help of Fig. 2 (in such a way that no part of the complex system is overlooked). Each corner of the triangle corresponds to a system in which only one of the three elements of the system is present, and this situation does not present any risk at all for the DEF-related occurrence.

Each side of the equilateral triangle in Fig. 2 represents the system in which only two elements are present in the absence of the third. In such a case, again, no risk exists for the DEF-related deterioration. For example, in microcracked concrete, not exposed to water, DEF cannot occur even if there is a potential late sulfate release due, for instance, to an excessive amount of unavailable clinker sulfate. In fact, in the absence of water, sulfate and other reactant ions (Al^{+3} and Ca^{+2}) cannot diffuse through the pore aqueous phase and migrate towards the existing microcracks to form ettringite.

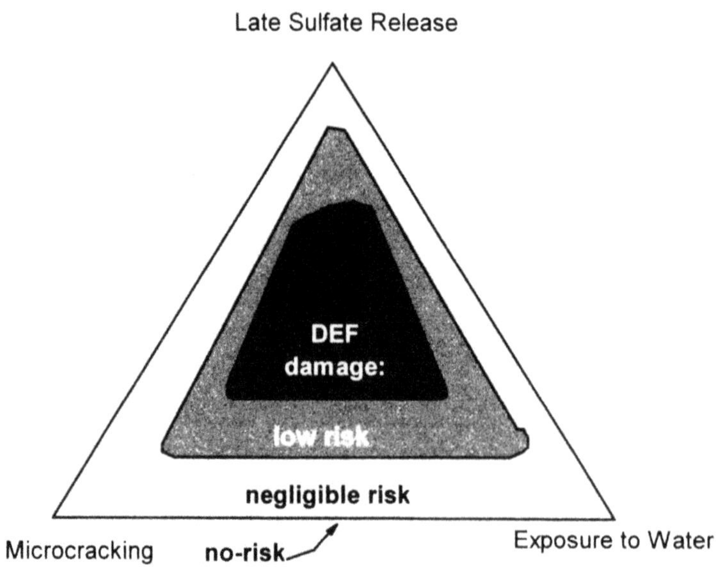

Fig. 2 - A holistic model showing DEF-damage as a function of late sulfate release, microcracking, and exposure to water.

The area in the middle of the triangle corresponds to situations of serious risk for DEF-related deterioration since all the three needed elements of the system are present: **late sulfate release** caused by cement with high sulfate content (especially that with high content of clinker sulfate in less available form) can feed the delayed deposition of ettringite in the pre-existing microcracks after diffusing through the pore solution in concrete, either intermittently, or continuously **exposed to environmental water**. Diamond and Ong (16) found that the total amount of the ettringite deposited in ASR-induced cracks of mortar specimens was significantly increased as compared with uncracked companion specimens. The author of the present feels that this statement is true regardless of the specific cause responsible for microcracking.

Each of the three needed elements of the system can be related to numerous causes. For instance, concrete microcracking can be promoted by one or more of the following causes:

- Curing at high temperatures (> 65°C), excessive heating/cooling rate or too short preliminary curing at room temperature
- ASR with microcracks around aggregate particles
- Weathering effects cycles including wetting/drying and heating/cooling changes
- Dynamic loads in service
- Plastic shrinkage in poorly cured slab structures
- Freezing/thawing cycles
- Excessively high porosity in aggregate particles
- Weak transition zone at the interface aggregate- or steel-cement matrix
- Localized high stress in prestressed structures.

Everything else being the same, steam-curing at high temperatures causes additional microcracking with respect to room temperature curing, and therefore there is a higher frequency of occurrence of DEF-induced damage in precast steam-cured concretes when cements with high sulfate content in the clinker phase are used. This also explains why in many laboratory studies (3-5, 7) only overheated steam cured specimens were subjected to DEF-related damage in contrast to what occured in the corresponding concrete specimens cured at room temperature.

The last cause of microcracking in the above list - that is localized high stress induced by the prestressing process itself - can occur in all the precast concrete ties, particularly in those manufactured by the *long-line method*. This provides a clear justification of why with this type of structure there is a higher frequency of occurrence of DEF-related damage than in any other concrete element. However, since this deterioration did not occur in concrete ties manufactured in the late 1970s by using the same method as that presently employed (24), an additional event must be identified to explain why the distress to some of these ties began to occur in the late 1980s. This additional event seems to be the late sulfate release, which, in turn, is attributable to a number of possible causes:

- High cement sulfate content is related to the clinker sulfate increase with the use of high-sulfur fuel or organic residues - such as tires - burned in cement kilns to destroy environmentally harmful products in a safe and cost-effective way (22).
- Other sources of slowly soluble sulfate from artificial lightweight aggregates or gypsum-contaminated natural aggregates, and shrinkage-compensating cements based on sulfo-aluminate products.
- Slowly released sulfate ions from that adsorbed on the *C-S-H* phase in high temperature (> 65°C) steam-cured concretes (6).

  Thermal decomposition of ettringite at high temperature in steam-cured concrete (3,4).

Only the first cause of late sulfate release in the above list can justify why there was an increase in the DEF-related distress-incidence, from the 1970s to the 1980s, particularly in concrete structures - such as ties - more prone to microcracking for the manufacturing process itself. On the other hand, in present-day manufactured concrete ties the DEF-induced damage seems to be an exceptional and discontinuous phenomenon rather than a general and continuous occurrence, although many ties per each process-day are microcracked. This erratic occurrence of the DEF-induced damage can be related either to intermittent use of sulfur-rich organic residues in the clinker kiln or to change in the sulfur content of ordinary fuels. In many present-day clinker kilns there are poly-functional burning systems which are capable of using either gaseous or liquid hydrocarbons, as well as solid small-particle coals, depending on the cheapest source of available fuels. The sulfur content of these different fuels can change and cause unwitting variations from one day to another of sulfate incorporated in the clinker phase.

Chemical interactions - causing a decrease in the pH of the pore solution phase - can favor normal formation of ettringite from environmental sulfate attack (28). Therefore, one cannot exclude that carbonation as well as ASR can interact with the DEF process, creating more favorable conditions for ettringite deposition in the existing microcracks. In particular, since ASR consumes alkalies and OH^- from pore solution (16), this process can reduce the pH and presumably the ettringite deposit in the aqueous phase of pores and microcracks.

The other potential causes of late sulfate release - related with the sulfate contamination of aggregates, or the adsorption of sulfate on the *C-S-H* phase at high temperatures or thermal decomposition of ettringite in steam-cured concrete - can occur, but cannot explain why there was a significant change in the DEF-induced damage-incidence during the last decade or so. Neither they can explain why microcracked concrete ties showed DEF-related distress regardless of the use of steam curing.

The third element - that is intermittent or continuous exposure to air or humid air - seems to be the one on which all the researchers agree. There is experiential knowledge that microcracked ties, with the same cement source causing DEF-related damage in concrete structures, do not show such a type of deterioration when they are protected from the contact with water. This field experience includes:

- Concrete ties in railroad tunnels (24)
- Stock-piled concrete ties below the outdoor storage stacks (24)
- Concrete ties in service when protected by hydrophobic treatment (13).

5 Recommendations to Prevent DEF-Induced Damage

On the basis of the proposed holistic model, DEF-related deterioration of concrete can be avoided provided that one or two of the above mentioned essential elements is precluded. The most easy preclusion apparently would relate to water exposure. However, in practice, permanent protection of concrete structures from exposure to water is very expensive by using the present-day available impermeable coatings or hydrophobizing products. This technique, may be considered for use with concrete structures in service showing initial DEF-related damage in order to reduce or block the deterioration process.

Therefore, the other two essential elements for the DEF-induced concrete damage should be taken into account first. One of these - microcracking - can be significantly reduced to harmless levels by adopting adequate measurements during the design and the execution phases. For instance, precast concrete ties with lower and more homogeneous stress distribution, deriving from the prestressing method itself, should be taken into careful consideration by the design engineers. Steam-curing process with lower thermal gradients, from controlled heating and cooling rates, can also reduce microcracking, and consequently mitigate the DEF-induced deterioration risk. Use of sound aggregates to prevent ASR-related cracking is a mandatory requirement regardless of the DEF-related deterioration. However, pozzolanic materials can advantageously be used to avoid or to reduce the risk of microcracking promoted by alkali-reactive aggregates when unwittingly used.

The most important challenge in preventing the DEF-induced deterioration of concrete structures is related to cement manufacture. First, cement producers should take into careful consideration the fact that clinker with lower sulfate content can significantly reduce the risk of DEF-induced damage in concrete structures. Moreover, there is experiential evidence that the risk of DEF-related damage significantly increases by using high-strength (high C_3S and C_3A) portland cements, whereas concrete mixtures with portland-pozzolan and portland-slag cements are much less prone to the DEF risk in steam-cured concretes (4,21,30). The behavior seems to be related with the lower amount of gypsum required for setting-time regulation and/or the lower amount of clinker sulfate content in blended cements. Pozzolanic materials, including silica fume at a dosage level as low as 10% by cement mass (29), can be additionally advantageous also in reducing the risk of DEF-induced damage through the pore-size refinement effect with consequent reduced diffusion rate of ions, in particular of sulfate ions, through the pore solution.

6 Conclusion

Concrete damage induced by delayed ettringite formation is one of the most complex problems in the science of concrete durability. A holistic model was adopted to study this problem. According to this model three elements are essential for the occurrence of the DEF-related damage: microcracking, late sulfate release, and exposure to water.

Each of these elements, in turn, is influenced by a number of causes. In this perspective ASR and steam-curing are two of the possible causes of microcracking. Therefore, DEF-related distress could occur even in the absence of ASR and steam-curing.

Theoretically DEF-induced damage can be prevented by eliminating one of the three elements from the system. From a practical point of view, two approaches are possible. First, portland cements with high clinker sulfate content - those responsible for the late sulfate release - should be avoided and portland-pozzolan or pozzoland-slag cements should be preferred. Second, microcracking should be reduced, particularly with precast concrete ties, by adopting lower and more homogeneous stress distribution from the prestressing process itself.

7 Acknowledgments

Luigi Coppola, Roberto Troli, and Silvia Collepardi collected many of the field experience data used in this paper. Kumar Mehta reviewed an earlier draft of the manuscript, and his helpful comments were very valuable in the preparation of the final draft.

8 References

1. Mehta, P.K., "Concrete Technology at the Crossroads - Problems and Opportunities", Mohan Malhotra Symposium, pp. 1-30, Editor P.K. Mehta, ACI SP-144, 1994.
2. Collepardi, M., "A Holistic Approach to Concrete Damage Induced by Delayed Ettringite Formation" Proceedings of the "Mario Collepardi Symposium on Advances in Concrete Science and Technology", pp. 373-396, Editor P.K. Mehta, CANMET/ACI 1997.
3. Heinz, D. and Ludwig, U., "Mechanism of Secondary Ettringite Formation in Mortars and Concretes Subjected to Heat Treatment" in Concrete Durability, Katharine and Bryant Mather International Conference, Editor J.M. Scanlon, ACI SP-100, pp. 2059-2071, 1987.
4. Heinz, D., Ludwig, U. and Rüdiger, I., "Delayed Ettringite Formation in Heat Treated Concretes", Concrete Precasting Plant and Technology, Vol. 11, pp. 56-60, 1989.
5. Lawrence, C.D., "Mortar Expansions Due to Delayed Ettringite Formation. Effects of Curing Period and Temperature", Cement and Concrete Research, Vol. 25, pp. 903-914, 1995.
6. Fu, Y. and Beaudoin, J.J., "Mechanism of Delayed Ettringite Formation in Portland Cement System", ACI Materials Journal, pp. 327-333, 1996.
7. Fu, Y., Xie, P., Gu, P. and Beaudoin, J.J., "Significance of Pre-existing Cracks on Nucleation of Secondary Ettringite in Steam Cured Cement Paste", Cement and Concrete Research, Vol. 24, pp. 1015-1024, 1994.
8. Fu, Y., Gu, P., Xie, P. and Beaudoin, J.J., "A Kinetic Study of Delayed Ettringite Formation in Hydrated Portland Cement Past", Cement and Concrete Research, Vol. 25, pp. 63-70, 1995.
9. Scrivener, K.L. and Taylor, H.F.W., "Delayed Ettringite Formation: a Microstructural and Microanalytical Study", Advances in Cement Research, Vol. 5, pp. 139-145, 1993.
10. Johansen, V., Thaulow, N., Jakobsen, U.H. and Palbøl, N., "Heat Cured Induced Expansion", Proceedings of the 3rd Beijing International Symposium on Cement and Concrete, pp. 144-156, 1993.
11. Lewis, M.C., Scrivener, K.L. and Kelham, S., "Heat Curing and Delayed Etringite Formation, Materials Research Soc. Proc. 370, Editor S. Diamond et al., Materials Research Society, Pittsburgh, pp. 67-76, 1995.
12. Diamond, S. "Delayed Ettringite Formation-Process and Problems", Cement and Concrete Composites, Vol. 18, pp. 205-215, 1996.

13.Shayan, A., "Behaviour of Precast Prestressed Concrete Railway Sleepers Affected by AAR", in Real World Concrete, R. N. Swamy Symposium, Editor G. Singh, pp. 35-56, 1995.
14.Oberholster, R.E., Maree, H. and Brand J.H.B., "Cracked Prestressed Concrete Railway Sleepers: Alkali-Silica Reaction or Delayed Ettringite Formation", 9th International Conference in Alkali-Aggregate Reaction in Concrete, Vol. 2, pp. 739-749, The Concrete Society, Wexham, Slough, U.K., 1992.
15.Thaulow, N., Hjorth Tacobsen, U. and Clark, B., "Composition of Alkali-Silica Gel and Ettringite in Concrete Railroad Ties: SEM-EDX and X-Ray Diffraction Analysis", Cement and Concrete Research, Vol. 26, pp. 309-318.
16.Diamond, S. and Ong, S., "Combined Effects of Alkali Silica Reaction and Secondary Ettringite Deposition in Steam Cured Mortars", in Cement Technology Symposium, Editors E.M. Gartner and H. Uchikawa, Ceramic Transactions, Vol. 40, pp. 79-90, 1993.
17.Taylor, H.F.W., "Delayed Ettringite Formation", Advances in Cement and Concrete, pp. 122-131, Editor M.W. Grutzeck and S.L. Sarkar, ASCE, New York, 1994.
18.Tepponen, P. and Eriksson, B.E., "Damages in Concrete Railway Sleepers in Finland", Nordic Concrete Research, Vol. 6, pp. 109-209, 1987.
19.Sylla, H.M., "Reactions in Hardened Cement Paste through Heat Treatment", Beton Vol. 38, pp. 449-545, 1988.
20.Fu, Y. and Beaudoin, J.J., "Microcracking as Precursor to Delayed Ettringite Formation in Cement Systems", Cement and Concrete Research, Vol. 26, pp. 1493-1498, 1996.
21.Lawrence, C.D., "Laboratory Studies of Concrete Expansion Arising from Delayed Ettringite Formation", BCA Publication C/16, 147 pp., British Cement Association, Crowthorne, Berres. U.K., 1993.
22.Hime, W.G., "Clinker Sulfate: A Cause for Distress and a Need for Specification", Concrete in The Service of Mankind, Concrete for Environment, Enhancement and Protection, pp. 387-395, Editors: R.K. Dhir and T.D. Dyer, E. & F.N. Spon, 1996.
23.Mielenz, R.C., Marusin, S.L., Hime, W.G. and Jugovic, Z.T., "Investigation of Prestressed Concrete Railway Tie Distress", Concrete International, pp. 62-68, December 1995.
24.Miller, F.M.and Tang, F.J., "The Distribution of Sulfur in Present-Day Clinkers of Variable Sulfur Content", Cement and Concrete Research, Vol. 26, pp. 1821-1829, 1996.
25.Taylor, H.F.W., *Cement Chemistry*, Academic Press, London, 1990.
26.Odler, I. and Zhang, H., "Investigation on High SO_3 Portland Cement Clinkers", World Cement, pp. 73-77, February 1996.
27.Jøns, E., "Measuring the Sensibility of Sulphate in Cement", World Cement, pp. 65-68, June 1996.
28.Brown, P.W., "An Evaluation of the Sulfate Resistance of Cements in a Controlled Environment", Cement and Concrete Research, 11, pp. 719-727, 1981.
29.Ghorab, H.Y., Heinz, D., Ludwig, V. and Meskendahl, T., "On the Stability of Calcium Aluminate Sulphate Hydrates in Pure Systems and in Cements", VII International Congress on Chemistry of Cements, Vol. IV, pp. 496-503, Paris, (France), 1980.

THE VERY LONG TERM PERFORMANCE OF CEMENT AND CONCRETE

F.P. GLASSER
University of Aberdeen, Old Aberdeen
Scotland, U.K

Abstract

Cement and concrete technology in nuclear engineering shares goals with civil engineering: for example, how can we predict the future performance of materials and validate our predictive models? Differences in aspects of performance emerge: cement backfills may be expected to retain a chemical conditioning ability rather than physical strength. However, sufficient communality of approach exists so that methodologies used in nuclear technology may prove relevant to chemically-related aspects of the durability of conventional concretes.
Keywords: Alkalinity, cement chemistry, corrosion, durability, nuclear wastes.

1 Introduction

Electric power production from nuclear reactors leaves an environmental legacy of radioactive wastes. This legacy could be left to our descendants, but one school of thought holds that we should manage this waste within our own lifetimes so as to render it safe and achieve disposal.

Cement and concrete play a potentially important role in waste treatment and management [1] but the specialised nature of waste disposal have meant that civil engineers have had only peripheral involvement. Yet disposal concepts have implications for more conventional constructions; also, construction for durability holds important lessons for nuclear waste treatment.

This paper briefly reviews the nuclear industry and draws attention to the numerous parallels which exist. The paper deals entirely with civil applications, including medical wastes: defence wastes are not considered.

Concrete Technology for a Sustainable Development in the 21st Century edited by O.E. Gjørv and K. Sakai.
Published in 2000 by E & FN Spon, 11 New Fetter Lane, London EC4P 4EE, UK. ISBN: 0 419 25060 3

2 Nuclear waste

Much attention focuses on highly active wastes arising from reprocessing partially-burnt uranium oxide fuel. These highly active wastes are, however, not suitable for direct encapsulation in cement as they generate too much heat: typically on the order of 1 kW/kg even after a few years decay while in storage. Reprocessing does, however, generate a wide range of other wastes ranging from solids (graphite, metal chips) to fine particulates (sludges, flocs, ion exchangers) and soluble constituents (inactive process chemicals, decontaminates) as well as a large volume of low-active materials (cellulose, plastic, etc.,) which are often immobilised by cement. Even if reprocessing is not used, large quantities of miscellaneous wastes accumulate, e.g., ion exchangers. An engineer making high quality concrete would not choose to add these materials to cement but this is nevertheless what happens.

In general, cements are very tolerant of low to intermediate levels of radioactivity. Some radiolysis may occur, in which water is decomposed to H_2 and O_2, but the rate is not higher in cement pore water than in pure water. In practice, no problems have been encountered with internal pressurisation: H_2 and O_2 appear to diffuse with sufficient rapidity to avoid problems due to internal pressurisation of low-to-intermediate active waste in cement.

3 Encapsulation

A first priority is often conversion of liquids, sludges, flocs, particulates, etc., to stable monoliths. Cemented monoliths meet many of the requirements for formation by automated processes, as well as requirements for interim storage and transportation to disposal sites and for final emplacement into a repository.

Waste stream constituents may interfere with cement set, e.g. borates, but such interferences are commonly overcome by adding NaOH [2]. While cement passivates steel, it may accelerate corrosion of electropositive elements and alloys (e.g., alloys of Mg or Zr). Since corrosion is accompanied by gas evolution, these metal wastes may require oxidative pre-treatment prior to cementation.

The leaching of monoliths is a function of surface area: volume ratio and it is therefore desirable to minimise leaching by using large volume monoliths. Unfortunately, cement hydration leads to a large thermal excursion with the result that cracking may occur, the severity of which increases with volume. Thus a trade-off exists: minimising the surface: volume ratio while at the same time avoiding thermal cracking.

Conventional civil engineering avoids thermal cracking in part by using aggregate as a heat sink but this option is disliked by the nuclear industry because it greatly reduces the potential waste loading per unit volume. Instead, pozzolanic materials are used to reduce short -term heat evolution: for example, 60-70 % fly ash or 80-90 % slag replacement of cement may be specified.

4 Post encapsulation

The preferred ultimate disposal scenario for radioactive wastes is underground. The depth of burial is selected according to the amount and character of the waste and the host geology. For example, radioactive Cs and Sr are often concentrated in a separate waste stream. While these radioisotopes are quite hazardous, they also have short half-lives with the result that 300-500 years of confinement is often deemed adequate. Other species, especially actinides, may require much longer confinement. Hence short-lived species are often suitable for shallow disposal while longer-lived species may go into deeper disposal.

Movement of radioactive wastes across national boundaries is often politically impossible and it is apparent that a number of sites will be required for waste disposal, perhaps one or more in each autonomous country. Even countries without reactors may have significant medical and industrial wastes which will require disposal. Therefore, a range of geologic barriers are likely to be exploited for repositories (salt, clay, granite ...). It is not the purpose of this paper to explore these except to note that, in practice, the performance of most if not all geological barriers have proven very difficult to quantify. For example, granite may have very low permeability, but how will fissures affect its performance? If fissures link to form "short-circuit" hydrogeologic pathways, transit times back to the biosphere could be unexpectedly short. Problems of proving a safety case based on geology alone have led, and will probably continue to lead, to very long delays in licensing sites. These factors lead to four features of present day planning:

- Provisional of additional man-made barriers, making use (for example) of the chemical conditioning action of cements in the repository.
- Possibly prolonged periods of above-ground storage prior to final disposal.
- Designing repositories so that guarantees of subsequent retrievability could be given, in the event that research reveals better disposal options.
- Less emphasis on reprocessing, with its chemically and physically inhomogeneous waste arisings, and more emphasis on disposal of used fuel elements: so-called direct disposal.

Table 1. Application areas in nuclear engineering

Application	Requirement
Shielding	Dense, perhaps using high density aggregate.
Grouting	Fluid grouts to make monolithic waste forms.
Structures	Containers for storage and shipment. Underground repository structures. Long-term interim storage facilities.
Backfills	Chemical and physical barriers to the release and mobility of radionuclides.

Translation of these requirement into specific applications areas is shown in Table 1. Some of these requirements follow from the prolonged period of initial storage and the possible need to keep repositories accessible for retrieval. This table emphasises the physical uses of concrete and, as such, will perhaps seem within the remit of conventional engineering: Table 2 suggests some points of relevance. However, there are a number points of divergence between civil engineering practice and radioactive waste treatment. Table 3 summarises these: a preference for cement-rich formulations has already been noted.

Table 2. Physical barrier action of concrete

Performance aspect in a repository	Relevant non-nuclear applications
Hardening, strength gain, permeability and cracking of waste monoliths	Toxic and hazardous conditioning and wasteform properties
Storage, transport and accident scenario	Conventional codes, including fire risk assessment
Crack resistance and impermeability	Design of water-retaining structures
Seals, controlled permeability backfills to control gas migration and canister corrosion	Conventional shrink resistance: permeation and corrosion studies
Long term material incompatibility	Selection of aggregates, possible limitations on other barriers, e.g. clay

These requirements deserve comment. Nuclear engineers are attempting to achieve immobilisation on a time scale with little precedent. Performance assessment models have therefore to be applied for which experience of cement in the environment is a useful, if qualitative, guide. These models show that good physical qualities (e.g. impermeability) are desirable for storage and to achieve good performance in accident scenarios: also in shallow burial. But physical qualities count for less in the post-closure phase of deep repositories, especially in favourable geological terrains: chemical immobilisation dominates the assessment of deep or impermeable host formations. Hence study of the *chemical* properties at long ages is an inescapable aspect of performance assessment.

Table 3. Special features of concretes used in nuclear waste treatment

Feature	Comment
Possible exclusion of organic plasticisers, waterproofers, etc., in formulations	Concern that the degradation products of organics may complex and solubilise radionuclides
Maintenance of chemical properties, especially high pH	Chemical, rather than physical control of radioactive species becomes important as time scale is extended
Internal oxidation - reduction properties	Reducing conditions favour retention of some species, e.g. Tc.
Resistance to thermal degradation over prolonged periods	Inclusion of heat-generating wastes may generate a thermal pulse for *ca* 10^2-10^2 years.

5 Performance

5.1 Benchmarks

Engineers are not free to select a location for a repository; instead, sites are often chosen based on a combination of factors including geological conditions (low permeability of the host rock, slow or stagnant water flow, geological stability) geographic factors (minimisation of transport, but with site access not too difficult, although remote from major urban conurbations) and availability of the site, i.e., no legal or planning restrictions. Amongst the terrains which have been or are being investigated are (i) granites or tuffs, (ii) clay, (iii) disused ore mines (iron, gypsum) and (iv) salt. It follows, therefore, that any one national programme might need to benchmark only one environment but globally, quite different environments will couple with those of cement.

The repository may also contain, heat-generating wastes: for example, depending on inventory, operational temperatures as high as 160-180 °C have been envisaged. This thermal pulse would occur shortly after closure and last for several decades, perhaps centuries, before significant thermal decay occurred.

5.2 Internal environment

Because cements are normally formulated with water contents in excess of those required for chemical hydration, many concepts applicable to aqueous solutions are also applicable to cements. Table 4 summarises characterisation properties: strictly, several apply to the permeating aqueous phase although we colloquially refer, for example, to "cement pH".

Table 4. Internal properties of cement

Property	Characterisation data
Pore solution composition	Initially, Na and K dominate. As these are leached, Ca dominates.
pH and pH buffers	Initially high, 12.5 to 14. As Na and K leached, buffered to pH 12.5 (25 °C) by $Ca(OH)_2$, C-S-H etc.
E_h and E_h poising within cement	Neutral, ca 100 to 200 mV. Neither affected by cement components nor well poised; may be lowered by S^{2-} in slag, by embedded steel, etc.

The two relevant concepts are pH and E_h[3]. The acidity or alkalinity, measured on the pH scale is familiar. Perhaps less familiar is the concept of buffering. For example, while the pH of water is ~ 7, addition of even traces of dissolved constituents may shift the pH: thus the water system is not buffered. Cement systems are, however, buffered: their content of $Ca(OH)_2$ and/or C-S-H gel and minor phases buffer the system. Thus if an acidic substance such as dissolved CO_2 is added, some Ca reacts to precipitate $CaCO_3$ thereby maintaining the high pH. The pH does not depend on *amount* of substance but the buffering capacity does. It is a key parameter in performance assessment.

Cement pore fluids also have an E_h, or redox potential. This is expressed relative to a standard hydrogen electrode (0.00 V at 25 °C, and unit activity of H^+ and H_2). Portland cement is normally made under oxidising conditions and has a measured potential of +100 to +200 mV, i.e., it is slightly oxidising. However cements lack poising capacity (the E_h equivalent of buffering) with the result that added electroactive substances affect the E_h. Blast furnace slag, which typically contains ~ 1 % sulfide, markedly lowers the Eh to negative ~ 300 mV and provides a poising reserve.

The E_h and poising capacity affect the performance of cement for immobilisation. For example, TcO_4^-, which is somewhat soluble at high pH, is reduced by S^{2-} (sulfide) to the much less soluble Tc^{4+} state. Thus slag-rich blends, which contribute sulfide, may be preferred for Tc-containing wastes. Corroding steel may also lower the E_h.

6 The source term

Many performance assessments, in attempting to model the transport and diffusion of radioactive species from a repository make use of a "source term". This inputs a limiting solubility value for each radionuclide, the value of which is independent of species concentration. An analogous concept is used for cement conditioned toxic-wastes. Defining the internal state of cement in terms of pH, E_h, buffering and poising capacities has been done and these data are available for technology transfer.

One reason why cements are attractive in waste treatment is the reduction in solubility achieved for many species. For example, depending on the choice of anion, it is possible to prepare uranium (VI) solutions of *ca* 1 M concentration. But when mixed with cement, U (VI) is very insoluble: probably not exceeding 10^{-9} or 10^{-10} M. The effectiveness of cement conditioning, as indicated by limiting solubilities, varies enormously from one species to another. For example, cement has little impact on the solubility of Cs and Sr, but great impact on the solubilities of U, Th, actinides, etc.

The solubility value is not dependent on the physical state of the cement, but only on the continued presence of cement hydrate phases and their ability to condition water to a particular range of pH and E_h values.

Below the solubility threshold, high surface area C-S-H may also exhibit useful sorption properties. Thus Ca-rich C-S-H's tend to sorb anionic species: for example, C-S-H in conventional Portland cement sorbs sulfate. On the other hand, Ca - poor C-S-H's, occurring in formulations rich in fly ash, silica fume, etc., preferentially sorb cations. This sorption is, however generally weak but just as silica fume blends are used to reduce soluble sodium and potassium, such blends also exhibit enhanced Cs sorption. This also mirrors experience of inhibiting alkali-aggregate attack, where silica fume is used to lower soluble alkali contents.

7 Specific deterioration processes

7.1 Introduction

It has proven easier to model purely chemical deterioration processes than those which combine physical and chemical attack. This is because the latter involve reaction kinetics, a topic which until recently has not received much systematic attention.

7.2 Chemical deterioration

The term 'chemical dissolution' suggests attack by aggressive agents such as mineral acids. However, pure water also behaves towards cement as an acid. The rate of dissolution of a cement barrier in contact with initially pure water has been successfully modelled [4] and practical tests confirm in detail the model [5].

These models utilise (i) the known Ca solubility of $Ca(OH)_2$ and of C-S-H (or mixtures thereof) in pure water and (ii) follow the decline in pH as deterioration progresses. Fig 1 depicts the solubility relationships: pH values are marked at appropriate points. Various threshold pH values can be noted: for example, the threshold value at which high pH no longer passivates steel. There is some question about its exact numerical value, but steel probably becomes active to corrosion below pH 11 to 11.5. Blending agents (fly ash, silica fume) lower the mean Ca:Si ratio and decrease pH once the buffering reserve of $Ca(OH)_2$ is consumed.

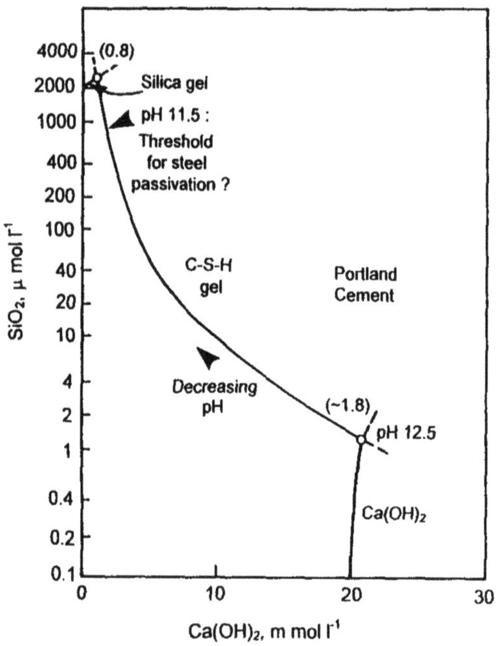

Fig1. Solubility of $Ca(OH)_2$ and SiO_2 gel at ~20 °C data are for well-cured preparations values in ( ) are the molar Ca:Si ratios of the C-S-H gels. Circles show solution invariant points.

In these models, which assume that permeating water becomes saturated with respect to $Ca(OH)_2$, the rate of deterioration per unit mass of cement is directly proportional to the volume of permeating water. This illustrates the importance of siting repositories in a low (preferably zero) ground water movement. The vast majority of groundwaters are not of course chemically pure. Nevertheless, the confidence generated by the evident success of these models has led to tackling the impact on cement of more complex groundwaters. Geochemists have much experience of calculating water-solid interactions, which has led to development of computer-based routines for calculation of mineral-solution equilibria. Extensive data bases have been compiled for minerals but, as most naturally-occurring waters are at pH 5 to 10, i.e., lower than in cements, the data bases do not include cement substances which are typically stable at high pH.

A series of papers describes the evolutionary development of a database [6-9]. Meantime, progress has continued in the assessment and compilation of databases and in the development and application of coupled reaction and transport models. These are described in the final report of CHEMVAL 2 [10]. Given its clear lead in accuracy and its recognition of the importance of transport, it has been decided to harmonise the cements database to CHEMVAL 2. This effort is supported partly by The European Commission under JETDEM and the cements database is expected to appear in its final report, early in 1999.

Earlier versions of the database have been used to calculate attack of single salts, e.g., of NaCl, on cement substances. The success of these calculations have led to modelling more complex interactions, for example of chloride, sulfate and carbonate attacking CaO-SiO_2-H_2O phases [11].

If the results of calculations are compared and contrasted with field studies on the deterioration of concretes, the following conclusions emerge:

- Dissolution of cement components occurs in natural waters and is an important mode of deterioration but precipitation of solid reaction products also affects and, in some instances, dominates the course of reaction.
- Reaction pathways lead to formation of the same minerals and mineral assemblages as are predicted from chemical thermodynamics, thereby providing corroboration of the validity of models.
- Experience of cement in the environment, by its nature, provides only qualitative data, whereas modelling can be made quantitative especially with respect to mass balances and solid mineralogy.
- Cement in the environment does provide evidence of the importance of kinetic controls over degradation pathways and rates of attack.

7.3 Physical deterioration

Some aspects of physical deterioration are relatively unimportant hazards in nuclear waste disposal, e.g. freeze-thaw in an underground disposal scenario. The impact of dissolution processes can be evaluated only insofar as porosity-permeability relationships are known. However, dissolution of concrete products is complicated by two factors: tortuosity introduced by the presence of aggregate and by incongruent dissolution of cement components. The former is well-known to engineers but the latter is perhaps less well appreciated. An example suffices to explain. If one compares the weathering resistance of $Ca(OH)_2$ - based mortars with those of Portland cement-based formulations, the superiority of the latter is evident. The reason lies in the low solubility of silica and alumina, present to ~ 25 % in Portland cement but largely absent from hydrated lime binders. The silica and alumina liberated during attack are resistant to dissolution and form semi-protective layers over undissolved cement: together with carbonate formation, they protect Portland cement with the result that it is much more durable than hydrated lime.

The above considerations apply without modification to dissolution in pure or slightly carbonated water, e.g., rain water. But cement is often exposed to salt solutions: for example, in the oceans. Reactions occurring in sea water are well characterised, and are marked by formation of a surface layer consisting mainly of brucite, $Mg(OH)_2$, silica gel and calcite, $CaCO_3$. Sulfate in sea water penetrates to greater depth than Mg, giving rise to a subsurface zone characterised by formation of ettringite and either (or both) gypsum and anhydrite. This zone plays an important role in the physical stability of the solid mass. If the net volume of products formed exceeds that of the cement substances which they replace, expansion and cracking occur. Cracking and spalling exposes fresh surface to attack and largely cancel the semi-protective action of the surface layer. The present situation is that we can predict the mineralogical zonation, because it occurs in response to pH and chemical gradients through the cement paste: each zone is in a local equilibrium. But the balance between precipitation and dissolution, with consequences for cracking, cannot be predicted.

8 Cracking

Physical barrier performance is much reduced, perhaps even cancelled, if barriers crack. Much depends on the spacing and width of cracks which affects their transmissibility. By and large, we know little about the long-term dimensional changes occurring in concrete except in a few extreme cases: of alkali - aggregate reaction or, as described previously, sulfate attack. However, field observations and laboratory experience shows that cracks, once formed, may reheal. The rehealing process may not ensure that strength is restored but it is known that even mechanically weak infillings can reduce flow, perhaps to the same order of magnitude as in uncracked concrete. The nuclear waste programme has contributed significantly to our understanding of the process by which cracks can heal [12]. Briefly, cracks are sealed partly by accumulation of debris and partly by precipitation of $CaCO_3$. The healing is assisted if part of the crack remains accessible to dissolution by percolating water: for example, branched cracks heal faster than unbranched because the branches are less affected by $CaCO_3$ precipitation and continue to contribute soluble Ca^{2+}. This soluble Ca migrates towards the zone where it mingles with incoming carbonate-containing water and precipitates, thereby blocking the crack. The potential relation of creep to cracking seems not to have been explored in the nuclear context.

9 Discussion

This presentation is, inevitably, somewhat telegraphic. But it is apparent that although the nuclear waste programme has contributed much knowledge of the long-term behaviour of cements and concrete, we have more to learn. Among the issues not discussed in detail in this presentation are:

- Advantages and disadvantages of blended cements: when do improved physical properties outweigh the reduction of pH buffering capacity resulting from blending agents?
- Can we estimate reliably long-term corrosion rates of embedded steel and of steel (or other alloy) canisters in concrete?
- Elevated temperatures may be encountered in deep repositories on account of geothermal gradients and heat generation by waste, coupled with poor thermal

conductivity of rocks. Can we predict the future of engineered concrete structures subject to prolonged thermal excursion in the range 55 ° to 160 °C?

- How do organics behave in cement in the long term? Can we use plasticisers, etc., without also solubilising radionuclides, many of which react with organics forming soluble complexes?
- Are we confident that aggregates which remain dimensionally stable for 50-100 years will remain stable for 1,000-10,000 years? And if so, with what confidence can predictions be extended to elevated temperatures?
- Are there geochemical environments unsuited to disposal of cement conditioned wastes? and if so, can we quantify these restrictions?
- What other factors, not yet assessed, could impact on cement and concrete properties over 10^4 years?

References

1. Conner, J.R. (1990) Chemical Fixation and Solidification of Hazardous Wastes. Van Nostrand-Reinhold New York.
2. Csetenyi, L.J. and Glasser, F.P. (1995) Borate Retardation of Cement and Phase Relations in The System Na_2O-CaO-B_2O_3-H_2O. Advances in Cement Res. **1** No. 25, 13-19.
3. Macphee, D.E. and Glasser, F.P. (1993) Immobilization Science of Cement Systems. Mater. Res. Soc. Bulletin, XVIII No. 3, 65-71.
4. Adenot, F. and Buil, M. (1992) Modelling of the Corrosion of Cement Paste by Deionised Water. Cement and Concr. Res., **22**, 489-496.
5. Faucon, P., Gerald, B., Jacquinot, J.F. and Marchand, J. (1998) Water Attack of a Cement Paste: Towards an Improved Accelerated Test? Advances in Cement Res. **10,** No.2, 67-73.
6. Atkins, M., Glasser, F.P. and Kindness, A. (1992) Cement Hydrate Phases: Solubility at 25 °C. Cement and Concr. Res. **22**, 241-246.
7. Atkins, M., Lachowski, E.E. and Glasser F.P. (1993) An Investigation of the Solid and Aqueous Chemistry of 10 Year Old Portland Cement Pastes: With and Without Silica Modifier. Advances in Cement. Res., **5**, No19, (97-102)
8. Damidot, D and Glasser, F.P. (1995) Investigation of The CaO-Al_2O_3-SiO_2-H_2O System at 25 °C by Thermodynamic Calculations. Cement and Concr. Res., **25**, 22-28.
9. Lachowski, E.E., Kindness, A., Glasser, F.P. and Luke, K. (1997) Compositional Development (Solid and Aqueous Phases) in Aged Slag and Fly Ash Blended Cement Paste. Proceedings of the X International Congress on the Chemistry of Cement, Vol 4, Paper 4 IV 066.
10. Read, D. and Falck, W.E. (Editors) (1996) CHEMVAL2: A coordinated Research Initiative for Evaluating and Enhancing Chemical Models in Radiological Risk Assessment. EUR 16648 EN (European Commission, Brussels and Luxembourg) ISBN 92-827-8988-8
11. Stronach, S.A. and Glasser, F.P. (1997) Modelling the Impact of Abundant Geochemical Components on Phase Stability and Solubility of the CaO-SiO_2-H_2O

system at 25 °C: Na^+, K^+, SO_4^{2-}, Cl^- and CO_3^{2-}. Advances in Cement Res. **9**, No. 36 167-182.

12.Harris, A.W., et.al (1998) The Performance of Cementitious Barriers in Repositories. EUR 17643 EN (European Commission, Brussels and Luxembourg) ISBN 92-828-2152-8

DEGRADATION MODEL FOR REINFORCED CONCRETE STRUCTURES UNDER SALT ATTACK ENVIRONMENT

Analytical process for degradation in the lifetime

K. MARUYAMA, T. SHIMOMURA
and H. HAMADA
Civil and Environmental Engineering
Nagaoka University of Technology
Nagaoka, Japan

Abstract

This paper discussed on the degradation model for reinforced concrete structures under salt attack environment. Reviewing the studies on corrosion problems of concrete so far, the authors constructed an analytical model which could treat the deterioration process of concrete structures with time elapse, from the beginning to the end of life time. The input of model consists of the environmental conditions, material properties and structural details, and the output shows the movement and accumulation of corrosion inducers (water, oxygen, chloride ion), the process of steel corrosion, the initiation and propagation of cracks as well as the reduction of flexural capacity of concrete structures. Finally, some analytical examples are presented.

Keywords: Salt attack, Corrosion of reinforcement, Degradation of capacity, Analytical model

1 Introduction

Corrosion of reinforcing bars in concrete has been a major concern for concrete engineers in a long time, and a lot of technical countermeasures have been developed widely in various aspects. Extensive research works have made it possible to evaluate the quantitative analyses of penetration and dispersion of corrosion inducers (water, oxygen, chloride ion), corrosion speed of steel and expectation of cracking of concrete due to corrosion of steel [1]. Moreover, studies are more actively conducted for the prediction of life time of existing concrete structures and the development of materials and repair methods to reduce the deterioration speed of structures [2,3].

This study aims at making a frame work to predict the time-depending changes of structural capacity of reinforced concrete structures. Combining the experimental

Concrete Technology for a Sustainable Development in the 21st Century edited by O.E. Gjørv and K. Sakai.
Published in 2000 by E & FN Spon, 11 New Fetter Lane, London EC4P 4EE, UK. ISBN: 0 419 25060 3

studies on the deterioration of mechanical properties of structures due to salt attack [9-11] with the studies on the penetration and dispersion of corrosion inducers into concrete [4-7] and on the mechanism of corrosion of steel [8], the authors proposed the overall simulation system on how concrete structures is deteriorating during a life time under the salt attack environment.

2 Outline

2.1 Deterioration process by salt attack

Deterioration process of reinforced concrete structures is generally illustrated as shown in Fig.1 [4]. At the first stage I, chloride ions are penetrating and dispersing into concrete, and the amount of the ions at the surface of reinforcing bars is increasing up to a certain level. The second stage II indicates the period from the beginning of generation of rust to the initiation of cracking due to the expansive force of rust. At the stages of III and IV, concrete structures display apparent defects produced by propagation of cracks.

2.2 Outline of analysis

The analysis at the first stage is to simulate the dispersion of corrosion inducers, such as water, oxygen and chloride ion into concrete. For simplicity, one dimensional analysis is adopted. Referring to the literature [5], a distribution of moisture in concrete was firstly analyzed. Based on the results, movements of oxygen and chloride ion were analyzed using the existent methods [5,7]. Maruya's model [7] for chloride ion movement can treat the influence of dry-wetting repetition condition. The movement of water was simulated using Shimomura's model [6] which takes account of the micro pore structures of concrete.

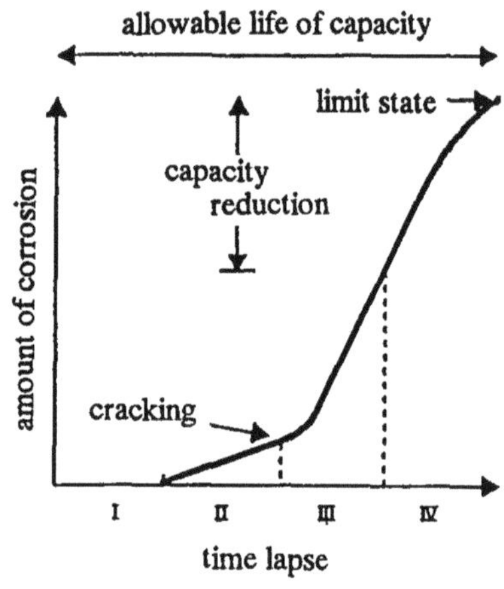

Fig.1 Deterioration process

At the stage II, corrosion rate was calculated based on the corrosion current density [5], and the prediction of cracking was done by the theory of thick shell structure. In the analysis of corrosion, reinforcing steel bars were assumed as uniformly corroded. A local corrosion was not taken into account. The corrosion rate at the stages III and IV was set as 10.2 $mg/cm^2/year$ after cracking under severe environmental condition [8]. Analysis was conducted assuming that the amount of rebar corrosion was linear to the crack width due to steel corrosion [9]. The details are discussed in Section 3. Finally, the reduction of flexural capacity was evaluated with regard to the longitudinal crack width due to corrosion of longitudinal reinforcing bar. More details are in Section 4.

The analytical flow diagram is shown in Fig.2.

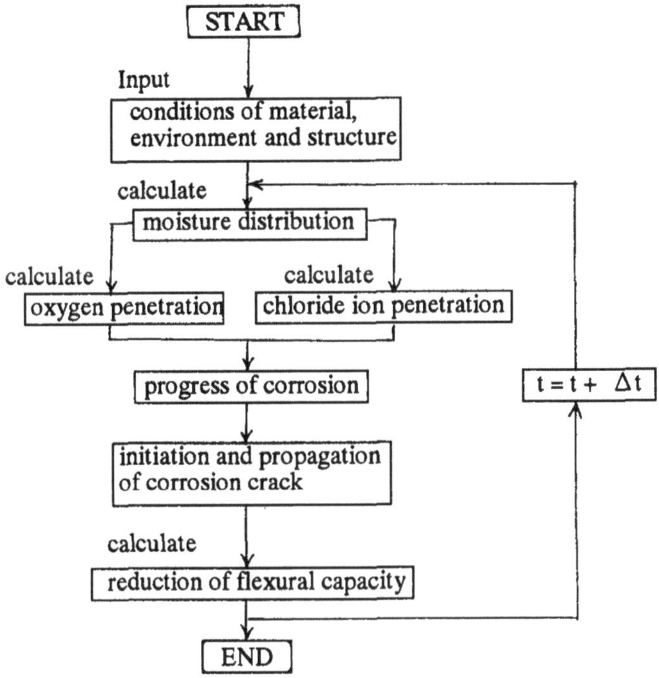

Fig.2 Flow chart of analysis

3 Cracking due to corrosion

3.1 Test specimens

Exposure test and electro-chemical corrosion test were conducted to examine how the amount of rebar corrosion in concrete and the corrosion crack widths increase with the lapse of time. Table 1 shows the mix proportion of concrete. For acceleration of corrosion of rebar, natrium chloride (NaCl) was added to water to make 3 % concentration solution. The test specimen is shown in Fig.3. Concrete cover thickness was 25 mm, and the rebars were D10 (deformed bar at 10 mm diameter) for the exposure test and $\phi 9$ (round bar at 9 mm diameter) for the electro-chemical corrosion test.

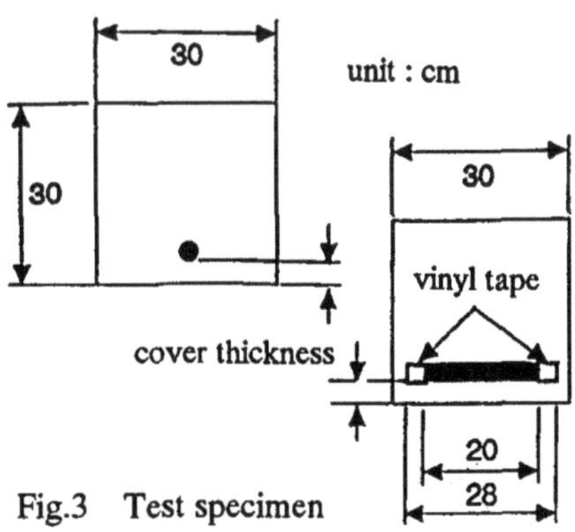

Fig.3 Test specimen

Table 1 Mix proportion of concrete

Type of test	W/C	s/a	unit content (kg/m^3)					
			W	C	S	G	ad.	NaCl
exposure	63	50	210	333	830	823	1.17	6.51
electro-chemical	63	42	165	260	799	1102	0.65	5.33

[Note] W/C : water to cement ratio (mass), s/a : sand to aggregate ratio (volume)
W : water, C : cement, S : sand, G : gravel, ad. : admixure

3.2 Details of tests

Concrete specimens for the exposure test were placed on the coast of Japan Sea, and the corrosion crack widths of specimens and the amount of rebar corrosion were measured at 2, 3, 4 and 6 years after exposure. The target length of rebar was 200 mm, and the ends of rebar were sealed by wrapping with vinyl tape.

For the electro-chemical corrosion test, specimens were submerged in water and direct current was charged so that a rebar was anode and two copper plates beside the specimen were cathode. 3 % concentration solution of NaCl was used for water to accelerate corrosion of rebar. The specimen was placed in water as the cover part being upward and out of water so that the measurement of crack widths of a specimen due to rebar corrosion was made easy. The crack widths were measured at 50 mm intervals and the average values were summarized. The amount of rebar corrosion was determined by weight difference before and after corrosion test.

3.3 Test results

The relationships between the amount of rebar corrosion and the average crack width are shown in Fig.4. The results from the exposure test and the electro-chemical corrosion test were expressed as linear regression curves. The calculated result in the figure was obtained by the corrosion crack model proposed by the authors [10]. The basic concept of the model is shown in Fig.5, and the relationship between the amount of corrosion and the surface crack width is formulated as below. The cracked part of concrete was assumed to behave itself as a rigid body.

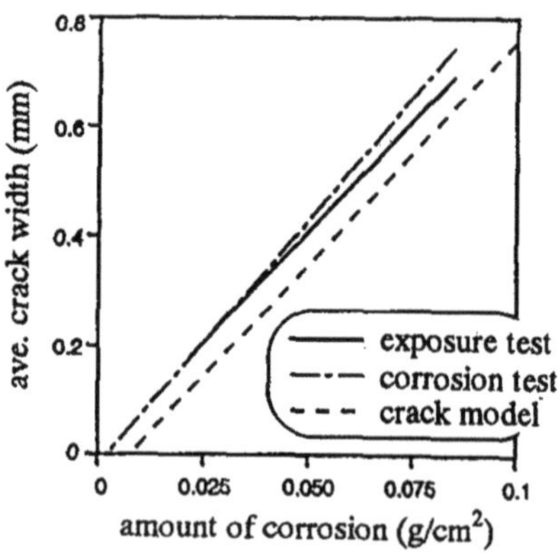

Fig.4 Crack width and amount of corrosion

$$\delta = (a\Delta\theta + u\cos\beta)$$
$$= 2u\left|\cos\beta + a(\cos\alpha + \cos\beta)/\{h + u(\sin\alpha \cdot \sin\beta)\}\right|$$

where, u: thickness of rust

The values for calculation were 3.2 of expansion rate of rust, 0 for angle α, 60 mm of crack length and 0.017 mm for thickness of rust at cracking. These values look good for the test results. Judging from Fig.4 the result of the electro-chemical corrosion test was quite similar to that of the exposure test while the corrosion product, kind of rust, by the electro-chemical test may be different from that of the exposure test. These results tell that the amount of bar corrosion in concrete can be estimated by measuring the average crack width. In other words, the reasonable prediction of bar corrosion in the life time of structures enables to predict the possible corrosion crack widths.

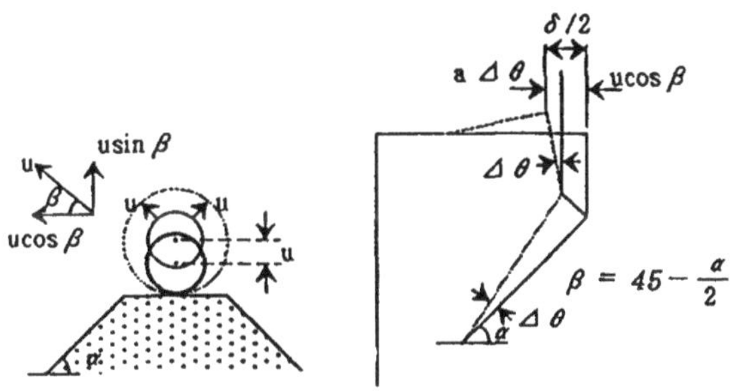

Fig.5 Crack model

4 Reduction of flexural capacity

The previous test results done by the authors [11] are summarized in Table 2. The test specimens were rectangular beams at an identical dimension of 100 mm (width) x 150 mm (height) x 1200 mm (length). Only one longitudinal reinforcing bar (D13) was placed with a cover thickness of 25 mm. The specimens were submerged in water and the direct current was charged by the electro-chemical corrosion test until the targeted corrosion crack widths in the longitudinal direction were obtained. With given longitudinal crack widths
the specimens were loaded statically up to failure. The loading test was done in a manner as the simply supported beams with 2 points loading.

All beams with/without corrosion cracks failed in flexure, and the ultimate flexural capacity of each beam is normalized by the reference beam without corrosion. The results are shown both in Table 2 and Fig.6.

Table 2 Test results

specimen	crack width (mm)	Max. load (kN)	capacity ratio
No.1	non	21.6	1.0
No.2	0.05	22.3	1.03
No.3	0.30	22.8	1.05
No.4	0.50	20.2	0.94
No.5	1.00	19.6	0.91

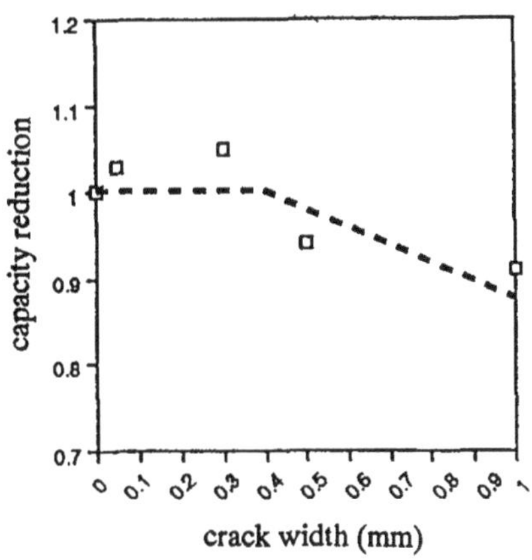

Fig.6 Reduction of capacity vs. crack width

Even if a corrosion crack is initiated, the flexural capacity may not decrease unless the crack width becomes large. For the analysis of flexural capacity reduction in the following section, the broken line in Fig.6 was adopted.

5 Analytical case study

5.1 Environmental conditions

The environmental conditions used in the analysis was such that the atmospheric temperature was 20°C, the relative humidity was 60 % and that the target structural member was on the coast having floating chloride ions in the air constantly [5]. Ten days drying with the above temperature and humidity and an instant saturation of the surface of the structural member with sea water were assumed to repeat.

5.2 Material properties

Material properties determined by the mix proportion of concrete is listed in Table 3. The coefficient of oxygen dispersion rate in concrete and that of chloride ion were determined referring to the literatures [1] and [7], respectively. 25 and 50 mm were chosen as the concrete cover thickness of member, and D13 rebar (deformed, 13 mm diameter) was assumed.

Table 3 Data for analysis

CONCRETE		
coefficient of dispersion (cm^2/sec)	oxygen, saturated	10^{-8}
	oxygen, unsaturated	10^{-6}
	chloride, surface	1.90×10^{-3}
	chloride, inside	1.59×10^{-3}
compressive strength	(N/mm^2)	30.4
tensile strength	(N/mm^2)	2.61
Young's modulus	(kN/mm^2)	26.6
Poisson's ratio		0.20
chloride solution in mixing water		yes or non
STEEL BAR		
Young's modulus	(kN/mm^2)	200
Poisson's ratio		0.3
Vulk expansion coeff. of rust (α)		3.2

5.3 Results

With the above mentioned conditions the simulation was conducted as discussed in Section 2. Table 4 shows the lapse of time at the initiation of cracking with the necessary amount of rebar corrosion.

Three cases are as follows;

① 25 mm of cover thickness, mixing water of 3 % solution of chloride ion

② 25 mm of cover thickness, normal mixing water

③ 50 mm of cover thickness, normal mixing water

Table 4 Analytical results on cracking

case	time lapse at cracking, (days)	amount of corrosion, (mg/cm^2)
①	721	5.34
②	1550	5.34
③	3497	16.8

Comparing the results of case ① with those of case ②, the existing of chloride ion accelerated the corrosion rate of rebar and the appearance of cracking was earlier more than two years. The results of case ③ proved the effectiveness of cover thickness. The 25 mm difference of cover thickness between cases ② and③ can delay the crack appearance by more than five years.

Fig.7 shows how the amount of rebar corrosion increases with lapse of time. The part of rapid change in curves should indicate the initiation of cracking. Once cracking happens, the corrosion rate increases. The rate of corrosion, however, looks similar in three cases under the same environmental condition.

The relationship between the amount of rebar corrosion and the average crack width is shown in Fig.8. As mentioned in Section 3, the relationship is influenced by the cover thickness. The rigid body movement of cracked part of concrete makes the surface crack width large as shown in Fig.5. The result of case ③ (50 mm cover thickness) represents this phenomenon clearly in Fig.8.

Fig.9 shows how the flexural capacity of member reduces with the lapse of time. The period when the capacity goes down to the 90 % of the initial stage is 15 years for case①, 18 years for cases ② and ③. As is expected, the existing of chloride ion in mixing water of concrete shortens the time lapse up to cracking and accelerates the reduction of flexural capacity. It may look strange that there is little difference

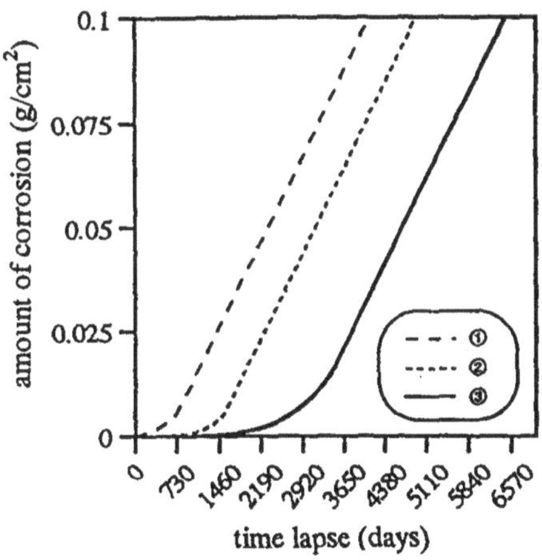

Fig.7 Corrosion vs. elapsed time

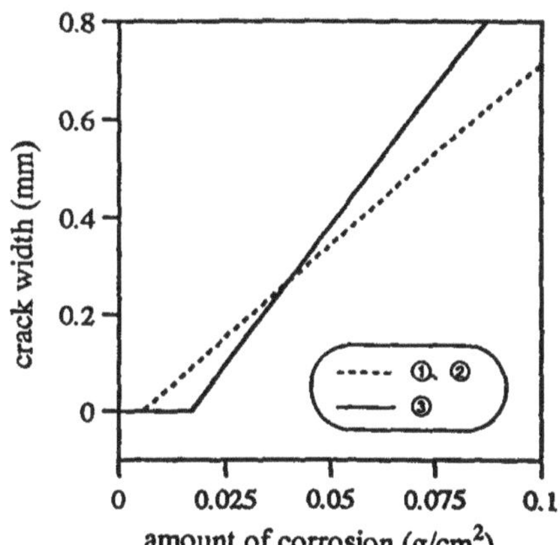

Fig.8 Crack width vs. amount of corrosion

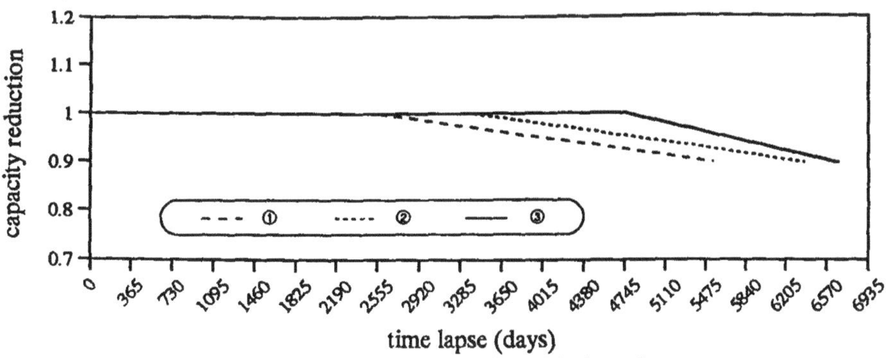

Fig.9 Reduction of capacity with time elapse

between case ② (25 mm cover thickness) and case ③ (50 mm cover thickness). This is because the rate of flexural capacity reduction is mainly related to the corrosion crack width in the analytical model.

6 Conclusion

There have been conducted so many studies on corrosion of steel in concrete and on mechanism of corrosion of steel in concrete. Those studies are, however, concerned with only some aspects of corrosion problem of reinforced concrete structures.

Although the discussion in this paper is just in the entrance of integrated approach of corrosion problem of reinforced concrete structures, the proposed approach shows the possibility of the life cycle assessment of reinforced concrete structures exposed to the salt attack environment.

Some analytical case studies show that increment of cover thickness should be effective to delay the penetration of chloride ion coming up to steel bar and to prevent the deterioration of capacity until corrosion cracks develop. The reduction of capacity, however, might be accelerated after cracking.

7 References

1. Japan Concrete Institute. (1994) *Technical Report on Repair Methods of Concrete Structures (II)*, JCI, Tokyo. (in Japanese)
2. Japan Concrete Institute. (1991) *Technical Report on Durability Design of Reinforced Concrete Structures*, JCI, Tokyo. (in Japanese)
3. Japan Concrete Institute. (1992) *Proc. of Symposium on Repair Methods of Concrete Structures*, JCI, Tokyo. (in Japanese)
4. Miyagawa, T. and et al. (1988) Prediction of Life Cycle of Concrete Structures exposed to Salty Atmosphere and Durability Design, *Proc. of Symposium on Prediction of Life Cycle of Concrete Structures and Durability Design*, pp.17-24. (in Japanese)
5. Japan Concrete Institute. (1996) *Technical Report on Repair Methods of Concrete Structures (III)*, JCI, Tokyo. (in Japanese)
6. Shimomura, T. (1993) *Drying Shrinkage Model of Concrete based on Pore Volume Distribution Density Function*, Doctoral Thesis at University of Tokyo.(in Japanese)
7. Maruya, T. (1996) *Development of Analytical Method on Chloride Ion Movement in Concrete*, Doctoral Thesis at University of Tokyo. (in Japanese)
8. Uhlig, H.H. and Revie, W. (1989) *Corrosion Reaction and Its Control - Third Version, Translation in Japanese*, Sangyo Tosho. (in Japanese)
9. Takaoka, U. and et al (1989) Cracking of Concrete due to Corrosion of Steel, *Proc. of JCI*, Vol.11, No.1, pp.591-596. (in Japanese)
10. Kimura, T. and et al (1996) Quantitative Evaluation of Corrosion of Steel in Reinforced Concrete Members, *Proc. of JCI*, Vol.18, No.1, pp.771-776. (in Japanese)
11. Nakata, Y. and et al (1990) Influence of Corrosion of Reinforcement on Capacity of RC Beams, *Proc. of JCI*, Vol.12, No.1, pp.551-556. (in Japanese)

INSTRUMENTED DURABILITY SURVEILLANCE OF CONCRETE STRUCTURES

Development of in service inspection systems

F. FLUGE
Norwegian Public Roads Administration, Oslo, Norway

Abstract

Instrumented surveillance of durability parameters has been performed on the pontoons of two Norwegian floating bridges constructed in 1991-92, data related to corrosion of reinforcement have been recorded 5 years after the pontoons were exposed to seawater, These registrations do not indicate corrosion of the reinforcement. This is supported by the visual inspections and the supplementary test on chloride ingress.

The reliability of the recorded data and the feasibility of this type of instrumented surveillance as part of a management system are discussed also taking economical factors into account.
Keywords: Management systems, durability surveillance, lightweight aggregate concrete, floating bridge.

1. Introduction

Efficient maintenance of highway structures is a high priority issue within the Norwegian Public Roads Administration. Efforts are made to establish reliable management systems for registration of the structures condition, taking decision as to the type of maintenance needed and when the repairwork should be performed.

In service inspection is an integrated part of a management system. Inspections of bridges are performed on a regular basis, comprising

- o daily/weekly observations of road conditions
- o annual visual inspections
- o principal inspection every fifth year, including both close visual inspection and collection of field data
- o special condition survey when deemed necessary.

Concrete Technology for a Sustainable Development in the 21st Century edited by O.E. Gjørv and K. Sakai.
Published in 2000 by E & FN Spon, 11 New Fetter Lane, London EC4P 4EE, UK. ISBN: 0 419 25060 3

During the last decades, instrumented surveillance of bridges has been investigated and different test set-ups have been evaluated. Up until now, monitoring of bridge-behaviour mainly has had relevance to static and dynamic response. Surveillance of durability parameters in order to record the durability status of the structure is however of great interest, but still in its infancy.

This paper will focus on methods for continuously measuring durability parameters related to corrosion of reinforcement, evaluate present findings, and discuss the usefulness of this type of instrumented surveillance.

Reference is made to the experience from two instrumented floating bridges located on the west coast of Norway. The objective is to correlate monitored effects with data from periodical inspections and discuss if these types of registrations can in the long-term decrease the magnitude of inspection work.

2. Field investigations

During the early nineties two floating bridges were constructed in Norway, the Bergsøysundet bridge in Møre & Romsdal County and the floating part of the Nordhordland bridge in Hordaland County.

Floating bridges had not previously been built in Norway and as such, they represented a new bridge type. Exposed to severe coastal environment it was essential to establish a relationship between loads acting on the bridge and its static and dynamic response. In addition, a system for durability surveillance of the pontoons of the two bridges was installed.

The pontoons of both bridges were constructed in Lightweight Aggregate Concrete. The structural reinforcement in the pontoons consists of prestressing steel and bonded reinforcement. The bonded reinforcement is protected by a sacrificial anode cathodic protection system.

This paper deals with concrete properties relevant to durability, instrumented corrosion surveillance of the reinforcement, inspections performed and the interconnection of these three items.

2.1 Description of the bridges

The Bergsøysundet bridge, figure 1, is a part of the primary road E 36 along the west coast of Norway, connecting the islands Bergsøy and Aspøy close to the town Kristiansund.

The floating bridge consists of a 830 meter long steel frame supported on seven pontoons constructed in high strength LWA concrete. The pontoons were constructed in the period from April 1991 to February 1992.

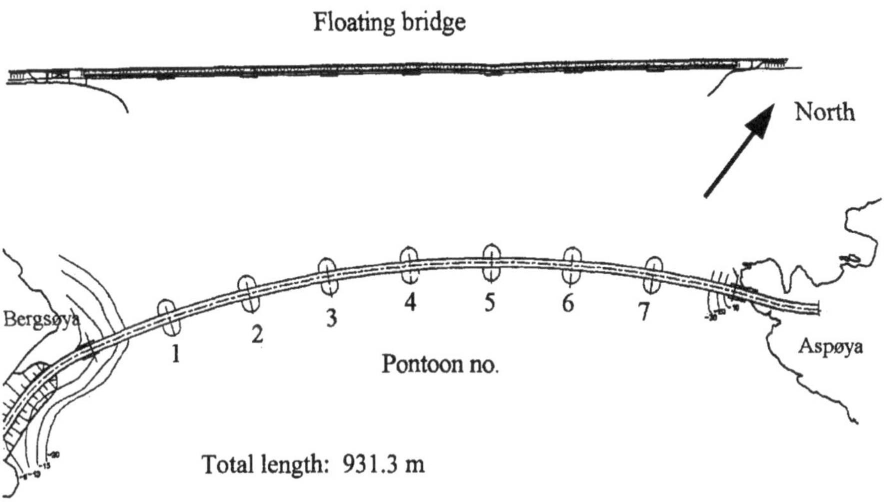

Fig. 1 Bergsøysundet bridge.

The elliptical pontoons are 34 x 20 meters with slipformed walls varying between 6.1 and 7.0 meters high. The pontoons were launched into seawater 7 days after completing the slipforming. Hence, the top slabs were constructed on the floating pontoons.

Also the Nordhordland bridge, figure 2, in the county of Hordaland is a part of the primary road E 36, crossing the Salhus fjord 20 km north of Bergen. The Nordhordland bridge consists of two separate bridges, the high level cable stayed bridge and the floating bridge.

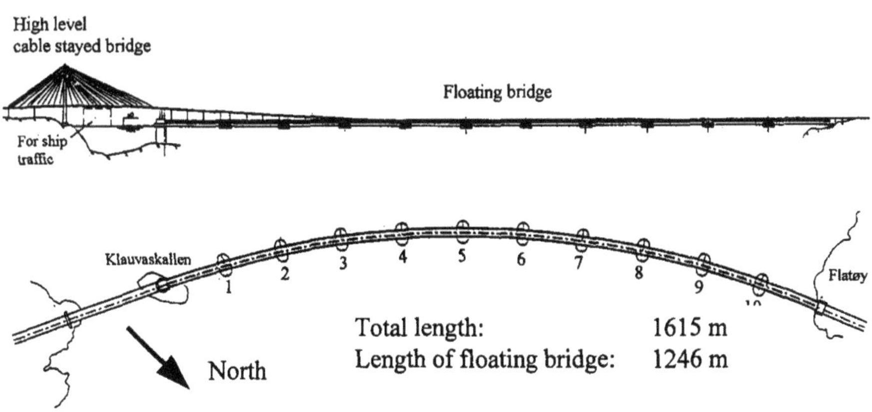

Fig. 2 Nordhordland bridge.

The floating bridge is a 1246 meter long steel box girder supported on 10 pontoons, constructed in high strength LWA concrete. The pontoons were constructed in dry dock from December 1991 to December 1992. The elliptical pontoons are 42 x 20.5 meters with slipformed walls varying between 7.2 and 8.9 meters high. Concrete cover on the outside of the bottom slab and walls is 60 mm, while 50 mm on the outside of the top slab.

The pontoons were launched into seawater at an age varying from 1 to 6 months.

2.2 Instrumentation for corrosion surveillance

The system for surveying the corrosion behaviour of the pontoons is based on three types of gauges:

- **Reference electrodes,** which record the electrical potential at the surface of the reinforcement through the concrete pore water environment. The embedded reference electrodes used were ERE-10, which give a potential 90 mV more positive than a standard $Cu/CuSO_4$ electrode.
  In the submerged zone, the readings might be very negative, i.e. below - 1000 mV. For a structure, which is cathodic protected by sacrificial anodes potential readings below - 600 mV indicate either that no corrosion is taking place or that the reinforcement are sufficient protected.

- **Resistivity** measurements are based on the four-pin, Wenner probe, which measure the electrical resistance in the concrete. Compared with wet concrete the resistivity is normally higher in dry concrete. However, when evaluating resistivity measurements, the relative change with time of the recorded values and the time dependent change in resistivity are more essential than the magnitude of the recorded values themselves.

- **Electrical resistance** gauges (ER-gauges) measure the electrical resistance in an exposed steel bar placed in the concrete cover.
  In the actual set-up, the ER-gauge is a 1-meter long reinforcement bar with 10 mm diameter. Half of the bar is coated with polyurethane, while the other half is exposed in the concrete cover approx. 20 mm outside the reinforcement. The bar is electrically isolated from the reinforcement and is therefore not cathodically protected.
  By measuring the electrical resistance in both the coated (reference) and the exposed part of the steel bar, increased recorded values in the exposed bar indicates that corrosion has started, hence chloride ions have reached the gauge.

According to the installation manuals /1/, the three types of gauges are arranged as show on figure 3. This configuration forms an instrumented area called a measuring point.

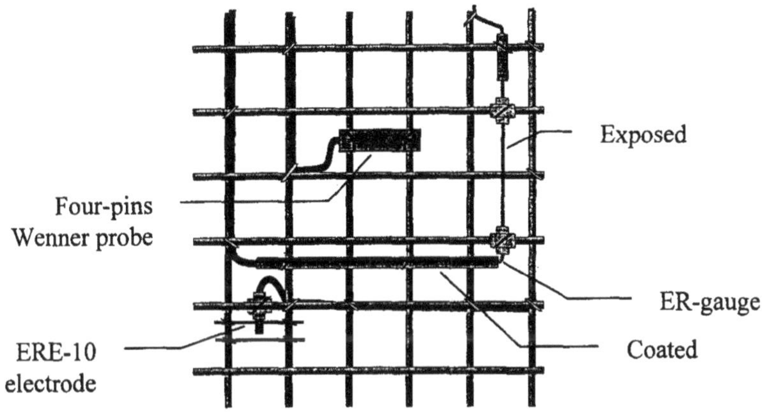

Fig. 3 Instrumented area – arrangement of sensors.

On Bergsøysundet bridge, all pontoons are instrumented. In total 14 measuring points were established on the pontoons 1 and 7 and 5 measuring points, No's 1,2,6,9 and 10, on each of the remaining other five pontoons. Location of instrumented areas is shown in figure 4.

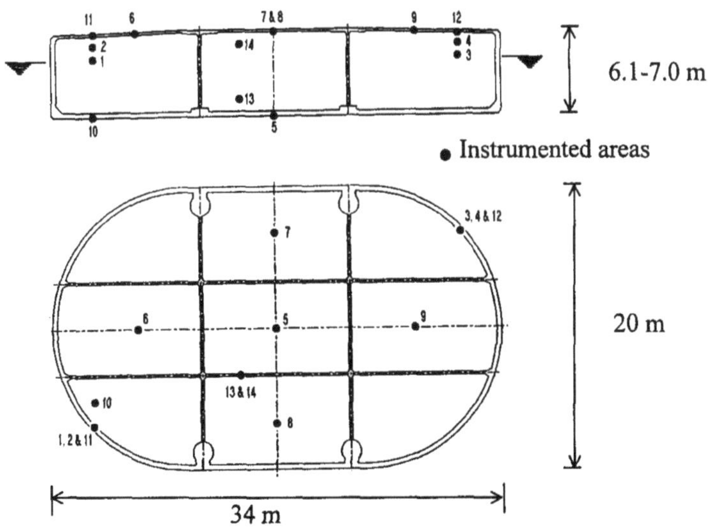

Fig. 4 Pontoons Bergsøysundet bridge.

Two of the pontoons at the Nordhordland bridge are instrumented. The locations of six measuring points on pontoon 1 and 5 measuring points on pontoon 10 are shown in figure 5.

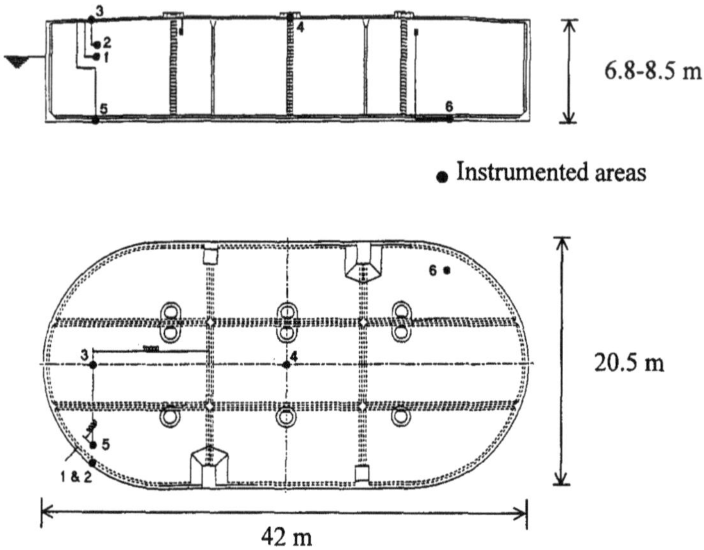

Fig. 5 Pontoons Nordhordland bridge.

3. Evaluation of measurements

Extensive measurements were performed on all the instrumented pontoons in December 1996, approx. 5 years after the exposure in seawater /2/. Visual inspection did not reveal any corrosion. Recorded electro-chemical data on both Bergsøysundet and Nordhordland bridges support this observation.

3.1 Recorded data

A summary of surveillance data monitored in December 1996 will be referred to and evaluated in the following.

3.1.1 Potential measurements

The potential readings at different locations are as follows:

Location	Bergsøysundet	Nordhordland
Submerged	- 1205 mV	- 1135 mV
Sea-level	- 1206 mV	- 1167 mV
Splash zone		
1 m above sea level	- 1100 mV	- 718 mV
Top slab, edge	- 607 mV	- 661 mV
Top slab, centre	- 576 mV	- 821 mV

The tabulated values are the average of recorded data from all the pontoons. Generally, there are small variations between data recorded at the same locations, e.g. the -1205 mV in the submerged zone is the average of nine readings varying from - 1180 mV to - 1223 mV.

The interpretation of the potential measurements says that there is no ongoing reinforcement corrosion.

3.1.2 Resistivity

Resistivity obtained at different locations:

Location	Bergsøysundet kohm cm	Nordhordland kohm cm
Sea-level	1003,5	467,4
Splash zone		
1 m above sea level	848,0	371,2
Top slab, edge	423,3	703,5
Top slab, centre	1290,6	628,5

The temperature in the seawater was not recorded and no temperature corrections of the resistivity values have therefore been made.

The recorded resistivity values are all above 20 kohm cm, which according to /3/ indicate a low corrosion rate. There are some variations between the recorded data, but all of them are very high. However, a tendency to increased resistivity in the areas with the driest concrete can be observed.

3.1.3 ER-gauges

The readings of the electrical resistance in both the coated and the uncoated part of the steel bar are consistent and no increase of electrical resistance in the exposed bars has so far been observed.

Hence, this indicates that no reinforcement corrosion would take place at locations 40 mm from the concrete surface, and that the chloride content at this location is below the threshold value.

3.2 Determination of chloride ingress

In order to evaluate the reliability of instrumented durability surveillance, supplementary investigations of chloride ingress in concrete exposed to seawater have been performed. The actual chloride ingress in the pontoons of Bergsøysundet bridge five years after exposure has been determined by J.E.Carlsen, Selmer, as part of a national research programme on lightweight concrete named "LETTKON" /4/.

The chloride ingress are measured at 10 locations on pontoons No's 6 and 7 at Bergsøysundet bridge, and consist of 20 separate chloride profiles in total. Nine of the investigated locations are in the splash zone at elevations 0.2, 0.7, and 1.2 meter above sea level on both the north and south side of the pontoon. In addition, one location on the top slab, 3 meters from the edge, has been tested.

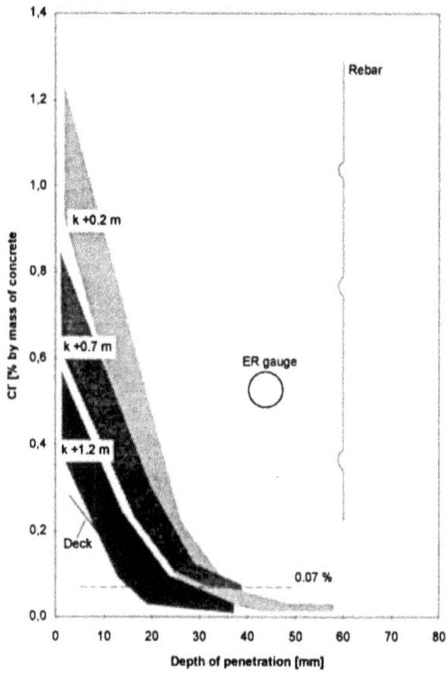

Fig. 6 **Range of chloride content at elevations 0.2, 0.7 and 1.2 meters above sea-level.**

The range of obtained chloride content, in per cent of concrete weight, for the three elevations on the walls investigated, is shown in figure 6. The scatters, show the variation in recorded chloride content. Using the upper limit for the recorded chloride profiles at elevation + 0.2 meter above sea level, the maximum chloride content at the depth of the ER-gauge, 40 mm from the concrete surface, is 0.06% by weight of concrete. This value is below the normally accepted threshold value of 0.07 - 0.09% Cl^- to initiate chloride corrosion.

3.3 Comparison between electro-chemical readings and chloride ingress

The evaluation of the recorded electro-chemical data shows very clearly that no corrosion of the rebars is taking place.

Potential readings over the pontoons, see figure 7, are consistent, and indicate either no corrosion of the rebars and sufficient cathodic protection of the reinforcement.

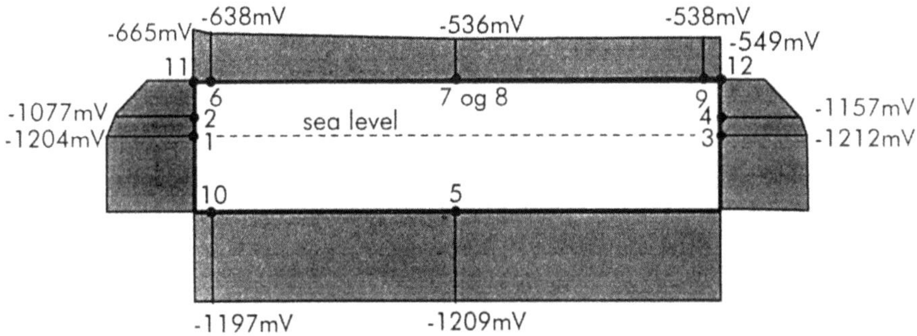

Fig. 7 **Potential measurements Bergsøysundet bridge. Average of recorded values from all seven pontoons.**

During the visual inspections performed no signs, which might indicate rebar corrosion, have been observed.

The chloride ingress altains the threshold value at a depth of 37 mm from the concrete surface. No chloride-initiated corrosion should therefore take place on the reinforcement or on the ER-gauges at depth 40 mm, see figure 6.

Hence, all the results are consistent with a **no corrosion** conclusion. This is promising with respect to the development of a reliable system for corrosion surveillance of concrete structures.

4 Conclusion

The investigations performed are still R&D.

We are not sure about the reliability of the data obtained, and we don't know how representative readings performed in some limited areas are for the whole structure.

The instrumentation used is relatively cheap. If costs increase, corrosion surveillance will be of less interest as a general part of a management system.

The investigations are in progress, final conclusions cannot be drawn, but we want to discuss the feasibility and the reliability of these types of systems.

However, most of the objectives mentioned have already been partially attained. They will undoubtedly remain only partially attained because as with any question concerning durability, the absolute answer giving the required 100 years protection cannot be given before it is obtained. Nevertheless, the project has taken a step in the right direction, and is promising as to establishing a sustainable system for corrosion surveillance.

Systems for durability surveillance represents a sustainable development within concrete technology, but before incorporated into a management system a number of topics, summarised below, have to be discussed:

- What parameters should be monitored?
- Where should sensors be located?
- What is the most cost effective sensor set-ups?
- Can recorded data easily be interpreted?

5. References

/1/ OCEANOR - manuals (in Norwegian)
OCN-BSB-105 (Bergsøysundet bridge) 17/6.1991
OCN-SAL-110 (Nordhordland bridge) 16/7.1992

/2/ DET NORSKE VERITAS
Reference measurement. Bergsøysundet bridge and Nordhordland bridge.
DNV-report No R 196423 11/2.1997 (in Norwegian)

/3/ Broomfield, John P.
Corrosion of steel in Concrete
E & FN Spon, London 1997

/4/ Carlsen, J.E.
Chloride ingress in Bergsøysundet floating bridge
Preliminary research report DP 3.7 of project LETTKON (in Norwegian)
April 1997.

SUSTAINABLE DEVELOPMENT OF THE CEMENT AND CONCRETE INDUSTRY

H. UCHIKAWA
Kanazawa Institute of Technology, Ishikawa, Japan

Abstract
Five problems to be solved for sustainable development of cement and concrete industry were described and discussed taking environmental compatibility in their manufacturing, use and disposal as well as development of high performance, multi-functional cement and concrete into consideration.
Keywords: Sustainable development, cement, concrete, intelligent function, materials design, environment, waste, recycling

1 Introduction

Cement and concrete have occupied a firm position in the modern society as low-priced and reliable basic material for construction through numerous improvements from long time ago. Accompanying with the recent remarkable advancement and diversification of the industrial and the social structure, projects for constructing the infra structure toward the information-oriented and creation-oriented age are extra-large ones such as ultra high rise buildings, great-depth underground structures, long bridges, offshore airports and underwater cities. The latest labor saving, rapid execution technology are putting into practice to correspond to the age of innovation and shortage of man power. In addition, the social request for creation of more comfortable living space harmonizing with the natural environment, facilitation and simplification in maintenance for structures and the like has been increasingly diversified. When we think of the social revolution and progress in technology from the second half of the 1990's toward the twenty-first century, the problems to be solved for sustainable development of cement and concrete industry may be concentrated on the following five items.

Concrete Technology for a Sustainable Development in the 21st Century edited by O.E. Gjørv and K. Sakai.
Published in 2000 by E & FN Spon, 11 New Fetter Lane, London EC4P 4EE, UK. ISBN: 0 419 25060 3

The first problem is to create "new type of cement" breaking through conventional cement though it may takes time to complete. It is not confined to minor improvement of the present cement, but provided with functionally gradient and intelligent function.

The second problem is to develop "improved cement". The users' needs for the quality and performance of concrete tend to be increasingly diversified. The users' needs for better workability, lower heat of hydration, rapider hardening, higher strength, less dimensional changes, and higher durability are also directed to cement, which is major material for concrete, and it is strongly requested to further improve the performance of cement soon. In order to cope with the request without delay, it is necessary to promote the development of "improved cement" which could be put to practical use in a comparatively short period of time by endowing the existing cement with higher performance.

The third problem is to develop "environmental load reducing type cement" less emitting CO_2, NOx, harmful material during manufacturing. Since it is manufactured by high temperature burning using a large amount of limestone as raw material, cement has an essential problem to generate a large amount of CO_2, and a considerable amount of NOx compared with the environmental level in the manufacturing process. For the reduction of the amount of CO_2 generated, it is useful to reduce the amount of fuel used, and the unit requirement of limestone. For the reduction of the amount of NOx generated, it is useful to lower the burning temperature. For the reduction of heat consumption, it is useful to reduce the unit requirement of limestone and to lower the burning temperature. It is of urgent necessity to develop such technology as to make the foregoing possible after securing the quality.

The fourth problem is to provide "concrete with higher performance". This can be achieved not only by analytical and inductive approach to carry out thorough characterization of concrete and concrete raw material, and to elucidate the mutual reaction mechanism for them, but also by synthetic and deductive approach to establish the criteria for material design for concrete based on these results, and to realize the optimization of mixing, compounding and placing technology.

The performance of concrete, which is composed of cement, mineral admixture (blending component), chemical admixture, aggregate, reinforcing material, water and the like, appears as the integration of character and physical properties of these individual materials, and various phenomena which occur in their mixing, compounding and reaction processes. If, therefore, the composition, structure and reaction mechanism of the respective materials, and the relationship between them and the physical properties are grasped, it becomes possible to implement material design of concrete, and to select optimum mixing and compounding operation.

The fifth problem is to preserve good, comfortable and high-quality living environment by not only restraining the amount of environmental loading material discharged in the manufacturing and discarding processes of cement and concrete, but also developing the technology of the recycling of industrial wastes and living wastes discharged in human social activities, to effectively utilize them as alternative raw material or fuel. To achieve this, it is indispensable to construct a social system for dividing and collecting wastes and to establish a "industrial system capable of coexisting with environment".

2. Creation of new type of cement

2.1 Conversion of conception for creation

The most popular, acceptable, safe method for the product development is need-oriented research and development. This is, however, hard to develop the state-of-the-art technologies breaking through present technologies and large technical and economic effects of it are not expected.

Further advanced research and development than the need-oriented ones are the want-oriented research and development. Although *'Want'* derived from the instinctive willing is unattainable at present, it is expected to be realized in the future. *'Need'* is the driving force of the development of improved technologies, while *'Want'* is the motive force of the creation for the technological breakthrough.

The creation requires a *'New thinking'* underlying it and the realization of the *'New thinking'* requires scientific and technical *'Seeds'* playing a role as the core.

New types of cement will be created by taking the measures for realizing the *'Seeds'* through considering what functions to be required for providing new types of cement with specified properties and what composition and structure to be required for providing them with those functions, based on not only good understanding of the known scientific facts or phenomena or the existing technologies but also *'Novel ideas'* independent of them.

Viewing from those standpoints, it is necessary to understand the concrete properties required from the *'Needs'* and *'Wants'*, and to bring out the functions required for developing those properties and the measures required for realizing those functions.

Purification and refining of substance is indispensable when we aim at high performance just as we have already experienced in developing the fine chemicals and advanced ceramics. This is because a pure single substance or that doped with a very small quantity of element gives superior performance to material with specified performance. Industrial materials are frequently required to be provided with several physical properties higher than constant levels. Mixing or compounding of several fine substances with specific properties is, therefore, generally effective for providing material with specially high performance.

2.2 Measure for creation

Cement is hardened by the reaction with water and the physical properties of the hardened paste, mortar or concrete are applied to various uses.

The performances required for cement in addition to the existing properties are classified as follows:

1. Performances regarding workability,
2. Performances regarding physical properties of hardened concrete,
3. Performances regarding maintenance of structures and
4. Performances regarding appearance of structures.

The performances regarding workability include those for freely controlling the handling time of concrete, those for keeping sufficient workability within the specified time, those for developing strength by rapidly beginning to accelerate the hardening at the set time, those for self-leveling and pumpably placing concrete with high fluidity without separation of materials, and those for requiring no vibrating compaction and forming no cold joints in hardened concrete. These properties are closely related to the rate of hydration reaction of cement together with the rapid development of high strength.

The performances regarding physical properties such as strength, durability and dimensional stability of hardened materials include those for increasing the final strength, those for increasing the tensile, bending and shearing strengths, those for increasing the resistance to freeze and thaw, those for decreasing the neutralizing rate, those for decreasing the rate of penetration of salts, and those for preventing from drying and cracking by heat. These physical properties are closely related to the structure of hardened paste, mortar and concrete.

The needs for appropriately maintaining structures by automatically detecting a part of concrete deteriorated by the natural and artificial factors including cracking, change of surface properties and lowering of strength caused by the neutralization, erosion by acidic rain, damage by briny air and alkaline aggregate reaction and successively repairing it in its early stages are related to the intelligent function of cement called self-diagnosis and self-restoration. Also the needs regarding the appearance of structure for beautiful surroundings and comfortable living space by maintaining the concrete buildings beautiful in harmony with surroundings keeping from getting moldy and by automatically switching the color of the wall surface of a building alternatively to a warm color in winter and a cold in summer are related to the intelligent function of cement for self-controlling by pH, temperature, atmosphere and light.

The functions required for creating cement satisfying the requirements mentioned above include the reaction-controlling and structure-controlling functions, and the intelligent functions implicating the former two. The key to the creation of new types of cement is, therefore, to find how to realize those functions.

To that end, it is necessary to research and create several kinds of substances other than the existing cement compounds having different hydration activity, hydration rate or temperature dependency, endothermic process during hydration, initiator or terminator of hydration and thermochromic and photochromic compounds. Also, it is necessary to establish reliable technology, capable of controlling as we like the morphology, size, crystal structure, flocculation structure and hardened structure of produced hydrate during hydration. Examples of the concrete measure were described in the literature [1].

3. Development of improved cement

With the advances of construction and civil engineering, the users' needs for the quality and performance of cement have also been diversified.

3.1 Ultra rapid-hardening cement

"Ultra rapid-hardening cement" was developed in the United States in 1970, and has been improved and commercialized in Japan in 1971 [2]. It is obtained by replacing $3CaO{\cdot}Al_2O_3$ solid solution in portland cement with another three-component compound with average composition of $11CaO{\cdot}7Al_2O_3{\cdot}CaF_2$, increasing content of it and alite, and mixing them with a large amount of anhydrous gypsum (II-$CaSO_4$) in place of gypsum ($CaSO_4{\cdot}2H_2O$). High initial strength within 1 to 3 hours is developed by a large amount of Ettringite($3CaO{\cdot}Al_2O_3{\cdot}3CaSO_4{\cdot}32H_2O$) produced during early stage of hydration and later strength is developed by C-S-H produced by the hydration of alite. Since it is possible to optionally control the setting time and the strength development by adjusting the components of cement and using retarder, the ultra rapid-hardening cement is widely used for urgent construction works, construction works in winter, spray and grouting works. In addition, many hydraulic products manufactured by mixing rapid-hardening cement with other cementing materials are also sold on the market.

3.2 High-strength cement

With the spread in construction of long large bridges, high rise buildings, great-depth underground-structures, huge underwater structures, it is requested to develop the cement capable of restraining the thermal stress in concrete structure generated by temperature rise after placing a massive concrete, and developing high strength exceeding 100 MPa which is three times as high as that of conventional concrete. Low-heat high-strength cement has been developed as cement meeting such needs. This cement is rich in belite and poor in interstitial material, and have wide particle size distribution by adjusting the particle size. It has both good fluidity and low heat of hydration. Multi-component blended cement has also been developed which is prepared by mixing normal portland cement with mineral admixtures such as blast-furnace slag, fly ash and silica fume at optimum rate considering those particle size distribution and hydration reactivity, and is used for the before-mentioned applications.

3.3 Ultra-fine particle cement

In order to perform dam construction, tunnel excavation and construction of underground structures, ultra-fine particle cement has been developed as cement for grouting to fill these underground splits and voids for reinforcement of foundation rock and stopping of water stream [3, 4]. It is also used for restoration of degraded portions and repair of cracks in concrete structures, solidification of peat bed ground in addition to the grouting.

With the progress of grinding and classifying technique in cement manufacturing and ultra-fine particle cement consisting of only particles of 2 to 10 μm required for grouting material can easily be manufactured. As a result, the grouting material is shifting from water-glass and organic resin to low-priced, pollution-free ultra-fine particle cement.

3.4 Shrinkage compensating cement (Expansive material)

Shrinkage compensating cement (expansive material) has been developed to prevent cracks in concrete structures caused by shrinkage. By using it, the number of joints of concrete pavement, which is the cause of bad running property, was resuced, and strength and durability of concrete structure was much improved. This cement is classified hauyne type, alumina cement type, lime type etc. depending upon the type of the expansive material contained in the cement.

Any one of these needs meeting improved cement has been developed by knowing the character (composition and structure) of cement which has close relationship to the developed physical properties of cement on the basis of the past fruits of research. It is, therefore, requested for cement industry to maintain and advance such high level capability of material design in the future.

4 Development of environmental load reducing cement

4.1 Belite cement

The reduction of the amounts of CO_2 and NOx gas emitted in the process of cement manufacturing is now an international matter of concern because of conservation of global environment. In order to reduce the emission of these gases which bring global warming, it is necessary to reduce the unit requirement of limestone and to lower the burning temperature. For this purpose, belite cement was developed [5]. Belite cement has low heat of hydration, and meets the users' needs for mass concrete of huge structures.

The low initial strength, which is the greatest disadvantage of belite cement, was overcome by making the concrete workable even at an exceedingly low water-cement ratio by the improvement of the belite hydration property, the adjustment of particle size distribution of the cement, and the use of organic admixture.

The belite cement is hardly form a transition zone at the interface of cement paste and aggregate both because it hardly produces $Ca(OH)_2$ during hydration and because it can prepare concrete with very low water-cement ratio. Therefore, the bond between aggregate and cement paste is also strong, thus providing stable, high final strength and good durability [6].

4.2 Modified belite cement

In addition to the improvement of the hydration reactivity of belite for increasing the early strength of belite cement as mentioned before, the method for adding an early strength-developing minerals together with belite is being explored [5]. The materials to be added include C_4A_3S, $C_{11}A_7{\cdot}CaF_2$ and C_4AF which have high rate of hydration reaction and produce ettringite in the presence of calcium sulfate.

The modified belite cement containing those minerals together with belite can be burnt at a temperature 200 to 300 °C lower than normal portland cement. Since $CaSO_4$ and CaF_2 much contained in the raw meal of modified belite cement are converted into low melting point compounds in the burning process and these compounds are often deposited on the cyclone, precalciner and duct of the suspension preheater and sometimes clog up them. The heat consumption is, therefore, not necessarily reduced though the burning temperature is low and the amount of emitted CO_2 is decreased.

The facilities suitable for the raw material preparation and clinker-grinding processes should be developed and improved to reduce the quantity of heat consumed for manufacturing such a type of cement and the merits of manufacturing modified belite cement should be discussed comprehensively viewing from the standpoints of the power consumption, product quality and environmental load.

4.3 Multi-component blended cement

Blended cement was developed in order to effectively utilize the industrial by-product indicating blast-furnace slag, fly ash, silica fume from iron and steel industry, electric power industry and electrothermo-chemical industry. On the other hand, "multi-component blended cement" is developed by mixing two or more kinds of mineral admixtures with the aim of further improving the performance of blended cement by increasing the strength development per heat generated at hydration (strength developed/heat of hydration) to prevent cracking due to thermal stress of mass concrete and reducing the load on the environment during cement manufacturing through the reduction in the quantity of cement clinker used [7, 8]. This cement contributes to provide higher fluidity, higher strength, higher density, higher durability, lower heat evolution increased resistance to alkali aggregate reaction of concrete, and is mainly used for massive structures including the foundations for piers of long large bridges and dam.

5. Up-grading of performance of concrete

Concrete is composed of cement, mineral admixture (blending component), chemical admixture, aggregate and water etc. Higher performance of concrete can be provided as the result of integration of technologies of compounding, mixing, placing and curing of those materials. Even if limited to the material field, progress in not only cement but also other concrete materials is indispensable to up-grade the performance of concrete.

5.1 Development of concrete raw materials and molding and curing method

(1) Utilization of organic admixture

The organic admixture for concrete is called superplasticizer, water reducing agent, AE water reducing agent etc. depending upon the function it has, and any of them utilizes their surface active function. The organic admixture gives required fluidity with a small quantity of water and entrains air when necessary by well dispersing the cement, mineral admixture and hydration products of cement in concrete. It is, therefore, an indispensable additive to manufacture high performance concrete having good workability, high strength and excellent durability.

Although the organic admixture thus plays a very important role in the up-grading of the performance of concrete, there are many unknown points on its mechanism for water reducing action and fluidity improving action [9]. The organic admixture has been developed and modified rather by trial and error method than by being based on theory. Various troubles concerning setting and hardening of concrete caused on the use of organic admixture have been dealt with by a symptomatic measure while the cause is not elucidated.

In order to break through such state of affairs, it is effective and important to carry out cooperative basic research between organic admixture manufacturers and cement and concrete manufacturers and to support the development of high performance admixture suitable for the concrete because cement and concrete manufacturers have many extensive information and experience on the application of organic admixture to concrete as well as profound knowledge on the character-property relationship of cement and concrete themselves.

(2) Utilization of mineral admixture (blending component)

The spherical particle shape, particle size distributing to the smaller diameter side, pozzolanic reactivity of mineral admixture improve the workability, strength and durability of concrete, contributing to up-grading of the performance of concrete [10]. The mineral admixture is often used in combination with chemical admixture to make the effect more remarkable. For example, a large quantity of fly ash and water reducing agent are jointly used for high volume fly ash concrete to be described later. Blast-furnace slag, fly ash, water reducing agent, and thickening agent are used for self-placeable concrete. Silica fume, which is spherical ultra-fine particle with high pozzolanic reactivity having a high specific surface area, AE water reducing agent, reactive organic compound are jointly used for high-strength concrete and high-durability concrete. As for fine powder with pozzolanic reactivity other than silica fume used are rice husk ash, metakaolin, fly ash, blast-furnace slag and limestone etc. In order to develop the performance of mineral admixture at its maximum, it is important to promote studying the relationship between reactivity and composition and structure, effects of mineral admixture on the structure formation of concrete.

(3) Utilization of composite material

In order to improve the bending and tensile strength of concrete, various fibers such as glass, carbon, aramid, and steel are used as reinforcing materials. Recently, various bars obtained by banding together fibers of glass, carbon and aramid in a bundle shape individually or mixedly have been developed as a substitute for iron and steel reinforcement. They are used as reinforcing bars for structures requiring light-weighing of concrete members, and for concrete structures of roads and bridges in a cold district where deicing agents are often used, and in an area having strata containing much chloride and sulfonate.

Among them, CFGFRP (Carbon Fiber Glass Fiber Reinforced Plastic) is these fiber bundles covered with plastic. Since it has both strength, alkali resistivity and fragility of carbon fiber, and the high strength and tenacity of the glass fiber, there is a sign of being rapidly spread due to the lowering of the price of the carbon fiber, particularly from pitch.

(4) Optimization of aggregate

The rheological property and fluidity of fresh concrete and the strength development of hardened concrete are closely related to the shape and size distribution of aggregate used, and there exist such shape and distribution of size as to render them best.

The surface structure of aggregate influences the bonding strength of cement paste to aggregate, and the strength of aggregate itself becomes an important factor in high-strength concrete.

In order to provide with durability, alkali reactive aggregate and abrasive aggregate must be excluded. It is important to select aggregate in accordance with strict standard to develop desired performance of concrete.

(5) Selection of molding and curing method

In order to develop required performance of hardened concrete, it is important to select appropriate molding and curing method from among vibration molding, pressure and mold integral molding, and confined, high-temperature, autoclave and gas atmosphere curing.

5.2 Control of hardened concrete structure

The texture and structure of hardened concrete, especially inter-surface structure between cement paste and aggregate, pore structure, orientation and crystal habit of deposited $Ca(OH)_2$ of transition zone, give a significant effect on the strength, gas and water permeability, ion penetrability and durability of concrete.

(1) Composition and texture of transition zone

An enlarged photograph by high magnification observation in **Fig. 1** [6, 11]reveals that the zone around surrounding the aggregate is a porous region rich in CaO and discontinuous from other parts. This zone is called transition zone [12], interfacial zone

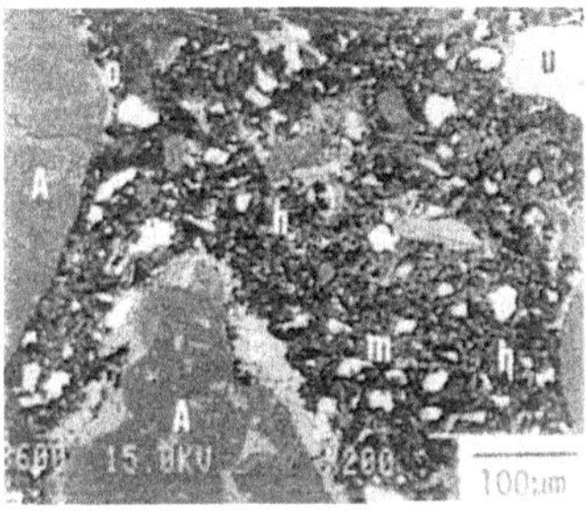

Back scattered electron image

Secondary electron image showing transition zone of concrete

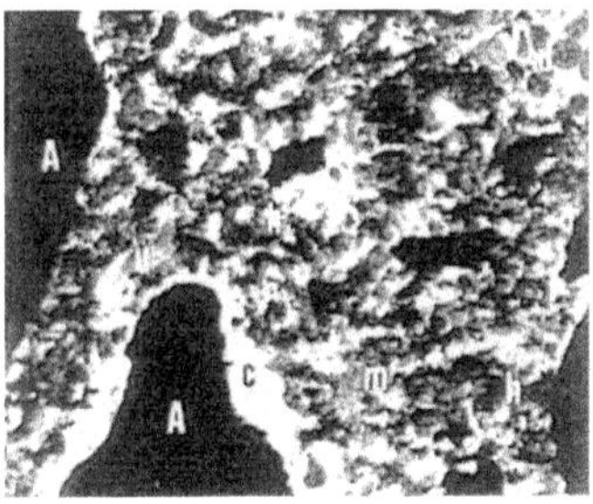

Molar ratio of Ca/Si

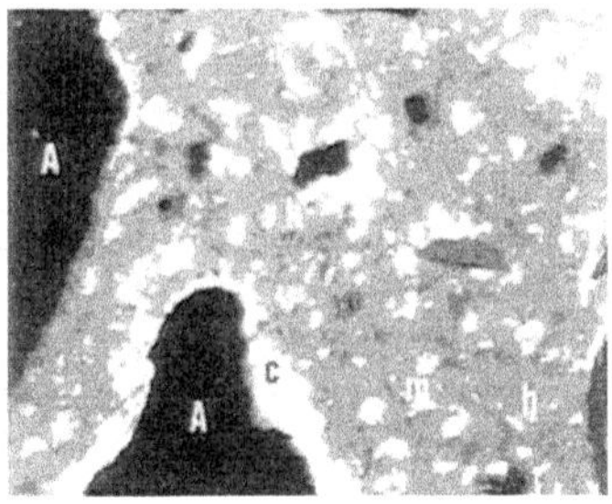

Distribution of Ca

Explanation of color representing the molar ratio and oxide content

The highest concentration of molar ratio calcium is indicated by the lightest part of the picture, graduating through to the lowest concentration being the darkest parts.

A: Aggregate, c: Ca(OH)2, h: C-S-H, m: Monosulfate, u: Unhydrated cement particle (alite, belite and interstitial phase)

Fig. 1. Back scattered electron image, secondary electron image and x-ray image map of Ca/Si molar ratio and Ca on the polished surface of concrete (W/C=0.6, 20°C, 28days) (11)

[13], transition ring [14] or transition aura [15].

The transition zone differs depending upon the mixture proportion and curing conditions of concrete, and is usually formed on the surface of aggregate with a thickness of 30 to 40 μm. The major component is a $Ca(OH)_2$ crystal of 10 to 20μm in diameter. The diameter of pore in the transition zone is mostly within a range of 50 nm to 2 μm. The existence of transition zone in concrete becomes clear at 1 to 3 days after mixing with water, and its thickness shows the maximum value at 7 days. The transition zone once formed is hardly densified much even if the age elapses. The porosity of the transition zone is affected by the orientation of $Ca(OH)_2$ crystal on the surface of aggregate, that is, the surface structure of aggregate.

The formation of the transition zone is restricted when the water-cement ratio of concrete is low, or when blended cement is used. The water-cement ratio at the boundary whether or not the transition zone is recognized is about 0.4.

(2) Formation mechanism of transition zone

Contacting cement with water, elements existing on the surface of cement particle are dissolved into the mixing water as ions. The concentrations of them are high near the cement particles or in pore water of cement paste and low on the surface of aggregate. The dissolved ions move from the cement paste part to the circumference of aggregate by diffusion or accompanied with water because there is a concentration gradient between particles and aggregate surfaces. Crystals of the hydrate are, therefore, deposited around the active site including structural defect and the depositing center on the surface of aggregate [11] with low potential and characteristic shape. Since Ca ion contained in a large quantity in cement clinker is water-soluble, single ion, has small ionic radius, it is easily movable without large resistance of diffusion. The hydrate deposited on the surface of aggregate is, therefore, mainly composed of calcium hydroxide. The water film covering the aggregate is generally thicker than the cement paste part. Since the hydrate is deposited in the wide space formed by the water film, the crystal of it is so large that the hydrate structure formed by entangling the large crystals of hydrate with each other is porous.

(3) Controlling method of formation of transition zone

The existence of transition zone strongly affect the physical and mechanical properties of concrete. The strength of concrete is reduced, drying shrinkage, gas and water permeability, deterioration of concrete by chloride attack and alkali aggregate reaction are increased. Then how can we control the formation of transition zone which give such unfabourable effects on the properties of hardened concrete?

Use of Low Lime (Belite) Cement

Since belite cement containing more than 50% of belite produces only a little amount of $Ca(OH)_2$ by hydration and the fluidity of it is superior to normal portland cement, it can be used at low W/C. Transition zone is hardly observed in concrete prepared from this cement in most cases.

Use of Blended Cement

As already described, transition zone with porous structure mainly consists of $Ca(OH)_2$ is formed in the early stage of cement hydration.

Diffusion of Ca ion to the aggregate and deposition of it on the surface of aggregate in the early stage of curing is depressed in the blended cement because blending components including blastfurnace slag, fly ash and silica fume consume $Ca(OH)_2$ by the pozzolanic reaction, adsorb much water to their surface having large specific surface area hence prevent Ca ion in water from diffusing, and fill the space with the cement particles with themselves or C-S-H produced by the pozzolanic reaction.

Silica fume which is sub-micron spherical powder with large specific surface area and high early pozzolanic reactivity is one of the most effective blending components to reduce the formation of transition zone [16].

Selective Use of Aggregate

It is important for densifying the transition zone to select a type of aggregate having few active sites for depositing $Ca(OH)_2$ crystal and having a composition and surface structure which does not form preferred orientation. An example is limestone based-aggregate.

Application of Water-reducing Agent

The textures, element distributions and pore structures of ordinary concrete prepared at a W/C of 0.5 and ultrahigh-strength concrete prepared at a W/C of 0.22 are illustrated in **Fig. 2** [6, 17]and **3**[6, 17]. No transition zone is observed at all in the ultrahigh-strength concrete with low W/C and the volume of pores between 50 nm and 2μm is much smaller than in the ordinary concrete. These facts suggest that the formation of the transition zone can be inhibited by reducing the W/C by adding a water-reducing agent without losing the fluidity of concrete.

Application of Multi-stage Mixing Method

The formation of the transition zone can be inhibited by reducing the bleeding by an improved mixing method of concrete. The method is carried out in such a way that cement is first mixed with aggregate at a W/C as low as 0.15 to 0.3 using water reducing agent to inhibit the formation of the transition zone by getting cement paste with a low W/C to stick onto the surface of aggregate, and then the rest of water is added to prepare concrete. The method is called the SEC method or the two stage mixing method [18]. The pore size distribution in hardened concrete prepared by the two stage mixing method compared with that in ordinary concrete is illustrated in **Fig. 4** [6]. The pore 50 to 100nm is significantly reduced by the two stage mixing. The aggregate covered with low W/C cement paste is so resistant to settling that the separation of materials after placing is reduced.

5.3 High performance concrete

The following high performance concrete has been developed by taking the measures described in 5.1 and 5.2 using the improved type and environmental load reducing type cement before described. These concretes have already been in the stage of practical use contributing to the up-grading of concrete performance and the simplifing the concrete placing.

(1) Self-placeable concrete

Self-placeable concrete is the concrete prepared by remarkably improving the fluidity and reducing the yield value using organic admixture having a high surface active function. Segregation of concrete raw materials is prevented by setting the water-cement ratio as low as 0.3 to 0.4, bringing the specific gravity of aggregate close to cement paste and increasing the plastic viscosity using thickener. Even if there is an obstacle such as reinforcement, placing can be performed by natural flow of concrete without using any auxiliary apparatus as vibrator. As regards mixture proportion, S/A is made to be lower than that of ordinary concrete and the volume ratios of cement paste or mortar existing between coarse aggregate are raised to facilitate deformation of fresh concrete due to the

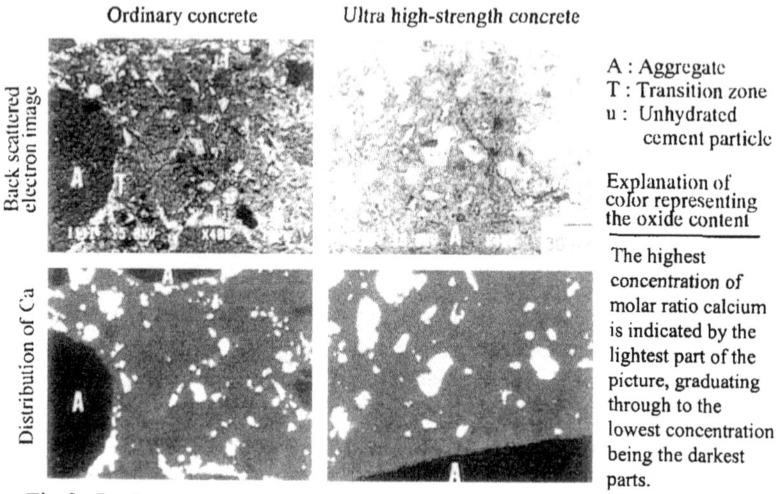

Fig.2. Back scattered electron image and distribution of Ca on the polished cross-section of hardened ordinary and ultra high-strength concrete (6)(17)

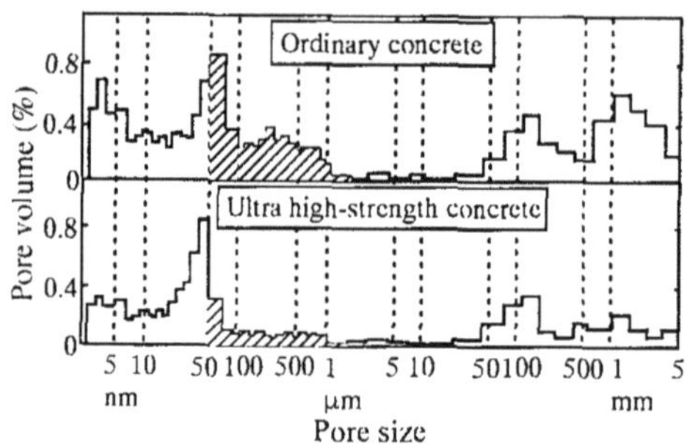

Fig.3. Over-all pore size distribution of hardened ordinary and ultra high-strength concrete (6)(17)

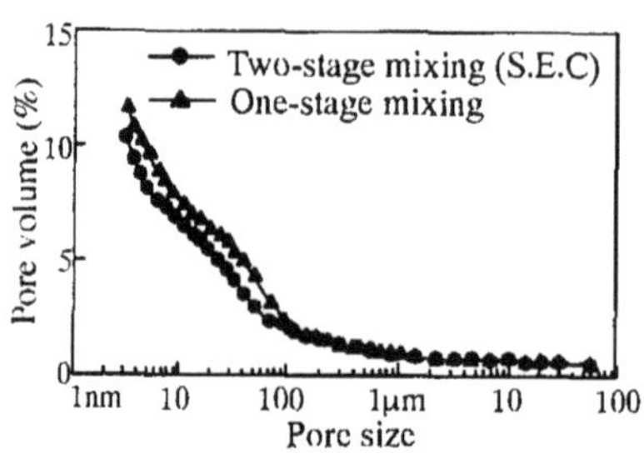

Fig.4. Pore size distribution of hardened concrete prepared by double mixing (W/C=0.5) (6)

self-weight. For the cement, three-component blended cement consisting of normal portland cement, blast-furnace slag and fly ash and belite cement, which adsorbs less amount of organic admixture and is excellent in fluidity, are often used to improve the fluidity and resistance to segregation.

(2) High-strength concrete

High-strength concrete is the concrete designed so as to retain required workability even at a low water-cement ratio and to form densified hardened structure. High strength cement, shape and size arranged non-reactive aggregate and high performance water reducing agent are usually used [17]. The closely-fillable particle shape and size arranged high-strength cement is consists of clinker powder containing reduced amount of interstitial material and increased amount of belite, and ground mineral admixture. The aggregate to be used is non-reactive with suitable surface structure for bonding to cement paste, sufficient strength and appropriate shape and size distribution.

(3) High-durability concrete

High-durability concrete is the concrete having low gas and water permeability and ion penetrability. The basic concept for manufacturing is the same as that of the high-strength concrete. In order to provide it with sufficient resistance to freezing and thawing by the nearly same small quantity of air bubble as plain concrete, studies are being done on an organic admixture capable of remarkably reducing the surface tension of mixing water and making the diameter of entrained air bubble smaller, an organic admixture which reacts with CO_2 and Cl ion in concrete and consume them to increase the resistance to neutralization and salt damage. The former is, for example, glycol ether derivative, and the latter is amino alcohol derivative [19].

5.4 High-performance composite material

High performance composite materials based on cement have also been developed. The technical development and expansion of application toward their practical use are an important problem for the future development of cement and concrete industry.

(1)DSP (Densified system containing homogeneously arranged ultra-fine particles)

The strength and durability of a hardened cement correlate with its porosity and pore size distribution, and are strongly affected by capillary pores 50 to 100 nm in diameter. DSP is obtained by dispersing particles well using organic chemical admixture to reduce the amount of mixing water, that is to say, the amount of capillary pores [20]. Ultra-fine solid powder with 5 to 500 nm in diameter having pozzolanic reactivity is also mixed to fill the pores by itself or with C-S-H produced by the pozzolanic reaction. The hardened body has very dense structure, high strength of about 200 MPa, extremely low gas and ion permeability and improved durability.

(2)MDF (Macro defect free) cement

MDF cement was developed aiming at high strength by reducing the number and size of defect in a hardened cement to less than a limited value and reducing stress concentration [21]. In order to reduce the defect and to improve the rheological property of fresh cement paste, organic polymer is used as admixture. The matrix part for combining cement particles is made to take such structure that fine particles of $Ca_2Al_2O_5 \cdot H_2O$ of a nm level are dispersed in polymer bonded by the medium of Al, whereby increased strength and improved tenacity are obtained. As for cement and organic compound, alumina cement and PVA and phenolic resin are usually used. The phenolic resin is effective to improve the resistibility of the MDF cement to water. It has good tenacity because it is a composite with organic polymer and has a bending strength of 200 to 300 MPa.

(3)RPC (Reactive powder concrete)
RPC consists of powder of less than about 0.1 mm without containing coarse aggregate and steel fiber [22]. The particle size distribution of powder is adjusted so as to obtain closest packing. The powder usually consists of smaller part of cement and larger part of mineral powder such as reactive quartz powder though silica fume of 5 - 10% of cement are often used. Usually, about 2 vol% of steel fiber is used as reinforcement. The water-cement ratio is lowered to 0.12 by using chemical admixture. By placing this raw material into an iron mold for pressure molding and/or heat curing, a hardened member develops marvelous ultra-high strength as several hundreds Mpa as shown in **Fig. 5**.

6. Establishment of environmentally compatible system in manufacture and use of cement and concrete

6.1 Recycling of wastes as raw material and fuel for cement manufacturing

With the briskness and development of industrial activities and improvement in living level, the amount of wastes discharged by the society keeps on increasing, and their treatment and disposal are becoming a serious social problem.

In the cement industry, industrial wastes such as blast-furnace slag, fly ash, silica fume, rice husk ash, red mud, by-product gypsum, and mill tailings have been used in large quantities as raw material and mineral admixture from long ago. In industrial wastes, however, there exist many type of wastes including combustible materials such as waste tires, waste oil, waste sludge, residual dust after burning, waste rubber, waste plastic, small pieces of wood, and sugar cane strained lees, and incombustible materials such as coal ash, aluminum ash, sludge and recovered foundry sand. In non-industrial wastes, there exist many kinds of unused ones such as sewage sludge and incineration ash of urban refuse. Since disposal sites for wastes become scarce, treatment letting them harmless is complicated and a enormous expenditure cost is required for the treatment and disposal, the cement industry is expected to construct a recycling system for consuming these wastes as alternative raw material or fuel, and to play a role as an venous industry having an environment decontaminating function.

The cement industry has positively carried forward a extensive research and development concerning utilization of these wastes as alternative raw material and fuel so far. On using the wastes as alternative raw materials and fuel for manufacturing of cement, the influence of waste on the manufacturing process and actual operation, influence of minor and trace components in the wastes on the formation, composition and structure of cement clinker, and on various physical properties such as rheological property, setting, strength and durability of cement paste were studied[5, 23]. Matters to which consideration should be given to secure the cement quality were also examined. The necessary measures to reduce the containinants to the environment by restraining their generation in the cement manufacturing process were also studied. The criteria to prevent contaminant form diffusing to the environment, technology to improve the homogeneity of raw material and handling property, the technology for stable operation

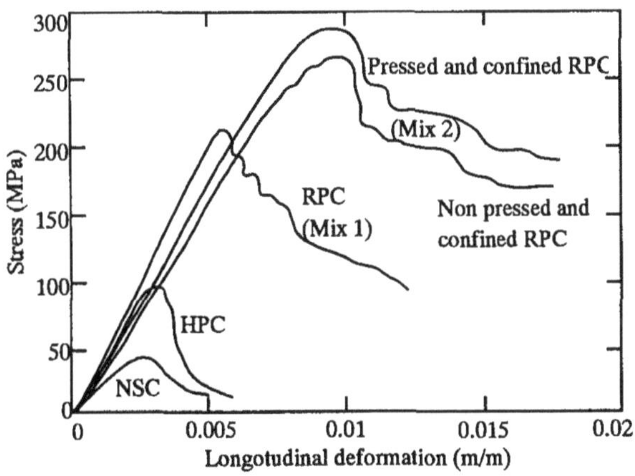

Fig. 5. Stress-strain curves for various type of concrete, including both bathes of RPC, NPC(normal strength concrete) and HPC (high performance concrete) (by courtesy of Prof. Aitcin)

were established and the investment on equipment for the expansion of wastes to be used were actively done. As a result, it became possible to continuously use wastes as alternative raw material and fuel for cement manufacturing[5, 23].

In order to further expand the kind and amount of wastes usable in the future, it is important to make efforts in technical development and to enhance the support system in aspects of taxation and finance.

6.2 Manufacture and spread of environmentally compatible cement (Ecocement) using urban refuses as raw material

The recycling of wastes as alternative raw material and fuel for manufacturing of cement described in 6.1 is to consume waste by substituting for a part of raw material and fuel. Carrying the conception a step further, an attempt has been made to partially add natural raw material to wastes for manufacturing cement having required performance. Thus disposing of the large quantities of wastes could be attained.

The incineration ash produced from the urban refuse contains heavy metals and harmful elements, all of which, apart from chlorine, phosphorus and zinc are within required tolerance limits as shown in **Table 1**. However, the chlorine content is as high as approximately 10%.

The dewatered sludge cake is a paste containing a large amount of water. It emits an offensive odor and contains bacteria. Although the exact composition varies according to the treatment, the water content is around 80 to 90%, organic matter and ash account for around 60 and 20%, respectively, of the solid matter as shown in **Table 2**. The heat value of dry material is approximately 5,000 cal/g, the level of harmful materials, such as heavy metals, are within the tolerance limits.

It is necessary to further dewater the dewatered sludge cake for easy weighing, mixing and grinding so that it can be used as the raw material for cement manufacturing. It is also necessary to sterilize and deodorize it to improve the working condition and safety. For this purpose, the dewatered sludge cake is further dewatered by the slaking reaction of quicklime added to it and sterilized and deodorized by the slaking heat generated by the addition of it through the decomposition of organic matter and alkalizing. A mixture of equal amount of the dewatered sludge cake and quicklime consists of grains and powders. The content of particles 0.15 mm or larger in diameter is approximately 50%. The water content is several percent and the heat value is approximately 700 cal/g.

Cement produced from those raw materials includes alinite cement containing calcium chlorosilicate including alinite $2C_3S \cdot CaCl_2$: $2[Ca_{9.9}Mg_{0.8}\square_{0.3}(SiO_4)_{3.4}(AlO_4)_{0.6}O_{1.9}Cl$, where $\square$ is vacancy] and belinite $C_2S \cdot CaCl_2$ and calcium chloroaluminate $C_{11}A_7 \cdot CaCl_2$ as a hydraulic material, and calcium chloroaluminate cement composed of calcium silicate and calcium chloroaluminate produced by substituting $C_{11}A_7 \cdot CaF_2$ of Jet Cement for $C_{11}A_7 \cdot CaCl_2$. The optimization of the composition of cement is indispensable for utilizing the characteristics of delayed-hardening, low strength type alinite cement and ultra rapid-hardening, high strength type calcium chloroaluminate cement. Industrial wastes including aluminum ash are often used for adjusting the composition of cement besides natural resources.

Raw materials of such a type of cement is composed of approximately 60% of wastes and the burning temperature of cement is as low as 1,000 to 1,200℃. Thus the fuel consumption and the CO_2 emission are also low. This type of cement is, therefore,

Table 1 Chemical composition of incineration ash of urban refuse (24)

Major and minor element (%)

	SiO_2	Al_2O_3	Fe_2O_3	CaO	MgO	SO_3	Na_2O	K_2O	TiO_2	P_2O_5	ZnO	CuO	Cl
Average	22.9	19.7	5.6	30.4	4.8	2.1	3.34	2.55	0.93	1.80	0.60	0.64	8.47
Range	13.7 – 29.1	16.4 – 24.9	4.0 – 6.8	24.3 – 39.8	2.2 – 12.1	1.4 – 2.4	2.40 – 3.9	2.00 – 2.93	0.87 – 1.00	0.88 – 2.3	0.546 – 0.657	0.12 – 1.29	5.02 – 10.6

Trace element (ppm-mg/kg)

	V	Cr	Mn	Co	Ni	As	Cd	Hg	Pb	F	CN	PCB
Average	67	538	438	33	345	55	11	3.5	311	120	ND	ND
Range	1-200	22-2000	80-2500	1-400	20-500	1-90	1-30	0.1-70	100-9200	0-250	-	-

Table 2. Chemical composition of sewage sludge and dewatered cake (Sludge blended with same amount of lime.) (1)

	Chemical composition	Sweage sluge			Dewatered cake of sweage sluge by blending lime		
		Average	Range		Average	Range	
	Amount of water (wt%)	84.0	80.7	87.0	7.9	2.5	16.1
Harmful element	Cadmium (mg/kg)	0.72	<0.05	1.1	1.55	1.0	1.9
	Cyanogen (mg/kg)	0.7	ND	<1	<1	<1	<1
	Organic phosphorus (mg/kg)	0.7	ND	<1	<1	<1	<1
	Lead (mg/kg)	13.4	<0.2	30	27.9	18.7	37
	Total chromium (mg/kg)	153.3	12	810	7.52	4.78	11
	Chromium^{+6} (mg/kg)	—	—	—	0.83	0.83	0.83
	Arsenic (mg/kg)	3.9	1.9	8.9	3.5	2.7	5.2
	Total mercury (mg/kg)	0.775	0.047	1.9	0.067	<0.005	0.15
	Alkyl mercury (mg/kg)	0.003	ND	<0.005	<0.005	<0.005	<0.005
	PCB (mg/kg)	0.068	ND	0.20	<0.005	<0.005	<0.005
Other element	Total phosphorus (wt%)	1.97	1.2	3.5(P_2O_5)	0.225	0.19	0.26
	Total sulfur (wt%)	0.97	0.73	1.2	0.24	0.16	0.29
	Total nitrogen (wt%)	5.50	4.6	6.9	0.45	0.16	0.53
	Ammonium nitrogen (wt%)	1.71	1.26	2.1	0.02	0.02	0.03
	Kjeldahl nitrogen (wt%)	5.05	4.75	5.34	0.53	0.53	0.53
	Organic nitrogen (wt%)	3.55	3.05	4.04	0.51	0.51	0.51
	Nitric nitrogen (wt%)	0.05	<0.01	0.09	<0.01	<0.01	<0.01
	Nitrous nitrogen (wt%)	0.01	ND	<0.01	<0.01	<0.01	<0.01
	Chlorine (wt%)	0.239	0.054	0.52	0.1046	0.0083	0.20
	Fluorine (mg/kg)	191.0	26.4	480	90.5	11	210
	Sodium (wt%)	0.07	0.05	0.08	0.02	0.02	0.02
	Calcium (wt%)	0.69	0.59	0.78	39.0	34.2	43.7
	Carbonic ion (wt%)	1.01	0.51	1.5	1.5	1.5	1.5
	Sulfuric ion (wt%)	0.75	0.40	1.1	0.52	0.52	0.52
	Total organic carbon (wt%)	41.9	41.9	41.9	3.6	3.6	3.6
Industrial analysis	Ash (wt%)	21.1	17.5	25.4	84.3	70.7	91.9
	Volatile matter (wt%)	77.06	72.41	81.7	29.3	29.3	29.3
	Fixed carbon (wt%)	5.05	<0.01	10.09	<0.01	<0.01	<0.01
	Heat value (cal/g)	4935	4710	5160	720	720	720

called "Ecocement" because it help to preserve the natural environment [24].

The setting and mortar strength of Ecocement manufactured in the pilot plant are illustrated in **Fig. 6** compared with those of ordinary portland cement and high early strength cement. The variation of slump, air content and compressive strength of concrete, the mixture proportion of which is illustrated in **Table 3** prepared by adding citric acid are illustrated in **Fig. 7**.

These results indicate that Ecocement with the setting and strength-developing properties lying between ordinary portland cement and high early strength cement can be manufactured by combining the delayed hardening, low-strength type alinite cement with the ultra rapid hardening, high-strength type calcium chloroaluminate cement and by controlling those compositions. Ecocement, with the composition developing higher early strength than that of high early strength cement, can be manufactured by using a retarder such as citric acid. Various physical properties including dimensional stability, creep, deflection and durability of Ecocement concrete are required to be determined by a future long-term test.

Since high chlorine-content wastes are used as the raw material, the quality of Ecocement is inevitably limited.

Ecocement is inapplicable to such uses as prestressed concrete, reinforced concrete and metallic fibre-reinforced concrete.

Promising applications other than concrete include soil stabilizer and grouting material.

It is necessary to optimize the composition of Ecocement and rationally combine mineral and chemical admixtures for keeping the specified quality level of Ecocement mortar and concrete. Although the amounts of heavy metals eluted from hardened mortar and concrete prepared from Ecocement are below the emission criteria, it is important to develop the applications suitable for the quality level of Ecocement taking various physical properties including strength and durability into consideration, because a small amount of chlorine is eluted and the contents of heavy metals vary according the type of raw material.

6.3 Utilization of waste for concrete raw materials

(1) High volume fly ash concrete

Since disposal land can be hardly secured for coal ash discharged from newly built or expanded thermal electric power plants, the use of significantly large quantity of fly ash to cement or concrete is required. There are following two technical concepts for increasing the addition of fly ash to concrete.

In the first case, fly ash cement containing 50% or more of fly ash is used. Although the strength of hydrated cement is decreased with the increase of the fly ash content, the strength of concrete is prevented from decreasing by using a water-reducing agent to decrease the unit quantity of water and water/fly ash cement ratio. The researches on high-volume fly ash concrete conducted in Electric Power Research Institute of Canada, Canada Centre for Mineral and Energy Technology and American Concrete Institute are based on this concept [25].

In the second case, fly ash is used as an alternative to fine aggregate. The unit quantity of portland cement in this concrete is equal to that in ordinary concrete for keeping the durability, preventing the neutralization, reducing bleeding of concrete, and supplementing fine aggregate lost in the washing process of sea sand to remove salts,

Table 3 Mixture proportion, slump and air content of concrete (24)

Cement	W/C (%)	s/a (%)	Unit quantity: Water (kg/m³)	Cement (kg/m³)	Sand (kg/m³)	Coarse aggregate (kg/m³)	Admixture *1 (% to cement)	Citric acid (% to cement)	Na_2SO_4 (% to cement)	Slump (cm)	Air (%)	Temperature *2 (°C)
Ecocement	55	44	152	276	821	1057	0.25	0.8	1.0	11.6	4.4	21.7
Normal portland cement	55	44	160	291	808	1040	0.25	-	-	10.5	4.8	18.5

*1: Lignin sulfonic acid-based admixture, *2: Temperature immediately after mixing.

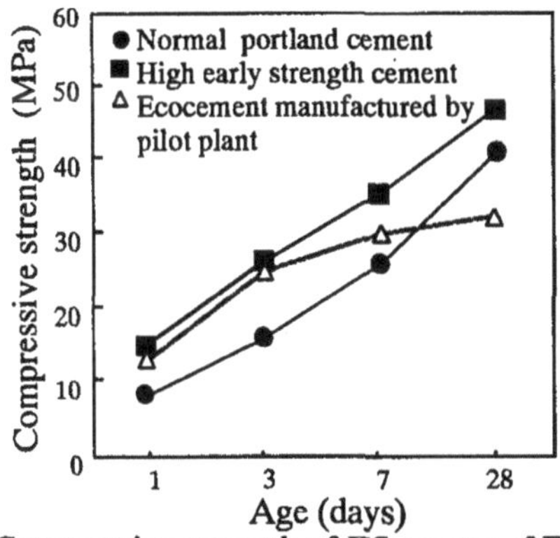

Fig.6. Compressive strength of JIS mortar of Ecocement Manufactured by the pilot plant (24)

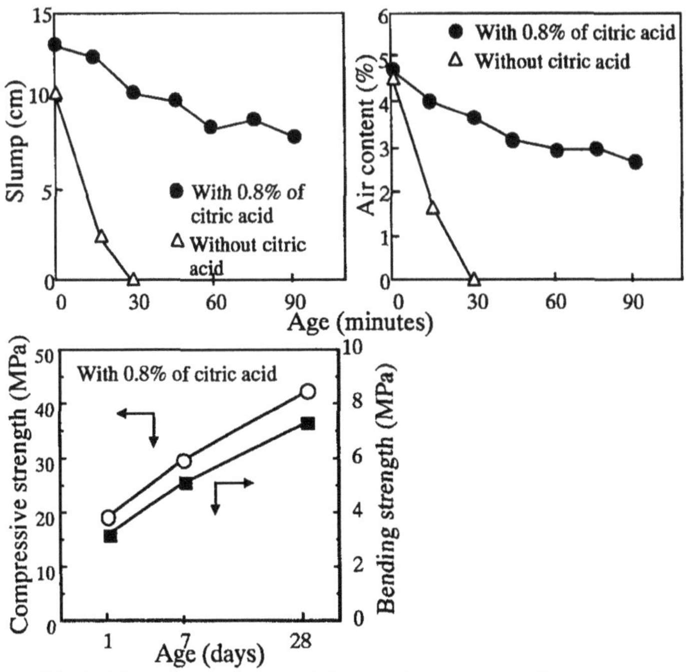

Fig.7. Time dependency of slump, air content and strength of Ecocement concrete (24)

which is increasingly consumed instead of fine aggregate in short supply.

Test results of the first case on the slump and strength of concrete in the case of high-strength concrete prepared using fly ash cement containing 50 to 70% of fly ash aiming at the 90 day- strength of 50MPa are listed in **Table 4** [1]. Since fly ash is substituted for tens percent of portland cement keeping the unit quantity of fly ash cement in concrete constant, the fluidity of concrete prepared at a constant W/C+F ratio can be maintained at a specified value though the addition of admixture is less. The portland cement/ water ratio is, however, as high as 0.64 to 0.74. The initial strength of concrete is, therefore, is low.

In the second case for preparing ordinary concrete with slump of approximately 12cm at the W/C ratio of 0.6 keeping the unit quantity of portland cement at a constant of $300kg/m^3$, the addition of admixture remarkably increases when the total content of portland cement and fly ash exceeds $450kg/m^3$. When the total content of cement and fly ash (C+F) reaches $600kg/m^3$, W/C+F become 0.3. Though the slump of concrete is increased by the increase of addition of admixture, the plastic viscosity is extremely increased and the workability becomes markedly poor. On the contrary, the concrete prepared at W/C+F of approximately 0.3 in the first case exhibits enough fluidity. Since the unit quantity of portland cement in concrete in the second case is large, the early strength is a little bit higher though the strength at 7 day-age or later is nearly equal to that of concrete in the first case. Though the quantity of fly ash in the unit volume of concrete is slightly larger than that in the first case, fluidity is reduced, plastic viscosity is extremely increased, creep of hardened concrete becomes larger and the cost of concrete is raised because the unit quantity of portland cement and admixture are larger.

High-volume fly ash concrete will lead to the mass consumption of fly ash. Although the high-volume fly ash concrete is on the way to development and there are some problems to be solved by investigating the durability, cost performance and others, attention is paid to this concrete viewing from the environmental problem including recycling of coal ash as resources and reduction of emissions of CO_2 and NO_x into environment. Research and development for improving the performance of high volume fly ash concrete have been actively promoted by combining the improvement of the quality of fly ash cement with the new concrete technologies for optimizing the composition and particle size distribution of cement, selecting the optimum admixture, selecting the aggregate considering the surface structure, and optimizing the particle shape and particle size distribution. These measures, however, increase manufacturing cost, so we will have problem of how to share this cost in the society.

(2) High volume waste concrete

Although the application of large quantities of waste in the concrete industry is the needs of the times, the waste used as the raw materials for cement and concrete are at present as little as 10 to 100kg per ton of finished cement and 10 to 200kg per cubic meter of concrete, respectively, and at most 300 kg per m^3 of concrete even in the case of the HVFC.

Recently, the research on high volume waste concrete (HVWC) has been reported [27] and attracted our interest. High volume waste concrete is the concrete containing as much as hundreds kilograms or more of waste per cubic meter of concrete by using a combination of several types of waste-derived materials and properly adjusting the partcile size distribution aiming at the usage of a large quantity of waste-derived

Table 4. Mixing proportion and properties of high volume fly ash concrete (1)

Mixing proportion (kg/m³) Water (W)	Cement (C)	Fly ash (F)	Sand (S)	Gravel (G)	Admixture (%)	Mixing ratio F/C+F	W/C+F	W/C	Slump (cm)	Compresive strength (MPa) 1d	3d	7d	28d	90d	Setting time (hr. : min.) Initial	final
The first case (26)																
115	428	0	682	1211	1.14	0	0.27	0.27	18.0	37.2	51.2	61.3	71.8	83.7	5:15	6:50
116	180	248	659	1171	0.86	0.58	0.64	0.27	24.0	8.6	20.5	33.7	44.1	51.9	7:45	10:30
	428															
115	156	215	685	1218	0.75	0.58	0.74	0.31	19.0	6.8	16.0	28.0	37.1	47.1	8:25	11:25
	371															
The second case																
160	300	0	789	1044	0.25	0	0.53	0.53	12.5	7.4	18.4	27.5	40.8	44.7	5:16	7:16
180	300	180	528	1044	1.0	0.38	0.6	0.38	6.5	10.3	19.9	27.1	42.9	58.1	7:55	10:35
	480	708														
180	300	300	339	1044	2.0	0.5	0.6	0.3	24.0 *	16.0	26.1	34.4	56.9	75.5	14:20	18:05
	600	639														

*: Plastic viscosity of concrete is extremely high

Table 5. Mixing proportion of high volume waste concrete (27)

	W/B (%)	s/a (%)	Unit amount (kg/m³) Water	Binder Cement	FA	SL	Fine aggregate Sand	IA	CA	*1 FM	Coarse aggregate Gravel	SS	GC	*1 FM	Admixture (to binder, %) *2	*3	Unit amount of wastes (kg/m³)	Slump (cm)	Flow (cm)	Air (%)
HVWC(I)	37.0	43.0	185	200	100	200	—	500	131	3.7	—	296	592	6.2	0.8	1.6	1,519	24.0	65	6.5
HVWC(II)	37.0	43.0	185	200	100	200	—	500	131	3.7	897	—	—	6.8	0.3	0.8	631	24.0	58	3.7
Reference concrete	37.0	43.0	185	200	100	200	661	—	—	2.7	897	—	—	6.8	0.4	0.4	—	23.5	—	1.5
Ordinary concrete	60.0	41.5	180	300	—	—	738	—	—	2.7	1059	—	—	6.8	0.1	—	—	24.5	—	4.4

*1: Fineness modulus of aggregate, *2: AE water reducing agent, *3: AE intensifier

materials not only as the mineral admixture for cement but also the substitute for aggregate for concrete.

In the example of concrete shown in **Table 5** prepared from the binder consists of large quantities of blastfurnace slag and fly ash besides portland cement, fine aggregate consist of incineration ash of urban refuse and sintered coal ash and coarse aggregate consists of sintered sewage sludge and glass cullet arranging the material size distribution, fresh concrete having predetermined fluidity without material separation could be obtained by using chemical admixture.

The hydration reaction of cement proceeds normally without being affected by the dissolution of minor and trace elements contained in water-derived materials. The use of slag and fly ash made it possible to increase the produced amount of C-S-H and to reduce that of $Ca(OH)_2$. The densification of hardened structure of the high volume waste concrete with time was proceeded as same degree as that of the concrete prepared form natural aggregate. Although developed strength was lower than that of concrete prepared from natural aggregate as shown in **Fig. 8**, the compressive strength exceeded 40 MPa at 91 days in the case when waste derived aggregate was used only as fine aggregate. The kinds and quantities of harmful elements dissolved from hardened high volume waste concrete were almost the same as those of ordinary concrete, and the values met the emission criteria of water. These results indicate that some types of waste are possible to be safely used as aggregate for concrete in large quantities.

6.4 Recycling of Concrete Waste

The annual consumption of cement in Japan totals approximately 80 million tons in 1993, made up ready-mixed concrete of approximately 54 million tons or 70%, concrete products of approximately 12 million tons or 15%, roadbed-improving materials of approximately 4.7 million tons or 6%, approximately 7.1 million tons or 9% for plaster work and others. Assuming that the unit quantity of cement used for concrete is 250 kg/m^3, the consumption of ready-mixed concrete is 216 million cubic meters. Assuming that the unit quantity of cement used for concrete products is 300 kg/m^3, the total quantity of concrete is 40 million cubic meters. Turning those cubic meters into tons, the annual consumption of concrete is as much as approximately 540 million tons.

Concrete wastes discharged in the process of manufacturing process, utilization of concrete and the dismantling of obsolete concrete structures are roughly classified into concrete sludge-containing water produced from water-washing of concrete batching facilities and ready-mixed concrete lorries, surplus fresh concrete remaining over from placing and crushed hardened waste concrete produced from dismantling of obsolete concrete structures. The kind and amount of concrete waste in Japan are illustrated in **Table 6**.

Surplus fresh concrete remaining over from placing accounts for approximately 2% of the total fresh concrete or as much as 10 million tons in a year. Concrete sludge produced from it and that separated from the concrete sludge-containing water produced from the water-washing of concrete-batching facilities and ready-mixed concrete lorries total as much as 2 million tons based on 50%-water content in a year.

The quantity of construction waste materials discharged in Japan in 1992 was as much as 86 million tons and 29 million tons of it corresponding to 34% of it was hardened waste concrete. Although 12 million tons of it was used as the roadbed materials, remaining 17 million tons was disposed.

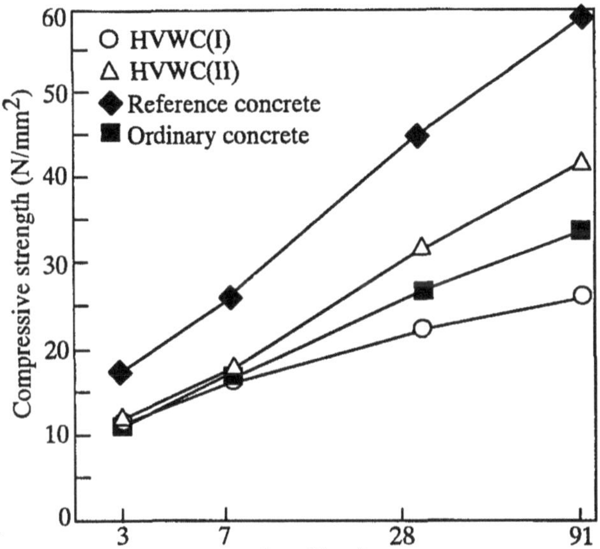

Fig.8. Compressive strength of high volume waste concrete (27)

Table. 6. Kinds and Amounts of Concrete Wastes in Japan (1992) (27)

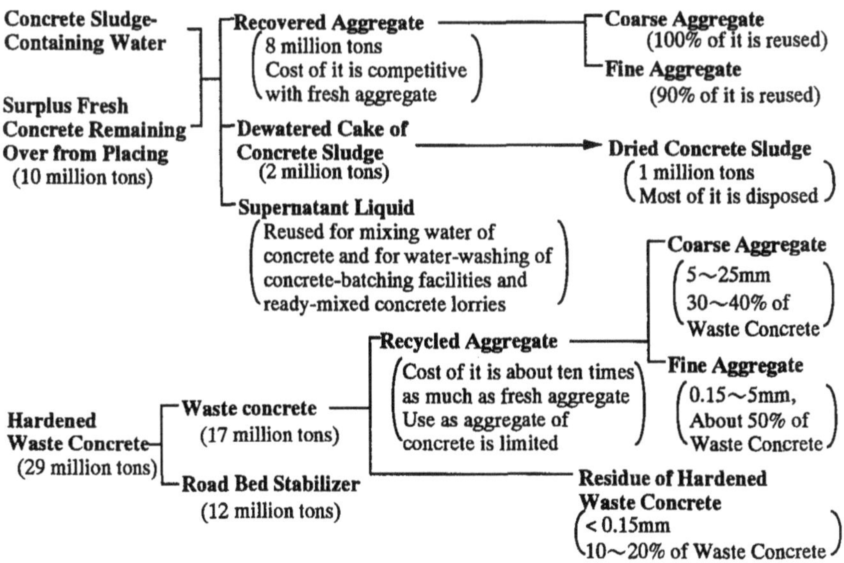

The quantity of concrete wastes to be disposed is 19 million tons. Although this quantity accounts for only 3.5% of approximately 540 million tons of the annual production of concrete at present, the percentage will increase remarkably in the near future considering from the high rate of the consumption of concrete and the shortening of the life-cycle of concrete structure in recent years. Recycling of concrete sludge and hardened waste concrete is, therefore, an important, inevitable problem taking shortage of disposal lands, increase of disposal cost, depletion of aggregate resources and protection of environment of aggregate-producing districts into account. Following measures are necessary to solve these problems [28].

(1) Improvement of performance of chemical admixture

In preparing concrete using recycled aggregate, it is necessary to increase the unit quantity of water for attaining specified workability because the hardened cement paste cohering to the aggregate absorbs the mixing water in concrete. Accordingly, the structure on the interface between the recycled aggregate and the newly produced cement paste becomes so porous that the bonding strength is insufficient, thereby more decreasing the compressive strength of concrete with increasing the addition of recycled aggregate.

It is required to increase the bonding strength between the surface of recycled aggregate, especially the surface free from previously hardened cement paste and the newly produced cement paste for preventing the strength from lowering. It is, therefore, necessary to reduce the W/C ratio for fresh cement paste compared with that for fresh cement paste using fresh aggregate alone. In order to still more reduce the W/C ratio without increasing the addition of admixtures causing cost increasing in the presence of remaining previously hardened cement paste on the surface of recycled aggregate, an chemical admixture with superior water-reducing, fluidizing and retarding functions to those of existing high-performance AE water-reducing agent should be developed.

(2) Development of terminator and initiator of hydration

More concrete than anticipated is routinely prepared in concrete manufacturing plants and transported to the sites to be placed with ready-mixed concrete lorries to smoothly progress construction works according to the requirements of the constructor. Accordingly, it is required to develop hydration terminator which discontinues the hydration reaction of surplus fresh concrete remaining and hydration initiator which initiates it again as occasion of demands for making efficient use of concrete inevitably remaining over from the shipment based on the production schedule mentioned above.

The hydration terminator has functions keeping cement particles apart from water for a certain period after mixing by adsorbing to cement particles, preventing the concentration of calcium ion in the solution from increasing by rapidly taking up Ca^{2+} dissolved from cement in the reaction of cement with water, and repressing the growth of hydrates including C-S-H and $Ca(OH)_2$ by adsorbing to the growing hydrates.

The hydration accelerator has functions breaking the chemical bond of unhydrated substances by promoting the attack of H^+ and H_3O^+ ions on cement particles in the presence of water, breaking the protecting film by the hydrates produced on the surface of cement particle and rapidly increasing the concentration of calcium ion in the solution after mixing.

The existing retarder having the function comparatively close to the hydration terminator include an admixture mainly composed of a composite material containing oxycarboxylic acid and alkylaminophosphate [29]. The performance of it do not, however, still meet the requirements of the hydration terminator judging form the rising inclination of the time of penetration resistance curve of fresh concrete illustrated in **Fig. 9** [1].

In addition to those chemical compounds, light, energy, ultrasonic wave energy and kinetic energy can be used for initiating and terminating the reaction.

(3) Development of uses of concrete sludge and residue of hardened waste concrete

Dewatered cake of concrete sludge and dried concrete sludge separated from the concrete sludge-containing water produced by water-washing the concrete-manufacturing facility and ready-mixed concrete lorries and surplus fresh concrete remaining over from placing are being investigated to use as a dried concrete sludge-blast furnace slag-based binder [30], dry concrete sludge-blast furnace slag-portland cement-based blended cement, and an alternative to part of fine aggregate for concrete. Since the basic physical properties of sludge has some problems described before and the fluctuation in quality is large, it is, however, quite difficult to reuse them for concrete.

The residue remaining after separating the aggregate from the hardened waste concrete has also the same problems as them. It is required to consider the reuse of them in the fields other than concrete. It is being investigated to use it as a soil neutralizer, roadbed materials and a neutralizer of hydrogen chloride contained in the exit gas from municipal refuse incinerators.

(4) Spread of concrete using limestone as the whole part of aggregate

It is now being actively developed to use limestone as the whole part of the aggregate including fine aggregate as well as coarse aggregate which has been used for concrete so far, for making the average chemical composition of concrete usable as a raw material for cement in such a way as crushing hardened waste concrete from obsolete concrete structures for using the whole crushed product as the raw material of cement. Thus, a closed cycle of materials from cement to concrete will be made by the process above mentioned. It has been already proved that the quality of cement trially manufactured by using a raw material produced by crushing and grinding the hardened concrete prepared by using limestone as the whole part of aggregate is almost the same as that of ordinary cement manufactured by using ordinary raw materials and the quality of concrete manufactured by using the trially prepared cement is also almost the same as that of ordinary concrete [31]. When this process will be developed, it will contribute to the material savings and environmental protection including the reductions of consumption of limestone used for manufacturing cement and emissions of carbon dioxide, because all of the concrete waste will be able to be used and $Ca(OH)_2$ and cement hydrates produced by the hydration of cement contained in hardened waste concrete will be reused. Taking the term of one hundred or more years for completely substituting the concrete using limestone as the whole part of aggregate for ordinary concrete used for all of the structures into consideration, it is necessary to pay a close attention to secure the limestone resources for aggregate and supply it stably viewing from a long-term standpoint.

7. Summary and conclusion

In order to meet the diversifying and advancing social needs, satisfy the wants and

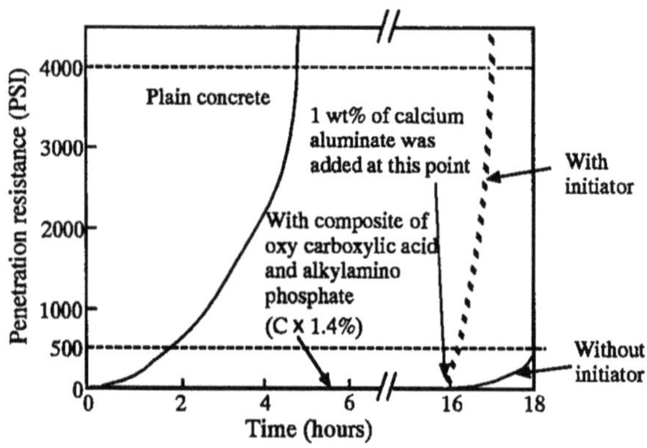

Fig. 9. An example of penetration resistance of concrete with or without retarder (Terminator) and accelerator (initiator) (1)

further continue the prosperity toward the twenty-first century surviving the age of keen competition, it is indispensable for the cement and concrete industry to establish an international production and business system considering the protection of environment as much as possible.

In order to correspond to the diversified users' needs without delay, it is necessary to create new type of cement having diversified functions and high performance broken through those at present though it may takes long time to complete. At the same time, it is necessary to actively develop the "improved cement" by providing the existing cement with higher performance and the "environmental load reducing type cement" less emitting CO_2 and NOx.

The needs for construction of huge concrete structures and labor-saving construction require higher performance of concrete. The improvement of concrete performance cannot be achieved by the improvement of cement performance alone, but it is indispensable to improve the performance of the other concrete materials and to establish control technique of the structure of concrete. The key materials for this purpose are chemical and mineral admixtures. It is strongly requested to advance such basic and theoretical research as to enable the material design of high performance organic admixture. For the enhancement of the performance of concrete, a new way of thinking that transcends conventional common sense is required also in any field of material development, design of mixture proportion, mixing technique, and transportation and placing of concrete.

In order to solve various problems concerning treatment and disposal of a large quantity of wastes discharged by the modern society as well as resources and energy-saving in the cement and concrete industry, it is necessary to develop the technology and to construct the system of wastes recycling as resources. Although wastes have already been used as alternative raw material and fuel for cement manufacturing, carrying a step further, the following problem have to be solved including development and spread of cement manufacturing using wastes as principal raw material, development of technology capable of utilizing wastes as raw material for concrete, reuse of concrete sludge, surplus fresh concrete remaining over from placing and waste water which are discharged during manufacture of concrete, and recirculation of hardened waste concrete as concrete material. It was not until these problems are solved that the importance of the role of the cement and concrete industry as the venous industry is recognized. Thus, closed circuit of material circulation of cement to concrete is achieved and a cement and concrete production and supply system capable of coexisting with the environment is established.

From the above-mentioned viewpoint, this paper has described the results of research and technical development performed until now and the future possibility based thereon. Any of these is a problem that the cement and concrete industry must clear by all means for their sustainable development. Time, capable men, enormous expenditure on the research and development and heavy investment in the cement manufacturing facilities are required for solving the problems and completing the technologies. Breaking through conventional framework, it is expected to promptly established a new international joint research and technical cooperation system toward the 21st century.

References

1. Uchikawa, H. (1994) Management strategy of cement industry in technology toward the next century, *Advances in Cement and Concrete*, ASCE, NewYork, pp.524-561.
2. Uchikawa, H. and Kohno, K. (1983) Ultra-rapid hardening cement (Jet cement), *New Concrete Materials,* University Press (England), pp.70-136.
3. Reinhardt, H.W. (1993) Ultra-fine cements for special applications -European developments, *Advanced Cement Based Materials*, Vol.1, pp.106-107.
4. Uchikawa, H. (1993) Ultra-fine cements for special applications -Asian developments, *Advanced Cement Based Materials*, Vol.1, pp.150-154.
5. Uchikawa, H. (1992) Present problem in cement manufacturing, Proc. 9th International Congress on the Chemistry of Cement- New Delhi (India), Vol. VII, pp.23-57.
6. Uchikawa, H. (1997) Role and regulation of the transition zone in realizing high performance concrete, Proc. 3rd CANMET/ACI International Symposium on Advances in Concrete Technology (Auckland), Vol.1, pp.109-129
7. Uchikawa H., Uchida, S. and Okamura, T. (1991) Role of fly ash on the hydration and structure formation of ternary components blended cement composed of blastfurnace slag, fly ash and portland cement, Proc. Ash Utilization Conference, Shanghai (China), Volume 2: Ash Use in Cement and Concrete, pp.65-1 - 69-9.
8. Uchikawa, H. and Okamura, T. (1993) Binary and ternary components blended cements, *Mineral Admixtures in Cement and Concrete*, Vol. 4, ABI Book Pvt. Ltd, India.
9. Uchikawa, H. (1994) Hydration of cement and structure formation and properties of cement paste in the presence of organic admixture, Proc. Conference in Tribute to Micheline Moranville Regourd 'Importance of Recent Microstructural Developments in Cement and Concrete' -Sherbrooke, Concrete Canada pp.63-117.
10. Uchikawa, H. (1986) Effect of blending component on hydration and structure formation, Proc. 8th International Congress on the Chemistry of Cement - Rio de Janeiro (Brazil), Vol. I, pp.249-280.
11. Uchikawa, H. (1989) Similarities and discrepancies of hardened cement paste, mortar and concrete from the standpoint of composition and structure, *Advances in Cement Manufacture and Use*, Engineering Foundation, N. Y., pp.271-294.
12. Mehta, P.K. (1986) *Concrete - Structure Properties and Material*, Prentice-Hall Inc. N.J.
13. Grandet, J. and Olliveier, J.P. (1980), Orientation of hydration products near aggregate surface, Proc. 7th International Congress on the Chemistry of Cement (Paris), Vol.III, pp.VII-63-68.
14. Maso, J.C. (1980), The bond between aggregate and hydrated cement paste, Proc. 7th International Congress on the Chemistry of Cement (Paris), Vol.I, pp.VII-1/3-1/563-68.
15. Massazza, F. and Costa, U. (1986) Bond : Paste-aggregate, paste-reinforcement and paste-fibers, Proc. 8th International Congress on the Chemistry of Cement (Rio de Janeiro), Vol.I, pp.158-180.
16. Malhotra, V.M., Carette, G.G. And Sivasundaram. V. (1992) Role of silica fume

in concrete: a review, proc. International Conference on Advances in Concrete Technology; Athenes, Vol.1, pp.925-991.

17. Uchikawa, H. (1994) Characterization and material design of high-strength concrete with superior workability, *Ceramic Transaction Vol. 40, Cement Technology*, American Ceramic Society, pp.143-186.
18. Itoh Y., Tsuji, M, Kaga S.and Yamamoto, Y. (1981), Properties of S.E.C. concrete and its prospects, Cement and Concrete (in Japanese), Japan Cement Association, No.410, pp.20-29.
19. Sakuda, M., Saito, T., Yanagibashi, K., Tanaka, T.and Sugiyama M (1988), Development of high durable concrete, *Ceramic data book* (in Japanese), Kogyou-seihin-gijyutsu-kyokai, pp.401-407
20. Bache (1981) Densitied cement/ultra-fine particle-based materials presented at 2nd International Conference on Superplasticizer in Concrete.
21. Birchall, J.D., Howard, A.J. and Kendall, K (1982) New cements - inorganic plastics of the future, *Chemistry in Britain*, December, pp.860-892.
22. Richard, P. and Cheyrezy, M.H. (1994) Reactive powder concrete with high ductility and 200-800MPa compressive strength, *Concrete Technology SP-144*, ACI, pp.507-518
23. Uchikawa, H. and Hanehara, S. (1997) Recycling of waste as alternative raw material and fuel in cement manufacturing, *Waste Materials and Industrial By-Product in Concrete Manufacturing*, Noyer Publications U.S. pp.430-553
24. Uchikawa, H. and Obana, H. (1994) Ecocement - Frontier of Recycling of Urban Composite Waste, *World Cement*, Vol.26, No.11, pp.33-40.
25. Radian Canada Inc. and CANMET (1992) ERPI TR-100577, Supplemental proceeding, 4th International Conference on Fly Ash, Silica fume, Slag and Natural Pozzolans in Concrete - High Volume Fly Ash Concrete Section
26. Sivasundarm, V., Carette, G.G. and Malhotora, V.M. (1989) Properties of concrete incorporating high-volume of ASTM class F fly ash, *ACI Material Journal*, Vol.86, No.5, pp.504-514
27. Uchikawa, H., Hanehara, S. and Hirao, H. (1998) High Volume Waste Concrete (HVWC), Proc. 4th CANMET/ACI International Symposium on Advances in Concrete Technology (Tokushima, Japan)
28. Uchikawa, H. and Hanehara, S. (1996) Recycling of concrete waste, Proc. International Congress - Concrete in The Service of Mankind (Dundee, U.K.), Volume: Concrete for environment enhancement and protection, pp.163-172.
29. Yoshida, S., Nishijo, O., Tsukuda, S. and Endo, Y. (1992) Delvo system, Transaction of NMB Central Research Laboratory, No.9, pp.105-113.
30. Yoshikane, T. (1992) Present state of technical development on the reuse of cement waste in Daiyu construction company, *Ready Mixed Concrete* (in Japanese) Vol.11, No.5, pp.55-63.
31. Tomosawa, F. (1995) Towards completely recyclable concrete - development of eco-concrete-, Cement and Concrete (Japanese), No.578, pp.1-8.

CEMENT AND CONCRETE DEVELOPMENT FROM AN ENVIRONMENTAL PERSPECTIVE

P.-C. AÏTCIN
Department of Civil Engineering
Université de Sherbrooke
Sherbrooke, Canada

Abstract
Concrete is the most widely used construction materials: more than 2 tonnes of concrete were produced for every person on the planet. As the use of concrete will increase in the future due to the development do urbanization in developing countries, time has come to look at cement and concrete to environmental point of view; cement and concrete manufacturing is producing 8 percent of the annual amount of CO_2 release in the atmosphere. The challenge is not to limit or forbid the use of concrete but rather to make more concrete having a lower impact on the environment. Different technological tools are already available but they have not yet received the attention they deserve. The three more promising avenues that can be implemented very rapidly are:

- a more efficient use of portland cement involving the use of chemical admixtures and supplementary cementitious materials;
- the production of low energy-intensive cement and/or cement with a lower lime content;
- recycling concrete.

Special effort should be made to transfer very rapidly this technology to developing countries.
Keywords: Admixtures, belitic cements, concrete, environment, pozzolans, Roman concrete, supplementary cementitious materials, water/cement ratio.

1 Introduction

Concrete is the most widely used construction material because it helps fulfill a basic human need: shelter against adverse climatic conditions, enemies, natural calamities, and the such (1). Concrete helps also societies to develop trade and business both within and between communities sometimes at great distances which requires roads, bridges, ports, and airports.

Concrete's advantage as a building material derives from a number of its intrinsic qualities. To illustrate, concrete is moisture resistant, does not rot, does not rust, is not susceptible to insect attack, and is fire resistant. Eighty-five percent of its constituents (water and aggregates) are available locally almost everywhere and the technology involved is very simple (superficially).

Concrete Technology for a Sustainable Development in the 21st Century edited by O.E. Gjørv and K. Sakai.
Published in 2000 by E & FN Spon, 11 New Fetter Lane, London EC4P 4EE, UK. ISBN: 0 419 25060 3

If it is true that, among living species, only human beings, and their parasites, have been able to colonize the whole planet from the deserts to the poles, concrete is equally ubiquitous.

This generalized use of concrete, as with any construction material (2), has some environmental consequences which should not be ignored. It is and will remain the responsibility of concrete producers and users to minimize the negative environmental impact of concrete. But before discussing this point, let's look at how concrete became the most widely used construction material.

2 Historical perspective

2.1 Natural Materials

The first building materials used were of natural origin: timber, stones, rocks, clay, soil, leaves, reefs, and even packed snow. The first binders used to improve the tightness of primitive constructions were also natural: clay and bitumen. These binders served to eliminate the necessity of dressing stone. Fairly rapidly, however, it became clear that natural materials and binders offered limited possibilities for societal development. Consequently, the need to find alternative construction materials better suited to the ambitions of organized societies came to the fore.

One of the first processed materials which could possibly be considered a composite and has been in use since biblical times is straw-reinforced clay brick (3). Its potential, however, was also limited.

The next major step toward a more durable construction material came about with the firing of clay bricks. Its two advantages were greater water resistance and no need for dressing. Once fired clay bricks were invented, however, the need arose to find a binder to make them easier to use. Concurrently, the development of the 2-D and 3-D arch, which can be used to convert tensile and shear stresses into compressive stress, drastically lengthened the limited spans that could be achieved with natural materials using post-and-beam construction.

2.2 Processed Binders

Gypsum and lime were the first two processed binders developed, probably because they are made from readily available natural materials, they involve a relatively simple production process (see Figures 1 and 2), and they require little energy to produce.

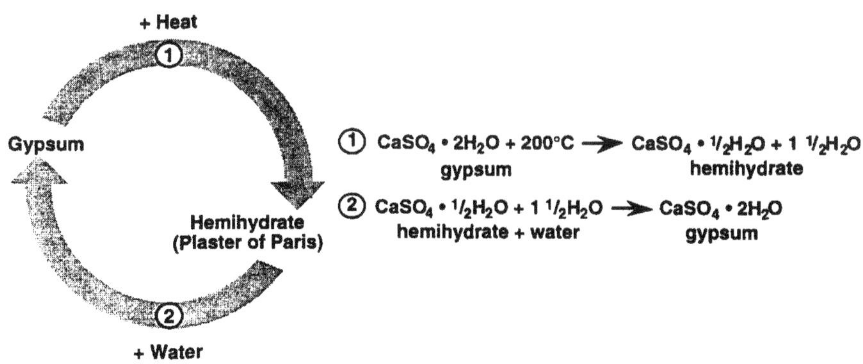

Fig. 1. The gypsum cycle

The thermodynamic process by which gypsum converts into calcium hemihydrate (plaster of Paris) requires an operating temperature of 120 °C, just slightly higher than the boiling point of water. Conversion of limestone into lime requires a temperature of 800 °C, which can easily be reached by a wood fire.

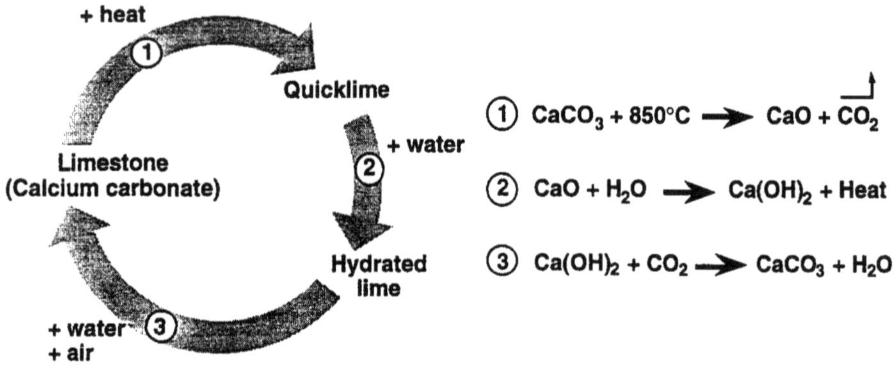

Fig. 2. The calcium carbonate cycle

It is interesting to note that, when gypsum and lime are used as artificial binders with construction materials, they revert back to their initial state (gypsum and calcium carbonate) after hardening. From a chemical and environmental point of view, in this regard, they can be considered environmentally friendly. Their only by-product is a small amount of carbon dioxide produced during firing.

Since gypsum is not moisture resistant, it offered low durability in the climates needed to develop agriculture, which were dependent on the availability of water. Lime performs somewhat better in this respect, but does not harden under water; to set, lime must react with carbon dioxide in the air.

Very early on, it was discovered that these artificial binders could be diluted with sand to make mortar. The development of mortar technology had a positive impact from an environmental standpoint, because more binding material could be produced from a specific volume of binder.

2.3 Roman Concrete

Technology took a leap forward when it was found that lime combined with a particular natural sand could be used to make a hydraulic mortar offering greater strength and durability than conventional lime mortars. This technology was in common use around Mount Vesuvius (and elsewhere), as evidenced by its discovery in the ruins of Herculaneum and Pompeii. The modern name of these specific natural sands, pozzolan, derives from Puzzoli, the name of the town where the Romans dug the volcanic tuff on the slopes of Mount Vesuvius.

Roman engineers all over the Empire learned also to took advantage of local reactive sands, such as Rhineland trass. Romans used this discovery throughout the Empire to build roads, bridges, ports, and buildings. Burnt clay dust also went into lime mortars, representing the first use of metakaolin as a supplementary cementitious material.

The discovery of a strong, durable binder requiring only water to harden opened the way to the development of concrete. In concrete, the binder is diluted not just with sand, but also with coarser particles, such as pea gravel, pebbles, and even stones. Lightweight pumice was also used to reduce the deadweight in some cases. The upper part of the Pantheon's dome was built with a lightweight concrete having a unit mass of only 1350 kg/m^3 (4).

At the close of the 20th century, we can still witness the culmination of Roman know-how in mastering natural pozzolans and concrete in the Pantheon in Rome, built from 118 to 128, and which remains one of the world's outstanding structures. The clear span of the Pantheon is 43.3 m; 1320 and 1460 years later, the domes of Brunelleschi's

Cathedral of Florence and Michelangelo's dome in St. Peter's Cathedral had clear spans 1 m shorter than that of the Pantheon. It took more than 1850 years before the Pantheon's record could be bested using metal-frame structures (4).

Moreover, Roman concrete was ecological because it required the burning of a small amount of lime and use of natural products.

2.4 Portland Cement and Concrete

Centuries passed before the development of a new generation of building materials: portland cement and steel which remain the mainstays of modern construction. Combining these two materials into a composite, referred to as reinforced concrete, resulted in a product that provided the intrinsic characteristics of both: the good compressive strength of portland-cement concrete and the high tensile strength of steel. Moreover, the alkalinity of portland-cement concrete was found to protect the steel from rusting. In later years, the concept of prestressing let designers use concrete and steel even more efficiently in building more slender or huge concrete structures. The tremendous technological potential of concrete is exemplified by the 7500-tonne precast concrete elements in Canada's Confederation Bridge.

From an ecological standpoint, the manufacture of portland cement and steel is energy intensive due to the high temperatures involved. Moreover, producing portland cement releases carbon dioxide over and above that which is produced during heating in the kiln to 1450 °C, the temperature at which clinker is formed (see Figure 3). A modern, efficient cement kiln produces just less than 1 tonne of carbon dioxide for every tonne of portland cement produced.

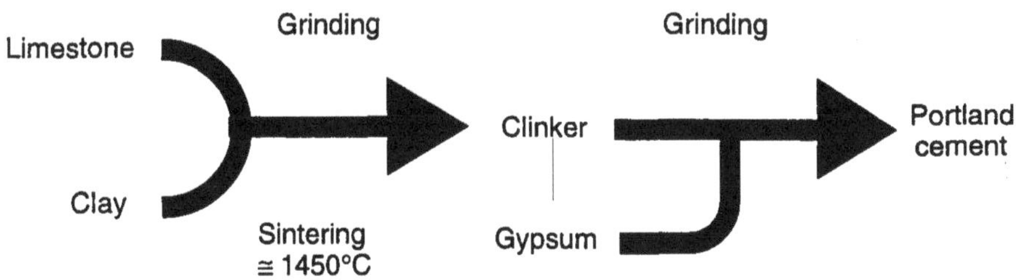

Fig. 3. Fabrication of Portland cement

Good concrete has a life cycle (in which the binder returns to its initial components) that largely exceeds the human time frame (see Figure 4). In fact, the span required for a durable concrete to revert to its initial components (calcium carbonate, clay, and silica) must be measured in geological time.

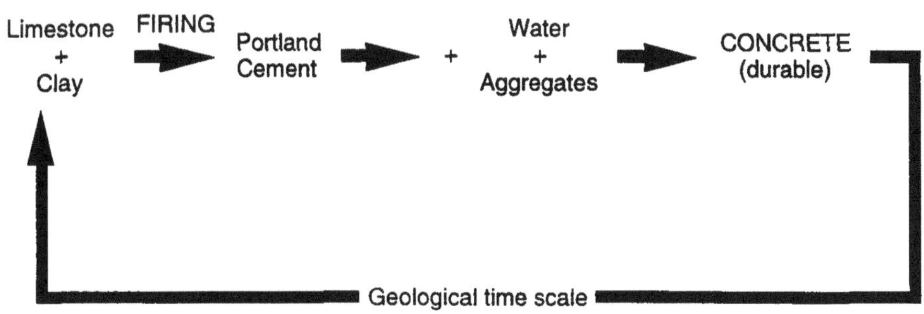

Fig. 4. Life cycle of durable concrete

3 Concrete the most widely used construction material

In the preceding, we touched on some of portland cement's inherent advantages as a construction material. Some of its success, however, derives from the cement industry's ingenuity in developing a very efficient manufacturing process that keeps the unit price of cement quite low (about US$60 per tonne in 1997).

Worldwide cement production reached 1.3 billion tonnes in 1996, which, based on an average cement dosage of 250 kilograms, corresponds to 5.3 billion cubic meters of concrete. Presuming that 1 cubic meter of concrete has a mass equal to 2500 kilograms, or 2.5 tonnes, a total of 13 billion tonnes of concrete was produced in 1996, which comes to more than 2 tonnes for every person on the planet. It is therefore not presumptuous to claim that modern society has been built, at least in part, on concrete. Through the years, concrete has appeared in our homes, cities, and countries in many different forms, ensuring our comfort and protection, and serving to develop trade and commerce that generate wealth.

While the use of concrete has beneficial aspects, just as any other human activity, its production has an environmental impact that should not be ignored. As mentioned above, manufacturing of 1 tonne of cement releases roughly 1 tonne of carbon dioxide into the atmosphere. At current production rates, the cement industry is releasing roughly 1.3 billion tonnes of carbon dioxide into the atmosphere every year, and very little of it recombines with concrete through carbonation. The time has come to look at cement and concrete from an environmental perspective.

4 Decreasing the environmental impact of concrete

While the carbon dioxide produced by the cement industry represents just a third of that released by the transportation industry, we have a responsibility to try to minimize this environmental burden. Several approaches, briefly reviewed below, have already been implemented. Until now, however, they have not received all the attention and promotion they warrant (2).

The three most promising approaches are:

- a more efficient use of portland cement involving chemical admixtures and supplementary cementitious materials;
- the production of low energy-intensive cement and/or cement with a lower lime content;
- recycling concrete.

Each will be briefly reviewed to assess how quickly they could be implemented to lessen the environmental burden created by the construction industry.

4.1 Developing a More Efficient Use of Portland Cement

Since concrete is the basic material for the development of modern society, its consumption will drastically increase in the future in developing countries. One immediate action that can be taken is to teach developing countries how to make high-quality, durable concrete with less cement. This does not imply that developing countries should be forced to use less cement, resulting in poorer quality concrete. Rather, they should receive the technology that will allow them to produce higher-quality concrete that contains less cement and water. This is achieved by reducing the mixing water content with chemical admixtures and supplementary cementitious materials.

Since concrete compressive strength relates to water/cement ratio not to cement content the same strength can be achieved with less cement if a water reducer or, even better, superplasticizer is used. Moreover, the use of superplasticizers with supplementary cementitious materials must be promoted from the standpoint of decreasing the water/binder ratio and obtaining concrete of higher quality, not just to reduce the cement requirement.

4.1.1 Decreasing the Water/Cement Ratio

If using a water reducer or a superplasticizer reduces the amount of mixing water by ΔW in a concrete having a water/cement ratio W/C, the cement saved ΔC will be equal to

$$\Delta C = \frac{1}{W/C}\Delta W$$

Note that savings increase as the water/cement ratio drops. Since the water/cement ratio of conventional concrete varies from 0.50 to 0.80, the practical savings in cement when using a water reducer or a superplasticizer is $1.25\ \Delta W \leq \Delta C \leq 2\ \Delta W$ or, on a mass basis, 1.25 to 2 times the mass of water saved.

Therefore, the amount of cement required and the CO_2 released into the atmosphere are inversely proportional to the efficiency of the water reducer in lowering the amount of mixing water. Consequently, if we accept that the worldwide development of the construction industry should not result in any increase of its present environmental load, the worldwide promotion and use of water reducer could result in the production of 8 percent more concrete without any increase of the amount of CO_2 released in the atmosphere.

Reduction of the CO_2 released when making concrete can be taken a step further when using high performance concrete with a lower water/cement ratio, since a decrease in the water/cement ratio increases concrete compressive strength and durability. It is easy to demonstrate that the use of a higher-strength concrete in structural elements working essentially in compression results in a significant decrease in cement consumption.

The increase in the compressive strength of high performance concrete also results in significant savings in sand and coarse aggregate consumption, because, when using three times less concrete to build the same column, the aggregate consumption is also reduced approximately by 3.

4.1.2 Increasing the Life Cycle of Concrete Structures

Concrete use can also be reduced by increasing the life cycle of concrete structures through enhanced durability (5). The durability of a concrete structure depends on the environmental conditions to which it is exposed as well as on the quality and curing of the concrete used. Since modifying environmental conditions is usually costly or difficult, adapting concrete quality and curing are preferred.

Concrete is still too frequently specified on the basis of its compressive strength and not its environmental conditions. As a result, the specified concrete is often partially or totally inadequate for its environment. For example, an outdoor parking garage can be designed and built in Canada like an office building with 20-MPa concrete. It won't last very long since low-strength concrete with a very high water/cement ratio affords no reinforcement-steel protection against the chloride from deicing salts brought in on cars.

National codes need to put more emphasis on the environment in which concrete structures have to perform rather than on concrete compressive strength. If the water/cement ratio required for adequate performance in a given environment must be low, the designer has to figure out a way to take advantage of the extra strength provided by the concrete to make the structure more durable but with less concrete.

Decreasing the water/cement ratio is just one requisite for durable concrete. Improper curing and lack of curing can significantly affect the performance of a potentially durable concrete, leaving the quality of the structure in question. Concrete must be properly cured to achieve its potential. Unfortunately, this is not always the case in the field, because contractors are not used to curing concrete and they want to cut their costs.

The lowest-bid philosophy that prevails worldwide in the construction industry has its drawbacks; a major one is the lack of effort and money devoted to concrete curing. Contractors are asked to cure concrete, but are not paid to do so. There is an urgent need to include paid items related to curing concrete in tenders. If the contractor is specifically paid to cure the concrete, he will. If all the items related to concrete curing are clearly identified and have a unit price, it will be easy for the lab in charge of the field inspection to enforce proper curing. If the contractor is caught cutting corners, he will lose money, and there's nothing he likes less.

Although there's no easy way to predict the increase in the life cycle of a concrete structure as a result of proper curing, the increase is real. The longer life cycle will decrease concrete consumption by reducing the frequency with which concrete structures need to be repaired, rehabilitated, or demolished.

4.2 Using More Supplementary Cementitious Materials

Several industrial by-products or natural pozzolans are already used for cement replacement in concrete, but not to the desired degree (6, 7). The cement industry has long claimed that supplementary cementitious materials offered no advantage due to their great variability in quality and because of a significantly lower early strength in the concrete. Although supplementary cementitious materials are not as reactive as portland cement during the early stage of cement hydration, their lower reactivity can be easily compensated for by getting the fine cementitious particles closer to each other by decreasing the water/binder ratio through the use of a superplasticizer. Since the binding particles are closer together, less "glue" is necessary to reach the same strength. The strength, even the early strength, and the porosity of a concrete depend as much as on the amount of "glue" formed as on the packing of the binding particles: closely packed cementitious particles need fewer hydrates to "glue" them together. Consequently, the lower the water/binder ratio, the closer the binding particles, resulting in a concrete that is stronger and less pervious (5).

The concrete industry will have to learn how to make stronger concrete than that produced with portland cement through the use of more supplementary cementitious materials.

This applies equally to natural pozzolans. We should learn how to make stronger and more durable pozzolanic concrete by improving the technology used by the Romans more than 20 centuries ago. There is no more environmentally sound concrete than a pozzolanic one containing a natural pozzolan. In a modern pozzolanic concrete, the cement and superplasticizer replace the lime and the hard work of the slaves, who painfully but successfully compacted the dry mixes they used in order to get strength from a very low reactive binder system.

If the 400 million tonnes of fly ash produced every year worldwide by coal-fired power plants were used to make concrete, about 30 percent more concrete could be produced without an increase in worldwide cement production. This also applies to concrete production with natural pozzolans, if properly promoted.

4.3 Making More Environmentally Friendly Cement

Producing 1 tonne of portland cement currently releases 1 tonne of CO_2 into the atmosphere. This figure can be reduced by decreasing the fuel consumption of cement kilns or by decreasing the amount of limestone used to make cement. Let us look at these two possibilities.

4.4 Making "Energy-Conscious" Cement

Over the last quarter century, the cement industry has been very successful in reducing the energy necessary to make cement. As things stand, further reductions appear problematic if traditional cements are to be produced.

The efforts currently under way in the cement industry to use all kind of alternative waste fuels are very valuable from an environmental point of view and help solve some critical environmental problems. On the other hand, this approach does nothing to decrease of the amount of CO_2 released into the atmosphere during cement production. Therefore, we must change the type of the cement produced today.

4.4.1 Mineralized Clinkers

The manufacturing of "mineralized clinkers," that is, clinkers doped with fluorine and sulfur, is very encouraging in this respect, because mineralized clinker only needs to be fired up to 1300 to 1350 °C, instead of 1450 to 1500 °C, as is now the case (8). It is almost certain that this new type of clinker will soon be the standard clinker due to the impact of fuel consumption on the production cost of cement and its environmental consequences. Some of these mineralized clinkers are already being produced for manufacturing less energy-intensive cement.

4.4.2 Belitic Cement

In response to contractor demands for concrete with a higher early strength to allow faster stripping of forms, the cement industry has gradually increased the C_3S content of its cements and ground them finer. Right now, portland cements with a C_3S content as high as 70% are not rare, while 50 years ago, less than 50% was common. This change in the phase composition of the cement has had two detrimental consequences:

- the higher C_3S content means firing more calcium carbonate to produce a tonne of cement and at a slightly higher temperature, which releases more CO_2;
- the concrete achieves its final strength earlier due to the higher C_3S content.

Specifically, modern concrete reaches a 28d compressive strength that is very close to its final strength, since there is no late strength increase due to the slow hydration of C_2S, which was common with the cement produced 50 years ago.

Fifty years ago, when a designer specified a concrete with a 28d compressive strength of 30 MPa, the concrete producer had to use more cement and a lower water/cement ratio than today for comparable strength, because the cement back then had a lower C_3S content (9). But this concrete gained strength after 28 days due to the slow rate of hydration of the C_2S. Therefore, the final strength of concrete from that period was usually much higher than that of a contemporary concrete with the same 28d compressive strength. When cored today, some kinds of old 30-MPa concrete record strengths of 40 to 50 MPa.

While both the older concrete and a modern concrete may be similar from a design standpoint according to our national codes, they are quite different with respect to durability. Without a doubt, today's 30-MPa concrete will not last as long as one cast 50 years ago. Designers should be aware of this change in the basic phase composition of modern cement and of its consequences on the durability of the concrete they specify.

Now that we know how to decrease the water/cement or water/binder ratio of a concrete, there is no need to use portland cements that are so rich in C_3S: it is possible to make concrete exhibiting the same strength development and 28 d strength using portland cement richer in C_2S. These cements are called belitic cements. Since they have a high C_2S content, their manufacture requires less calcium carbonate, so that less CO_2 is released into the atmosphere during production. Some types of high C_2S cement are already produced, but they represent a very low percentage of the worldwide cement production, despite their success.

The only case in which cement containing a high amount of C_3S is beneficial is when supplementary cementitious materials are used, because lime activates pozzolanic materials.

From a long-term perspective, sulfobelitic cements could well take over the market since they can be fired at around 1200 °C. The manufacturing process of sulfobelitic cement has already been mastered and this very ecological cement is just waiting to be used more (10).

4.5 New concrete concepts

The basic principles governing mix proportioning are well established and seemed to be "cast in the concrete" until recently. The increasing use of chemical admixtures, particularly water reducers and superplasticizers, has resulted in some changes in the composition of a concrete but they are not drastic enough to alter significantly the basic composition of concrete. However, quite recently new innovative concepts in mix proportioning have been developed and put successfully into application which, for different reasons, result in a significant decrease of the cement dosage or a more efficient use of the binding properties of portland cement expressed in terms of MPa per kg/m^3 of cement. Three of these new concepts will be briefly presented here.

4.5.1 High-volume fly ash and high-volume slag concretes

The efforts made by CANMET researchers (11, 12) to promote the concept of high-volume fly ash and high-volume slag concretes should deserve more attention, because it is definitely a very interesting way to decrease the environmental impact of concrete while producing a low cost, strong and durable construction material. In this type of concrete 50 to 60 percent of the total cementitious material is fly ash or slag. The water content is kept very low and the high workability is obtained by the use of an appropriate dosage of superplasticizer. Due to its low water-to-cementitious materials ratio this type of concrete has adequate early-age and long term strength as well as excellent durability characteristics.

4.5.2 Reactive Powder Concrete (RPC)

This new ultra high performance concrete developed by P. Richard (13) is made from different **powders** that are at one time or another **reacting** during RPC processing. The term **concrete** is used rather than mortar to describe RPC because of the mechanical behavior of RPC when steel fibres are added to enhance ductility. It is easy to demonstrate that adding 0.2 × 25 mm steel fibres to RPC in which the maximum size of the powder particles is 600 µm is like adding 8 × 1000 mm rebar to concrete made with 25 mm coarse aggregate.

The basic principles used to develop RPC concept are:

- Enhancement of homogeneity by elimination of coarse aggregates
- Enhancement of compactness by optimization of grain-size distribution and application of pressure before and during setting
- Enhancement of the microstructure by post-set heat treatment
- Enhancement of ductility by incorporating fine steel fibres or external confinement

In spite of the fact that in its present formulation RPC contains a very high amount of cement it is easy to show that due to its very high strength the use of RPC in a concrete structure result in a significant decrease of the total amount of cement necessary to build that structure (14). As an example the amount of cement used to build the Sherbrooke RPC pedestrian/bikeway bridge was half the amount of cement that would have been used to build it with a high performance concrete.

It is sure than in very few years RPC will be made with much less cement than presently making it even more environmental friendly than presently.

4.5.3 Low-Heat High Performance Concrete (LHHPC)

Researchers from Atomic Energy of Canada Limited (AECL) (15) have developed a new concept called low-heat high performance concrete (LHHPC) containing only about 100 kg/m^3 of Type V portland cement and the same amount of silica fume as binder and about 200 kg/m^3 of silica flour. This 0.50 W/B concrete (silica flour is not considered as a binder) experience a maximum temperature rise that does not exceed 20° C in quasi adiabatic conditions. In spite of their very low binder content these concretes exhibit quite high long term strength approaching 100 MPa.

It is the author opinion that RPC and LHHPC represent in some ways the two extremes of a new family of ecoconcrete that will be developed in the near future, as many powders will have to be recycled in concrete.

5 Recycling concrete

One final attempt to decrease the environmental burden created by the tremendous volume of concrete used worldwide is greater recycling of used concrete in order to save our natural aggregate resources and to lower the energy consumption in transporting aggregate to urban centers.

In large cities in industrialized countries, concrete is starting to be recycled on a significant scale, because:

- in-town or nearby natural aggregate quarries are unable to extend their operations because they are getting closed down or depleted;
- it is increasingly difficult to open new quarries because of environmental regulations, which are being enforced. Modern societies are more conscious of the impact of these quarries on the environment and on a sustainable development;
- it is increasingly difficult and costly to dispose of construction waste.

But one fact must be emphasized: recycling cannot produce a material that performs as well as the original. In fact, most of the recent breakthroughs with high-performance materials rely on the optimized mixing of a combination of highly pure raw materials in very specific proportions according to a complex process. Since these pure materials have to be mixed in very precise proportions at specific stages in the process, and, since they react with each other, they cannot be separated back when recycling the used material. Therefore, it is impossible to make a material from recycled components that performs as well as the original.

Moreover, the quality of recycled concrete can vary greatly; separating good concrete from bad isn't very easy, since they have the same ingredients only their proportions differ. Once separated from its principal contaminants (steel, gypsum, plastics, organic matter, and wood), recycled concrete is only suitable for low-strength concrete, a fairly low-quality granular material, or as a subbase in road construction. Such low-tech uses still benefit the environment because recycled concrete can replace virgin aggregates that will not have to be quarried and transported from far away. Any decrease in transportation distance has a direct impact on the CO_2 released.

6 Conclusion

Concrete is and will remain the most widely and intensively used construction material on the planet because of its intrinsic qualities. Its use will not increase significantly in developed countries, where a decrease in concrete consumption has already been noted. Its use will increase drastically in developing countries, where concrete technology is less advanced and where people are more concerned with their own development than with the global environment.

The urbanization process in developing countries is unavoidable, because it is the process that has created wealth in human societies since the most remote times. This urbanization process will inevitably lead to an increase in the use of concrete with a concomitant increased release of CO_2. There are already several means that can be implemented very rapidly to allow us to make more concrete with less cement. Unfortunately, these means are developed or under development in industrialized countries where the use of cement and concrete is stagnating or even already decreasing. Moreover, it must be admitted that cement companies do not promote such uses very strongly; in some cases, they deliberately ignore or fight it. We have the responsibility to teach developing countries how to make more durable concrete with less cement, how to make and use more environmentally friendly cements, and how to maximize recycling of old concrete.

7 References

1. Aïtcin, P.-C. (1995) Concrete the Most Widely Used Construction Material. *Adam Neville Tribute*, (ed. Mohan Malhotra), Las Vegas, pp. 257-66.

2. Kreijger, P.-C. (1987) Ecological Properties of Building Materials. *Materials and Structures*, Vol. 20, pp. 248-254.

3. The Bible "Genesis The Babel Tower," Verse XI, The Babel Tower and Exodus Verse V.

4. Mark, R.(1990) *Light, Wind and Structure - The Mystery of Master Builders.* McGraw-Hill, N.Y., pp. 49-74.

5. Aïtcin, P.-C. (1994) Durable Concrete - Current Practice and Future Trends. *ACI SP-144*, March 1994, pp. 85-104

6. Malhotra, V.M., Mehta, P.K. (1996) *Pozzolanic and Cementitious Materials,* Gordon and Breach Publishers, ISBN 2-88449-211-9, 191 p.

7. Mehta, P.K. (1993) *Concrete Technology at the Cross Roads - Problems and Opportunities*, ACI SP-144, pp. 1-30.

8. Li, P., Liu, C., Zhuo, H. (1997) Practical Application of Composite Mineralizer Technology in Precalcining Cement Kiln, *10th International Congress on Cement Chemistry*, Gothenburg, Sweden, Paper li028, 8 p.

9. Wischers, G., *The Impact of the Quality of Concrete Construction on the Cement Market*, Report to Holderbank Group, No. DIR 84/8448/4.

10. Zhang, L., Su, M. (1997) Control Properties of Sulpho- and Ferro-aluminate Cement Concrete with Special Admixtures, *10th International Congress on Cement Chemistry*, Gothenburg, Sweden, Paper 3iii026, 7 p.

11. Ramachandran, V.S., Malhotra, V.M., Jolicoeur, C., and Spiratos, N. (1998) *Superplasticizers: Properties and Applications in Concrete*, CANMET Publication MTL 97-14 (TR), March, pp. 291-294.

12. Malhotra, V.M. (1994) CANMET Investigations Dealing with High-Volume Fly Ash Concrete, *Advances in Concrete Technology*, CANMET Publication MSL 94-1 (IR), pp. 445-482.

13. Richard, P., Cheyrezy, M. (1994) *Reactive Powder Concretes with High Ductility and 200-800 MPa Compressive Strength*, ACI SP-144, pp. 507-518.

14. Bonneau, O., Poulin, C., Dugat, J., Richard, P., Aïtcin, P.-C. (1996) Reactive Powder: from Theory to Practice. Concrete International, Vol. 18, No. 4, pp.47-49.

15. Gray, M.N., Shenton, B.S. (1998) *For Better Concrete, Take Out Some of the Cement*, 6th ACI/CANMET Conference on Fly Ash, Silica Fume, Slag and Natural Pozzolans in Concrete, Bangkok, May, 28 p.

SUSTAINABLE CONCRETE FROM A CEMENT MANUFACTURERS POINT OF VIEW
Manufacturing process and modified cements

B. SØPLER and T.F. RØNNING
Norcem A.S / R&D Section, Brevik, Norway

Abstract
Sustainable concrete comprises careful selection of raw materials as well as qualified and advanced manufacturing techniques. Portland cement accounts for the main emissions related to concrete, but additional substituents and transport may play an important role.

However, one should focus on the right issues; the "environmental competitiveness" of the complete structure and compared to alternative solutions/materials. The cement and concrete industry should contribute to and influence the models for selections of sustainable construction and building products. In this perspective, one should not forget the ability of concrete, or concepts based on concrete, to protect the environment against human activity.

"Sustainable concrete" is a major Norwegian research project headed by Norcem A.S, in cooperation with several industrial partners and financial support from the Research Council of Norway. The main objective of the programme is to consider various components of the mix design and find solutions to reduce the impact on the environment.

One approach has been to apply fly ash cement, another to utilize "in-situ" or nearby available processed sand. The investigations comprise fresh concrete properties as well as durability issues.

Finally, with respect to "sustainable competitiveness", evaluation should follow an engineering design based on the actual application.
Keywords: Concept, emission control, product development, cooperation.

1 Significance of work

1.1 General
Sustainable building materials is about to become a public issue. The industry is worldwide facing governmental actions against wastes and emissions. In order to stay in business, the construction industry (among others) will have to adopt.

Concrete Technology for a Sustainable Development in the 21st Century edited by O.E. Gjørv and K. Sakai.
Published in 2000 by E & FN Spon, 11 New Fetter Lane, London EC4P 4EE, UK. ISBN: 0 419 25060 3

Moreover, maybe this industry - together with its allies - should take initiatives and responsi-bility themselves, doing their part to ensure a sustainable future. The aim should be developing and engineering concrete structures to serve mankind and environment, not human life on the terms of our industry.

1.2 Models for sustainable products

The cement industry tends to look at the cement only, the concrete manufacturers on the (economically) competability of the concrete, the constructor considers the structure as a complete product. The owner will (should) focus on the lifetime of the structure. However, it is essential not to sub-optimize on each level - or to generalize improperly. Concept models **are** being developed - with or without our contribution. The concrete industry should participate in order to create the right terms as far as possible.

1.3 Emissions from cement manufacturing

Apart from existing taxes on fuels, the Norwegian cement industry will shortly be facing some sort of CO_2-emission taxes. It is also obvious that emissions from cement clinker manufacturing varies significantly worldwide and, hence, represent important potentials.

Equally important in order to decrease emissions, per m^3 of concrete, is the utilization of cement substitutes. The optimal selection of these may vary with field of application and market structure. Norcem has chosen an approach with a CEN type II cement, containing 20 % PFA and reducing the emissions accordingly. On the domestic market, with many small concrete manufacturers, it is considered appropriate to offer a blended cement, in stead of another cement substitute (in addition to silica fume), to ensure the total quality.

1.4 Sustainable concrete engineering

In order to reduce the emissions per m^3 of concrete as far as possible, it is important to investigate the performance of concrete of different compositions for the relevant applications. Concrete standards (e.g. the future EN 206) provide information/ guide-lines, but will presently for blended cements and/or concrete containing cement sub-stitutes not sufficiently cover performance or explain such in the Nordic environment. It is necessary to establish experience with such materials and to carry out investigations on the concrete performance.

2 INVESTIGATIONS AND RESULTS

2.1 Concrete vs. environment

2.1.1 General

In the 90's, increasing interest has been focused on the environmental adoption of concrete. Several attempts have been made to describe these relations qualitatively

[1], [2]. Comparative simulations require a systematic, model-based approach comprising the main criteria for sustainable product development;

- Main environmental and health impacts
- Main customers' requirements and maintenance practice
- Life cycle costs (LCC)
- Life cycle assessment (LCA)

Principles for LCA studies are given in NS-EN ISO 14040 [3] and further examplified in [4]. It is a systematic mapping of health and environmental impacts during the complete life of a structure, including extraction of raw materials, maintenance of the structure, demolition and dispose of waste. For these kind of exercises, it is convenient to choose a "unit structure", e.g. 1 km road pavement [5].

Such analysis should start with a flow chart for the structure and alternative materials, as shown for concrete production in Fig. 1;

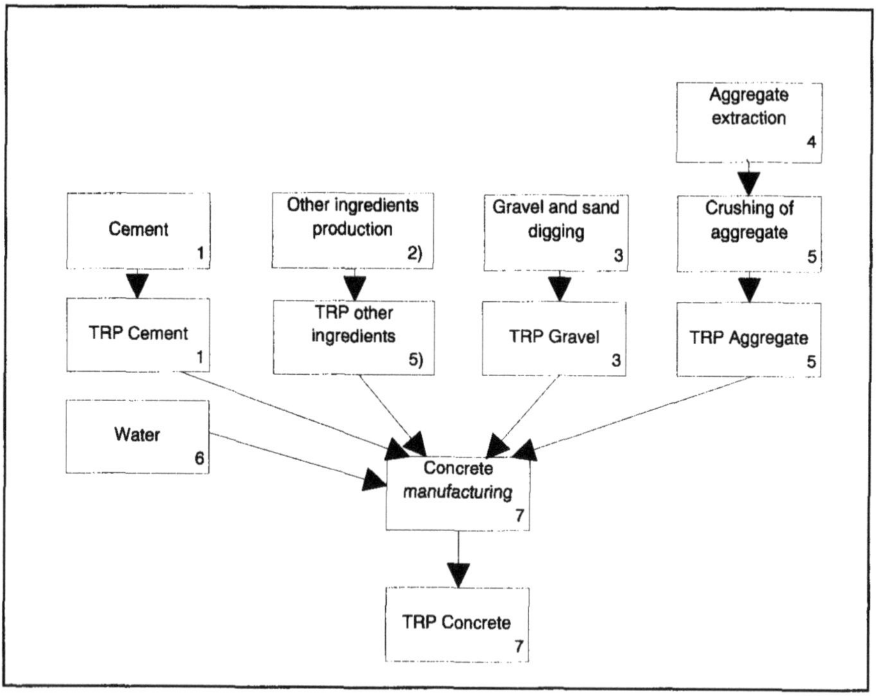

1. Environmental burdens from production and transportation of cement (Lappeenranta). Data of energy and material flows are from Finncement.
2. Silica fume is dealt with as by-product. No environmental burdens except those from transportation to the concrete manufacturing place are allocated on it.

Concrete additives (total weight < 1 % by weight of concrete) are excluded because lack of data.

3. Fuel consumption data from extraction and transportation of gravel and sand from Lohja Rudus.
4. Fuel consumption data from quarrying of crushed aggregate from Lemminkäinen.
5. Fuel consumption data from crushing and transportation of aggregate materials from Lohja Rudus.
6. No environmental burdens are allocated on water.
7. Energy and material flows in concrete manufacturing and transportation from Lohja Rudus.

Figure 1 Flowchart of concrete production (TRP = Transport /6/

2.1.2 Concrete pavement

One of our projects compared the "environmental costs" of concrete and asphalt pavements [6] with alternative scenarios; Table 1:

Table 1 Scenarios valuated according to different valuation methods. Emissions are given in kg/km, energy in GJ/km, land use in m^2. /6/.

	1 Asphalt Maint. A	2 Asphalt Maint. B	3 Concrete Maint. A	4 Concrete Maint. B	5 Concrete Maint.A Including carbonation
CO_2	590 000	670 000	940 000	940 000	880 000
SO_2	2 500	2 800	1 700	1 700	1 700
No_x	3 000	3 600	4 700	4 700	4 700
CO	610	670	2 000	2 000	2 000
VOC tot	1 900	2 100	1 000	1 000	1 000
CH_4	1 500	1 500	880	880	880
As	0.0028	0.0040	0.019	0.019	0.019
Hg	$42x10^{-6}$	$64x10^{-6}$	0.0076	0.0076	0.0076
Cd	0.00041	0.00081	0.0016	0.0016	0.0016
Cr	0.0012	0.0017	0.0066	0.0066	0.0066
Pb	0.028	0.040	0.042	0.042	0.042
Zn	0	0	0.015	0.015	0.015
Dust	1 200 000	1 200 000	650 000	650 000	650 000
N into water	0.91	1.3	0.13	0.13	0.13
COD	18	26	68	68	68
Fossil fuel	15 000[1)]	20 000[1)]	7 600	7 600	7 600
Nuclear fuel	4 600	4 600	2 930	2 900	2 900
Renewable energy	4 600	4 600	2 900	2 900	2 900

1) Including inherent energy.

Among the main conclusions from this study were:

- The environmental burdens of concrete pavements are 40-60 % higher for CO_2, 30-60 % higher for No_x.
- The environmental burdens of asphalt pavements are 40-60 % higher for SO_2, and approximately double for VOC_s, dust and non-renewable energy, compared to manufacture, maintenance and use of concrete pavements.

However, the study still is inflicted by some uncertainties, e.g. fuel consumption. A decrease in fuel consumption of traffic due to properties of concrete would bring significant "sawings" in emissions with this alternative and, hence, promote concrete.

2.1.3 Concrete abutment of a bridge

A study from Netherland examines the environmental cost for alternative materials concrete and steel, including maintenance, Fig. 2

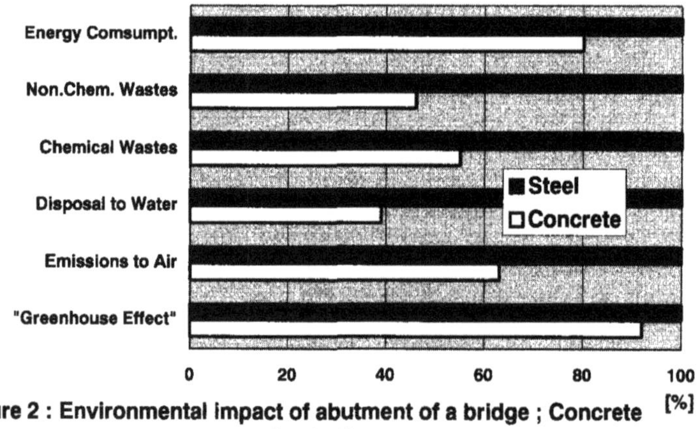

Figure 2 : Environmental impact of abutment of a bridge ; Concrete or Steel /7/.

2.1.4 Concrete pipe for sewage

Another study [8] illustrates the preferences of concrete pipes to pipes made of PVC concerning energy consumption, effluents and wastes.

2.2 Emissions from cement production

In 1996, Norcem completed its part "Cemenviron" of the national programme FORFOR [9]. This study enabled a systematic mapping and clarifying of the local potentials at the plant in Brevik, Norway.

Another study is completed as a joint project together with German and Italian cement industry. One example of these studies is outlined in [10]. Statistical analysis (PLS; Partial least square regression) of log (No_x emissions) gave an explained variance (with six factors included) of appr. 80 %. The feed of reburning fuels, i.e. supply of energy per hour [GJ/h], had a positive and significant influence of No_x reduction. An increase in the energy supply of 80 GJ/h (≈ 35 % of the total energy of the secondary fuel applied) reduced the No_x-emissions with appr. 15 %.

The emissions of CO_2 is of course very much linked to the decomposition of limestone and remains approximately the same per ton clinker. However, through different production technology, the emissions due to fuel consumption may vary 20-25 %. One way to "reduce" emissions may also be to apply "CO_2-neutral fuels".

2.3 Investigations for sustainable concrete engineering

Within "Sustainable Concrete Programme", a joint project with several domestic partners and governmental support, extensive studies have been made on utilization of the new blended cement ("Norcem Std FA"; 20 % Fly Ash) and processed (crushed) sand, with or without silica fume and various admixtures. All these materials may be considered "substitutes", and hence - the application of such - environmentally favourable with reduced emission from production or transport (e.g. transport of natural aggregates).

Just to examplify some of the work done, Fig. 3 shows the slump values of realistic ready mix concrete (of same strength class), containing 40 % (of fraction 0-8 mm) washed 0-2 mm crushed sand.

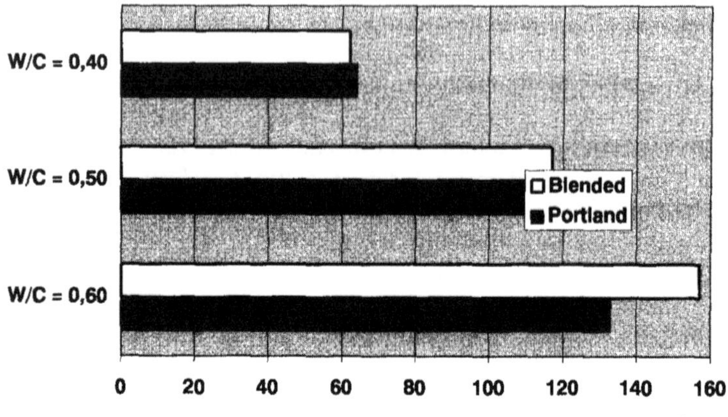

Figure 3 : Slump value with pure portland and blended cement /11/

These laboratory results concur quite well with the field experience with these materials. Especially in lean concrete mixes, the feedback has been very positive concerning these combinations - or these materials used separately. The investigations are continued with more sophisticated methods, e.g. a coaxial cylindrical viscometer.

The above mentioned results can also be directly applied in cost analysis, which is a dedicated project activity. However, to complete the picture, it is compulsory to investigate the performance of the concrete structure in the relevant field environment.

Parallell to the examinations on fresh concrete properties - but much more time consuming - durability issues are being considered. Partly together with "end-users", like the public road ministry - and with Swedish partners, studies are in progress on chloride ingress, freeze-thaw durability and resistance against alkali-aggregate reaction. The investigations comprise laboratory and comparative in-field studies, even methology. However, the results are "premature" to be presented at this stage.

3 CONCLUSIONS

In order to develop sustainable concrete, , it is essential to work jointly to establish the correct framework and not to sub-optimize.

Most of the emissions related to utilization of concrete originates from the manufacturing process of cement clinker. There are potentials to reduce these to some extent, depending on local conditions. Present work considers e.g. No_x-reduction and secondary fuels.

One of the main potentials is to utilize supplementary cementitious maerials. It is in all partners interest to promote this development, after careful examinations of concrete performance under relevant conditions. Even technical improvements are attainable with supplementary materials, but this will not be the rule. Hence, it is also essential to agree on a "sufficient quality level" and to develop "sustainable engineer-ing".

4 REFERENCES

1. Wischers, G.: Beton und Umwelt - Ökobilanz für Beton ("Concrete and the environment - ecological analysis for concrete") Beton- und Fertigteiltechnik, vol. 4 (1992), pp 50-61. German.
2. Torghele, K.: Beton im Wohnbau - Ökoloische Gesichtspunkte ("Concrete in housing - ecological point of view"), Beton-Materialen für eine neue baubio-logische und ökoloische Position, Vienna 1995, ISBN 3-900 403-25-2, pp 113-136. German.
3. EN ISO 14040 Environmental management - Life cycle assessment - Principles and framework (ISO 14040 : 1997).
4. Johnsen, T. et al: Sustainable product development in cement and concrete industry. A Cement Nordic initiated project. Fredrikstad, Norway, October 1996. ISBN 82-7520-265-5, 61 p.
5. Stoltenberg-Hansson, E.: Miljøregnskap - Betong, et konkurransedyktig materiale ("Environmental accounting - Concrete, a competitive construction material"). The Norwegian Concrete Association, Trondheim, October 1997, 9 p.
6. Häkkinen, T. et al: Environmental adaption of concrete - Environmental impact of concrete and asphalt pavements. A Cement Nordic initiated project. VTT, Espoo, Finland 1996, ISBN 951-38-4907-4, 97 p.
7. Fraanje, P. et al: Milieubelasting van twee aan bruggen, Amsterdam University, May 1992
8. Milieuprofiel en milieumaten van een buitenriolering van pre en gres in vergelijking met betong. Intron report no. 95195.
9. FORFOR; Forskning for bærekraftig prosessindustri 1992-1996 ("FORFOR; Research for a sustainable process industry 1992-1996"). Main report from the Research Council of Norway, ISBN 82-12-00773-1, 40 p, Norwegian.
10. Svinning, K. et al: Statistical analysis of the correlation between No_x emissions and production conditions in a cement kiln applying staged combustion. World Cement, Jan. 1998, pp. 68-75.
11. Mørtsell, E. et al: Betydning av varierende mengde vasket og uvasket maskinsand på betongens konsistens ("Significance of various amounts of washed and unwashed, crushed fine aggregates on concrete consistency"). Non-official, internal project report from the Norwegian "Sustainable Concrete Programme", Trondheim 1997.

ROLE OF SUPPLEMENTARY CEMENTING MATERIALS IN REDUCING GREENHOUSE GAS EMISSIONS

V.M. MALHOTRA
CANMET, Natural Resources Canada, Ottawa, Canada

Abstract
The World Earth Summits in Rio de Janeiro, Brazil, 1992 and in Kyoto, Japan in 1997 have in no uncertain terms, emphasized the dangers associated with the unchecked increase in CO_2 and other greenhouse gas emissions to the atmosphere, and their undesirable effects on climate change. As a result, most industrialized countries are in the process of formulating regulations to stabilize these emissions to 1990 level by the year 2010. This paper describes the amount of CO_2 being contributed by the portland cement industry, and discusses how these emissions an be reduced considerably by the increased use of large volumes of fly ash and other supplementary cementing materials in concrete industry.
Keywords: Cement, CO_2 emissions, concrete, fly ash, greenhouse gas, supplementary cementing materials.

1 Introduction

The environmental issues will play a leading role in the sustainable development of the cement and concrete industry in the 21st century. The World Earth Summits in Rio de Janeiro, Brazil in 1992 and Kyoto, Japan in 1997 have made it abundantly clear that unchecked increase in the emission of greenhouse gases to the atmosphere is environmentally and socially no longer acceptable for overall sustainable development. The primary greenhouse gas emissions discussed in the deliberation of the above conferences are CO_2 emissions. The other greenhouse gases such as NO_x and CH_4 are of serious concern, but the amount involved is relatively small compared with that of CO_2. As a consequence, the developed countries are considering regulations and mandatory quotas on the emission of these gases, and the main thrust is to stabilize these emissions to the 1990 level by the year 2010. As the manufacture of portland cement contributes significantly to the CO_2 emissions, this paper discusses the role the increased use of large volumes of fly ash and other supplementary cementing materials in the construction industry can play in reducing these emissions.

Concrete Technology for a Sustainable Development in the 21st Century edited by O.E. Gjørv and K. Sakai.
Published in 2000 by E & FN Spon, 11 New Fetter Lane, London EC4P 4EE, UK. ISBN: 0 419 25060 3

2 Portland cement production

Portland cement is a major construction material worldwide and will remain so for the foreseeable future. The net cement production in the world is expected to rise from about 1.4 billion ton in 1995 to almost about two billion ton in the year 2010 (Table 1). It is expected that the production will not increase significantly in North America and Western Europe but the major increases will take lace in China, India, Eastern Europe and Africa. In view of the huge tonnage involved, it is imperative that the manufacture of cement be made as environmentally friendly as possible. If this is not possible, then ways and means must be found to reduce significantly the consumption of portland cement by the increased use of supplementary cementing materials in concrete.

Table 1. Regional and World Cement Production to Year 2010*

Summary: regional and world cement consumption to 2010

	European Union*	Other Europe*	Former Soviet Union*	North America *	C/S America*	Africa	Middle East*	East Asia*	S/SE Asia*	Oceania*	World Total
1995	168.1	65.8	58.1	92.9	89.4	64.8	63.5	623.4	161.2	8.0	1396.1
2000	187.9	80.0	80.3	94.9	106.6	74.3	75.6	732.7	219.1	10.6	1662.1
2005	194.1	90.2	110.1	94.8	127.4	80.7	76.9	798.8	255.0	11.1	1839.1
2010	189.3	94.7	128.2	94.7	145.0	85.5	73.4	844.3	279.2	11.8	1946.1
% of total 1995	12.1	4.7	4.2	6.6	6.4	4.6	4.6	44.6	11.6	0.6	100.0
% of total 2010	9.7	4.9	6.6	4.9	7.5	4.4	3.8	43.4	14.4 ·	0.6	100.0

*million tonnes
Figures may not add due to rounding, source: Ocean Shipping Consultants Ltd.

*From World Cement Annual Review 1997. World Cement, Vol. 28, No. 7, July 1997, pp. 3-60.

3 Energy consumption, CO_2 emissions, associated costs in the manufacture of portland cement and situation in developing countries

3.1 Energy consumption

After aluminum and steel, the manufacture of portland cement is the most energy-intensive process. The manufacture of portland cement require about 4 Gigajoules (GJ) of energy per ton of the finished product. Over the past decades, the cement industry has made major strides in reducing the energy consumption. This has been achieved primarily by replacing wet production facilities with new modern dry-processing plants and by moving away from the petroleum-based fuels. For example, in the U.S.A. the cement industry has been able to reduce the average energy use per ton of cement from about 7 GJ in 1974 to about 5.5 GJ per ton in 1991 (Fig. 1). However, it has reached about the limit beyond which it is extremely difficult to reduce futher energy use in the cement production process. Obviously, the existing cement plant cannot be shutdown. This leaves only one option and that is to limit the installation of new plants, and phasing out of the old inefficient installations. The loss in capacity due to this change can and must be met by the increased use of supplementary cementing materials such as fly ash, blast-furnace slag, natural pozzolans, rice-husk ash, silica fume and metakaolin.

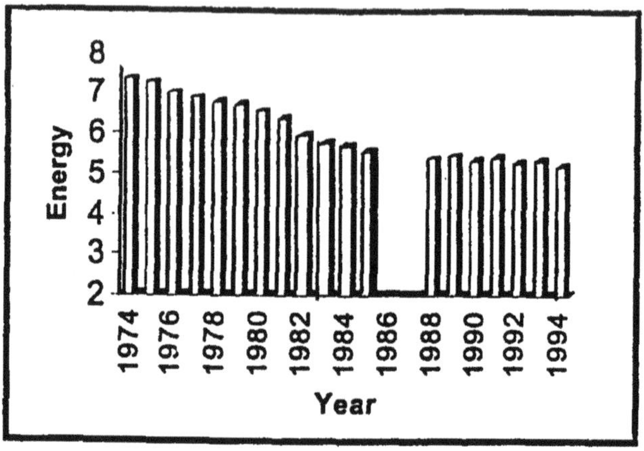

Fig. 1. Trend in energy consumption m Btu/t of cement.
Source: PCA Economic Research Dept.
US Cement Industry Fact Sheet, Fourteenth Ed.
No data available for 1986 or 1987.
From: World Cement, August 1997 p.65

3.2 Greenhouse gas emissions

Not only is the manufacture of portland cement highly energy intensive, it also is a significant contributor of the greenhouse gases. The production of every ton of portland cement contributes about one ton of CO_2 into the atmosphere. Minor amount of NO_x and CH_4 are also released into the atmosphere. The total CO_2 emissions per ton of cement range from about 1.1 ton of CO_2 from the wet process to 0.89 ton from a precalciner kiln. About half of the CO_2 emissions are due to the calcination of limestone (a major raw material) and the other half are due to the combustion of the fossil fuels. According to Cahn et al. [1], the emissions from the calcination of limestone are fairly constant at about 0.54 ton of CO_2 per ton of cement; the emissions from the combustion depend on the carbon content of the fuels being used and the efficiency of the fuel usage.

Globally, in 1995, the amount of production of cement was about 1.4 billion tons. This means that the manufacture of portland cement contributed about 1.4 billion tons of CO_2 to the atmosphere. According to IEA World Energy Outlook 1995, the worldwide CO_2 production from all sources was 21.6 billion tons. On the basis of the above figures, the worldwide production of cement accounts for almost 7% of the total world CO_2 production, and from the projections made by the cement companies, this proportion is expected to remain steady in the next decade [1]. Obviously, the cement companies are not expecting the emergence of environmentally-friendly cement manufacturing technologies in the near future.

3.3 Cost consideration

The cost of a new portland cement plant is in the order of 175 million dollars per million tons of installed capacity. Till recently, a one million tons capacity plant was considered to be the most efficient installation. However, lately huge cement plants with capacities of about 3 million tons have come on line in Thailand and South Korea. The above cost considerations, combined with the CO_2 emissions and energy consumption issues, cast doubts that many new installations will be brought on line in the developed world. Instead, the government regulations on greenhouse gas emissions would force the cement industry to look for supplementary cementing materials that could be used to produce blended cements or could be used as separate ingredients at concrete batch plants. The developed countries would also attempt to import more cement from the developing countries. The export of cement from Mexico to the U.S.A is one such example.

3.4 Situation in developing countries

The infrastructure needs of the developing countries have led to uge increases in demand for portland cement (Table 1). This has led to the installation of a large number of new cement plants in China, India and South America. Paradoxically, these are the same countries that are also installing huge coal-fired power stations to supply electricity to meet the growing needs of the population and the manufacturing industries. For example, it is anticipated that in the year 2002, India will produce about 109 million tons of portland cement and the coal ash by-products from thermal stations will reach about 130 million tons annually. Unfortunately, the much needed essential development in theses countries are adversely affecting the environment in two ways. The installation of the new cement plants is increasing tremendously the CO_2 emissions, and the construction of the monster thermal-power stations is resulting in huge amounts of by-products such as fly ash, bottom ash and boiler slag that are not being used in any significant manner. Most of the fly ash is being dumped in lagoons, land-fill sites, abandoned quarries, and in some instances being slurried directly to the open sea. Thus, potential valuable cementing resources are being wasted in precisely the countries that need it most to reduce the greenhouse gas emissions, and to make economical and durable concrete infrastructures.

4 Role of fly ash and other supplementary cementing materials in sustainable development of cement and concrete industry

In view of the energy and greenhouse gas emission concerns in the manufacturing of portland cement raised above, it is imperative either new environmentally friendly cement-manufacturing technologies be developed or substitute materials be found to replace a major part of portland cement for use in the concrete industry. At present, there are not many major new technologies on the horizon for manufacturing environmentally-friendly portland cements that are economically acceptable to the cement industry. This leads to the question whether other supplementary cementing materials and technologies are available to replace significant amounts of cement in concrete. These issues are discussed below.

4.1 Availability of fly ash

In 1989, the total coal-ash production in the world was in the order of 400 million tons annually [2]. The former Soviet Union had the largest production of 90 million tons followed by china at 55 million tons. Other major sources of coal ash production in the world were the United States at 48 million tons, and India at 36 million tons. Since then the fly ash production has increased considerably. The production data for 1995 are shown in Table 2. It is now estimated that both China and India will be producing more than 100 million tons of fly ash by the year 2000. There has been some increases in the other developed and developing countries but not to the same extent. Based on the fragmented information available, it can be predicted safely that by the year 2000, the availability of fly ash will be greater than 600 million tons in comparison with the estimated production of portland cement of 1.4 billion tons. The vast majority of the

Table 2. Coal-Ash Production and Utilization for Selected Countries for 1995*

Country	Production** Million Tonnes	Utilization Million Tonnes
Australia	8.7	0.76
China	91.1	13.8
Denmark	1.3	0.4
Germany (East)	16.8	3.5
Hong Kong	0.63	0.59
India	57.0†	2.0†
Japan	4.7	2.8
Russia	62.0	4.3
South Africa	28.0	N/A
Taiwan	1.5	0.89
U.S.A.	60.0	8.1

*The above data include fly ash, bottom ash, and slag. For every 100 tonnes of fly ash there are approximately 25 tonnes of bottom ash and slag.
**Adopted from: Worldwide Production and Utilization of Coal Ash in Concrete and Other Products--A Survey for the Period 1959-1995 by Oscar Manz. Paper to be presented at the Sixth CANMET/ACI International Conference on Fly Ash, Silica Fume, Slag and Natural Pozzolans in Concrete, Bangkok, Thailand, May-June 1998.
†Estimates only.

available fly ashes for use in concrete are low-calcium fly ashes (ASTM Class F), and are basically the by-product of burning anthracite or bituminous coal. These fly ashes, in themselves possess little or no cementitious value, but will in finely-derived form and in the presence of moisture, chemically react with calcium hydroxide at ordinary temperatures to form compounds possessing cementitious properties. In recent years, high-calcium (ASTM class C) fly ash has come on the market in the U:S.A., Canada, Poland, Greece and some other countries. These fly ashes are by-products of burning lignite or subbituminous coal, and have cementitious properties in addition to being pozzolanic.

4.2 Present and potential use of fly ash in concrete and blended cements

It is agreed that not all the available fly ash is suitable for use in concrete. However, it is believed that technologies are available that can beneficiate those fly ashes that fail to meet the fineness and carbon content requirements, the two most important parameters of a fly ash for use in concrete. These include separation technologies, floatation technologies and grinding methods. From the published information, the current worldwide utilization rate of fly ash in concrete is about 10 per cent. In China the utilization rate approaches about 15 per cent, whereas in India the utilization rate is less than 5 per cent. In North America, the utilization rate is in the vicinity of 10 per cent in concrete. The utilization of fly ash in Hong Kong and some European countries like the Netherlands and Denmark is very high, but these are exceptions. Thus, there is huge potential for the increased use of fly ash in concrete in the world. In this regard the following statement in 1984 from Ed Abdun-Nur, a noted U.S. concrete technologist, should be of interest [3].

> "In the real world of modern concrete, fly ash is as essential an ingredient of the mixture as are portland cements, aggregates, water, and chemical admixtures. In most concrete I use, it is no larger amounts (by volume) than portland cement, and therefore it is not admixture i.e. an addition to the mixture. Concrete without fly ash and chemical admixture should only be found in museum show cases."

The writer agrees totally with the above statement, and this prophetic observation is slowly coming true.

4.3 New technologies in the use of fly ash in concrete

One of the major developments in the area of fly ash utilization in concrete has been the technology of high-performance, high-volume fly ash concrete by Malhotra and his associates [3,4], and it is believed that in years to come, it will revolutionize the use of fly ash in concrete. Briefly, in the production of high-performance, high-volume fly ash concrete, about 60 per cent of portland cement is replaced by ASTM class F fly ash. The water-to-cementitious materials ratio is maintained at 0.30 ± 0.01 with portland cement and fly ash contents at about 150 and 225 kg/m^3 and flow slumps are achieved by large dosages of a superplasticizer. This type of concrete that was developed about 10 years ago has excellent long-term mechanical and durability properties, and this is supported by voluminous research data and some field data. The early-age strength (6-8 MPa at one day) is ample enough for formwork removal except in cold regions. The later-age strength of more than 110 MPa at 10 years has been achieved in demonstration blocks (see Tables 2,3). This type of concrete has been successfully used in several projects in Eastern Canada. Somewhat similar type of concretes can be produced using ASTM Class C fly ash.

Other fly ash technologies being investigated include alkali-activated fly ash concretes, ternary blends incorporating portland cement, fly ash and slag or silica fume, and high-volume fly ash blended cements.

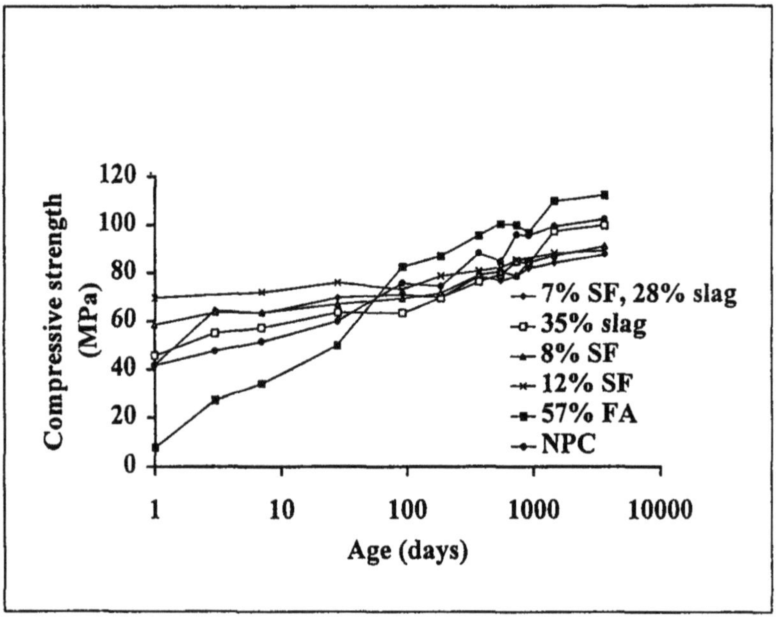

Fig. 2. Compressive strength versus age for various high-performance concretes. The strength is based on 100x200-mm cores taken from large blocks, 2.0x1.2x1.35-m in size.

5 Tradeable emission rights and utilization of fly ash

The "tradeable emissions" refers to the economic mechanisms that are expected to help the countries worldwide meet the stringent emission reduction targets established by the 1997 Kyoto Protocol. At present there are no regulations as to how this international trading of emissions is going to operate. It is being speculated that one ton of emission will have a trading value of about $100.00. Thus, for example, if a country can replace 20 per cent of cement utilization by fly ash or slag, the country would have saved about 20 per cent of the CO_2 emissions. For a country that produces 50 million tons of cement,

Table 3. Compressive Strength of 100x200-mm Cores Taken at the 10 Years from 2.0x1.2x1.35-m Thick Concrete Blocks Cast, Cured, and Left Exposed to Elements in Toronto, Canada

	Age (Days)	Mixtures					
		1	2	3	4	5	6
		Total Cementitious Content, kg/m³					
		485	484	488	486	350	485
Thick-Column Element C (2.0mx1.2mx1.35m)		cement - 65% slag = 28% silica fume = 7%	cement = 65% slag = 35%	cement = 92% silica fume = 8%	cement = 88% silica fume = 12%	cement = 43% fly ash (class F) = 57%	cement = 100%
		Compressive Strength of 100x200-mm Cores, MPa					
	1	41.7	45.6	58.4	69.4	7.8	41.5
	3	64.3	55.0	63.5	--	27.1	47.5
	7	63.2	57.2	63.4	71.8	34.0	51.4
	28	69.4	63.4	66.8	75.8	49.9	59.9
	91	70.8	63.3	69.1	72.9	82.5	75.6
	182	69.6	69.4	71.4	78.7	87.0	74.3
	365	78.5	76.2	79.2	80.8	95.6	88.2
	546	76.6	79.3	80.3	82.2	100.5	94.7
	730	78.6	84.7	85.3	85.3	99.9	95.9
	912	81.8	83.9	--	--	96.8	95.5
	1460	84.1	97.3	87.1	88.2	109.9	99.4
	3650	87.6	99.8	91.2	89.2	112.3	102.3

Table 4. Concrete Blocks, 2.0x1.2x1.35-m Thick, Cast, Cured and Left to the Elements in Toronto, Canada: Mixture Proportions and Resistance to the Chloride-Ion Penetration at 10 Years

Mix	Cementitious materials (kg/m³)						Charge passed (coulombs)
	Total	OPC	Silica Fume	Slag	Fly Ash	w/cm	
1	485	315	35 (7%)	135 (28%)	--	0.29	102
2	484	317	--	167 (35%)	--	0.28	237
3	488	449	39 (8%)	--	--	0.27	565
4	486	427	59 (12%)	--	--	0.27	118
5	350	150	--	--	200 (57%)	0.29	0
6	485	485	--	--	--	0.27	381

this 20% replacement by fly ash would amount to savings in CO_2 of 10 million tons. This, in turn, translates into a trading value of 100 million dollars. One must keep in mind that the value of the trading of emission rights is only an estimate because the international community has not yet reached a conclusion on the subject. But given the political and environmental pressures, the utilization of fly ash will pay rich dividends. The developed countries have a major stake in this issue. If the developed countries can transfer fly ash utilization technology to developing countries, and demonstrate actual slow down in the installation of new cement plants in these countries, the developed countries would rightly claim CO_2 emission credits. In order to give an idea of huge financial stakes involved, it is estimated that it will cost about $100 billion over 15 years for Canada to cut greenhouse gas emissions by as much as it promised at the Kyoto climate change conference. Canada made a legally binding pledge to cut greenhouse emissions 6 per cent from 1990 levels by 2008-2012 at the above conference.

6 Other supplementary cementing materials

In addition to fly ash, the other supplementary cementing materials that can be used to partially replace portland cement in concrete include granulated blast-furnace slag, silica fume, metakaolin and rice-husk ash. Compared to the quantities of fly ash, the availability of these materials is rather limited. The worldwide production of granulated blast-furnace slag is only about 20 million tons per year. The availability of silica fume and metakaolin is very limited, (combined tonnage is less than 5 million tons/year worldwide) and these highly pozzolanic materials are generally specified for specialized applications such as structures exposed to aggressive chemicals. The rice-husk ash is not yet available commercially, though the potential is of about 20 million tons worldwide. The use of blast-furnace slag has increased considerably in recent years and this trend is expected to continue. Rice-husk ash, when it becomes available commercially will along with fly ash and granulate blast-furnace slag be the most significant supplementary cementing material for use as partial replacement for portland cement in concrete.

7 Conclusions

Environmental issues associated with the CO_2 emissions from the production of portland cement, energy and resource conservation considerations and the high cost of portland cement plans demand that supplementary cementing materials in general and fly ash and ground granulated blast furnace slag in particular be used in increasing quantities to replace portland cement in concrete. Given the almost unlimited supply of good quality fly ash worldwide, and the development of technologies such as high-volume fly ash concrete, it is believed that the installation of new cement plants in developing countries should be avoided a far as possible. In addition, the ageing portland cement plants should be phased-out in both developed and developing countries, and the resulting loss in capacity should be compensated for by the use of supplementary cementing materials. In order to achieve the above goals, increased attention to the training of concrete engineers, and increased emphasis on quality control in the field will be needed. The combined use of superplasticizers and supplementary cementing materials can lead to economical high-performance concrete with enhanced durability. It is hoped that the concrete industry would show leadership and resolve, and make contributions to the sustainable development of the industry in the 21st century by adopting new technologies to reduce emission of the greenhouse gases, and thus contribute towards meeting the goals and objectives set at the 1997 Kyoto Protocol. If the above leadership and bold initiatives are not forthcoming, it is certain that the bureaucrats will impose unpleasant regulations and tax on the industries contributing significant amounts of greenhouse gases to the atmosphere. The manufacturing of portland cement is one such industry.

8 References

1. Cahn, D. et al., (1997) *Atmospheric CO_2 and the U.S Cement Industry,* World Cement, August, pp.64-68.
2. Manz, O.E. (1989) *Worldwide Production of Coal ash and Utilization in Concrete and Other Products,* Presented at CANMET/ACI Third International Conference on Fly Ash, Silica Fume, Slag and natural Pozzolans in Concrete, Trondheim, Norway, Supplementary Paper, 15 pp.
3. Malhotra, V.M. and Ramezanianpour, A.R. (1994) *Fly Ash in concrete,* 2nd Edition, CANMET. Energy, Mines and Resources Canada, Ottawa, Canada, 307pp.
4. Malhotra, V.M. *CANMET Investigations Dealing with High-Volume Fly Ash Concrete,* Advances in Concrete Technology, 2nd Edition, CANMET, Ottawa, pp. 445-482.

CLEAN CONCRETE CONSTRUCTION: AN AUSTRALIAN PERSPECTIVE

Clean concrete construction in Australia

D.W.S. HO, S.L. MAK and K.K. SAGOE-CRENTSIL
CSIRO Building, Construction and Engineering, Victoria, Australia

ABSTRACT

Ecologically sustainable development (ESD) is becoming one of the most important challenges for the construction industry in the new millennium. This issue is primarily driven by the need to conserve materials and energy resources so that future generations can have sufficient resources for their needs. Therefore, ESD strives to provide inter-generational equity. The construction activity is a major consumer of both materials and energy resources. With respect to construction with concrete, there are many opportunities to apply advanced concrete technology to enhance sustainability in construction. To ensure long-term sustainability, we need to adopt an holistic approach which views the entire life cycle of construction. This will include aspects of materials processing, product manufacture, transportation, construction, service life, demolition and disposal. Whilst the management of waste is one of the most important aspects in ensuring sustainable development, waste management alone is not sufficient. In taking a life-cycle approach, we need to adopt clean construction practices which encompass waste management as well as minimise the environmental impacts of precursor activities such as the manufacturing of construction materials. In this paper, a broad discussion of clean concrete construction will be discussed in the context of the Australian concrete construction industry. Examples of recent advances in concrete technology research and application in Australia which contribute to clean construction will be provided.
Keywords: Life cycle, clean construction, closed-loop recycling, sustainable development, waste management.

1 Introduction

Ecologically sustainable development (ESD) or durable development is an important issue for both developing and developed countries in view of the continued depletion of virgin material and energy resources and the emission of greenhouse gases. It has been

Concrete Technology for a Sustainable Development in the 21st Century edited by O.E. Gjørv and K. Sakai.
Published in 2000 by E & FN Spon, 11 New Fetter Lane, London EC4P 4EE, UK. ISBN: 0 419 25060 3

estimated that 40% of all materials used is related to the construction activities. In addition, 40% of all energy is consumed during the life cycle of a building, from the material production to construction and the in-service phases. Like many other construction materials, concrete may also extract an environmental cost. It is commonly accepted that when 1 tonne of Portland cement is produced, 1 tonne of CO_2 is also produced from the clinkering process and products of combustion. This constitutes approximately 5% of total global CO_2 emissions. In Australia, more than 60% of material used in construction is cement-based products.

Sustainable development can be defined in many ways but principally means that current needs should not be met at the expense of future generations. Therefore, to provide intergenerational equity, there is a need to conserve natural material and energy resources by increasing efficiency, reducing wastage and maximising reuse and recycling. Ideally, these efforts should be directed at all levels of activity. In the context of construction, these include planning and design, specifications, material selection, raw materials processing, manufacturing of construction materials and products, transportation, construction, in-service performance and, eventually, decommissioning or demolition of a structure or facility. The United Nations Environment Program (UNEP) uses an all-encompassing concept of 'cleaner production' to describe a comprehensive and sustained approach to achieving sustainability. Therefore, cleaner production implies an holistic approach, which needs to consider the life cycle of any activity. Using this principle, the term 'clean construction' may then be used to characterise construction processes which consider the environmental impact over the entire life cycle of *structures*.

Many recent advances in concrete technology provide solutions to not only reducing the environmental impact of construction, but also offer technological solutions with potential economic and ecological benefits. In order to understand the relevance of these technologies, we need to first understand the targets that we need to achieve in order to improve sustainability.

2 Principles of clean construction

2.1 The sustainability equation

Clean construction is aimed at producing environmentally sustainable development. At its most basic, clean construction with concrete should follow the following three tenets:

- managing the consumption rate of natural resources
- minimising ecological impact of construction-related activities
- ensuring that design life requirements for structures are met with minimal costs

USE LESS MATERIALS AND ENERGY	+	REDUCE WASTE AND ENVIRONMENTAL IMPACT	+	ENHANCE SERVICE LIFE PERFORMANCE	=	SUSTAINABILITY

In principle, clean construction aims to conserve the state of our environment in terms of its resources and habitability by improving efficiency, minimising wastage and maximising utility, whilst maintaining costs at an acceptable level. At the most fundamental level, clean construction calls for the adoption of efficient waste management practices in order to maximise the rate of material recycling and reuse. But more than that, clean construction also requires the minimisation of wastage in the first place so that no undue pressure is placed on the environment. Accompanying these broad aims, structures need to be constructed efficiently with sufficient durability so that the costs of maintenance are manageable and that premature replacement is avoidable.

2.2 Life cycle approach to clean construction

Efforts to improve sustainability must not be limited to piecemeal actions, but rather should take a consistent and holistic approach that considers the big picture in order to put in place continuous improvements in all aspects of the construction activity [1]. An holistic approach to achieving clean construction needs to consider the life cycle of a constructed facility, as shown in Fig. 1. At its simplest form, the life cycle encompasses the following:

- extraction of raw materials to produce a construction material
- conversion of the partially or fully processed material into a manufactured construction product
- transportation of materials or products to construction site
- placement or installation to build a facility
- usage of the facility in service
- repair, maintenance, rehabilitation and upgrading required to achieve the designed level of utility, and very often the prevailing 'fashion'
- decommissioning or demolition of the facility
- disposal of demolished materials

True sustainability requires that at each stage, we consider the impact on materials reserves, energy consumption and generation of wastes. Therefore, efficient waste management is the first step towards clean construction. If all demolished materials can be successfully reused, remanufactured or recycled, this demonstrates a successful closed-loop recycling process as shown in Fig. 1. Another more globalised view of how different activities within the service economy feed themselves into various loops is shown in Fig. 2.

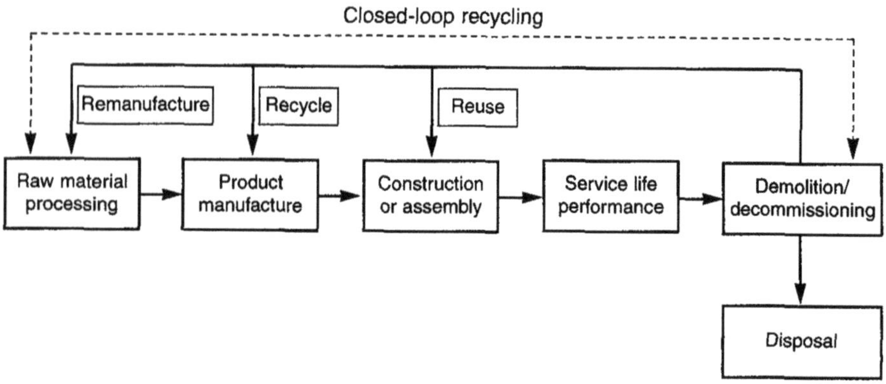

Fig. 1. Life cycle of a constructed facility.

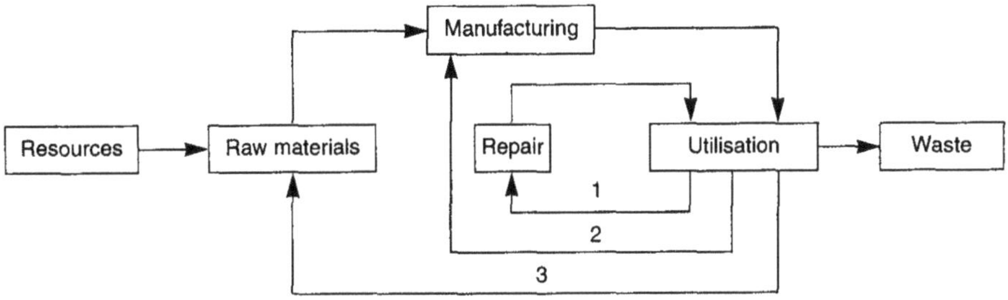

Fig. 2. The service economy (after Curwell and Cooper [2]).

2.3 Waste management

A primary requirement for clean construction is to manage wastes in such a way that the environmental impact and costs are acceptable. Australia, through the Australian and New Zealand Environment and Conservation Council (ANZECC), has set out a national overall target to reduce waste going to landfill by 50% by the year 2000, based on 1990 per capita levels. Whilst an intrinsically simple concept, waste management has many different implications. This is because there are many different types of waste streams that arise from the life cycle of a building or a structure, e.g. at the construction or the demolition stage. There are three major contributing sectors to the total waste stream: construction and demolition (60%), commercial and industry (25%), and council (15%). Currently, the total amount of waste going to landfill is approximately 17 million tonnes per year, with concrete contributing some 60% of this waste problem. Hence, waste management is a multifaceted concept, which can be operated at many different levels. A generic waste management hierarchy is shown in Fig. 3.

The obvious strategy to reduce the consumption rate of virgin material and energy resources is to use less, use more efficiently, or to reuse and recycle to a higher degree. In terms of wastes generated prior to or during construction, the need to reuse or recycle wastes becomes increasingly irrelevant if the quantities of wastes generated can be

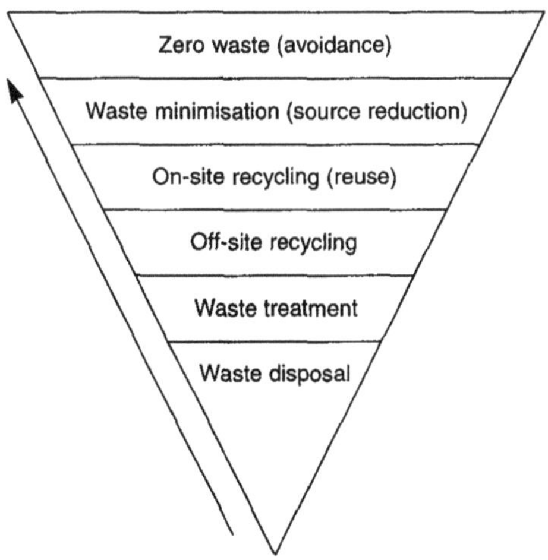

Figure 3. Waste management hierarchy

minimised in the first place. Hence, the waste management hierarchy in Fig. 3 shows that the ideal solution is waste avoidance in the first place. Successful waste avoidance requires the implementation of an optimal and coordinated chain of activities ranging from planning and design to procurement, transportation and construction. At the construction site, waste avoidance in terms of a 'zero waste' concept primarily means not ordering more than is needed. This minimises the generation of off-cuts, waste concrete etc. Therefore, everything that is transported to site is consumed directly or reused on site even if for marginal applications. At a more basic level, waste avoidance implies that design should be streamlined or optimised to use as little material as is required to satisfy structural, aesthetic or durability requirements.

In terms of concrete construction, waste avoidance on site also implies more extensive use of precast units rather than cast-in-situ concrete. Thus, builders become assemblers instead of builders in the traditional sense. To support this waste management strategy, researchers need to develop cost-effective, high-performance lightweight products as well as 'fashionable' architectural finishes to satisfy individual requirements.

For concrete production, waste minimisation also involves the reuse of waste water from truck and yard washing. In Australia, like many countries around the world, environment protection agencies are concerned with contaminations to waterways from this washing due to its high pH. Concrete producers are required, as good corporate citizens, to recycle this waste water. Systems have been well developed and implemented for this purpose. However, the potential of possible cement reduction through alkali activation from this cement-rich, high pH 'slurry' has not yet been fully realised.

When concrete buildings are demolished, the recyclable materials may be mixed and may not often yield a high value. This is due to the economic costs of separating and transporting the wastes and/or the diminished quality of the recycled product. For instance, recycled concrete aggregates (RCA) obtained from crushing demolished concrete is often considered a marginal waste product, which is generally suitable for comparatively low-grade applications such as road sub-base. This is due to a combination of factors, which are associated with the demolition and recycling techniques used, e.g. air classification or wet treatment [3]. Some of the physical factors associated with diminished performance of RCA may include bond reduction, secondary fracturing, excessive fines, the presence of contaminants, variable feedstock and high absorption. Improvements in feedstock selection, sorting, separation and crushing techniques can potentially bring about major improvements to the quality of the material.

The extent of utilisation of recycled materials is mainly driven by economics. The problem of mixed waste has been overcome to some extent by cost-incentive schemes through the cooperation between government and industry. In recent years, the cost of tip fees on waste disposal has risen sharply to a staggering $40 per tonne in certain city areas. While at the same time, recycling operators charge around $10 to $20 per tonne depending on the level of mixing. For 'clean' concrete that requires little or no separation, the saving in disposal costs could be as high as $30 per tonne. As a result, contractors on demolition sites have developed a culture of separating materials for disposal. On many demolition sites around Sydney and Melbourne, separation of wastes at source is becoming a norm rather than an exception.

Utilisation of recycled materials would be successful only if technology is available and standards are in place to ensure the quality of supply. While the supply of RCA as a material for road sub-base can fetch a modest $8 per tonne, suppliers could charge three times this price if RCA could be used to replace natural aggregates for concrete production. Comparing this with some $50 per tonne for natural aggregates, there are obvious environmental and economical benefits. To provide confidence to the potential user, CSIRO is currently working closely with industry and local governments to generate the necessary technical data from laboratory and field investigations. Research projects are also in place to develop technology and standards for RCA to be used for higher valued products other than as a material for road sub-base. In the longer term, the properties of RCA need to be modified for wider applications and markets.

For concrete elements to be effectively reused, an obvious starting point is the increased use of modular construction with simple deconstructible fixing technologies that allow structural components to be easily dismantled. The concept of deconstruction or disassembly also requires the availability of innovative joining and sealing systems.

3 Materials and processes for clean construction

Waste management is only one aspect of clean construction. Effective waste management strategies are fundamentally important yet insufficient to ensure long-term sustainability. For instance, efficient waste management practices will ease the pressure on landfill and reduce the consumption of natural resources. However, there are other more basic areas where the environmental impact of construction can be minimised through clean construction principles. Whilst waste avoidance practices on construction sites minimise waste generation, true sustainability needs to consider the material production and manufacturing phase which precedes construction. In this context, it is useful to examine the selection of materials and products with minimised or manageable impact on the environment with consideration of:

- type, source and reserves
- energy requirement for extraction, processing and transportation
- emissions and waste streams from extraction and processing
- potential for recycling and reprocessing

3.1 Embodied energy of materials

There are many ways to evaluate the energy efficiency of building materials. This depends on which aspect of construction is being considered. A common measure of embodied energy is the gross energy requirement (GER) which includes both direct and indirect components. The direct component of GER includes the energy used to transport building materials to the construction site, the construction activity itself and the energy consumption at the building site. The indirect component of GER includes the energy required for extraction of raw materials, processing it into a form that can be used to make building products, and that required in the actual manufacturing of the building product itself. The latter also includes the capital cost of manufacturing equipment. Estimates of GER do not include energy requirements for maintenance or demolition.

Concrete is intrinsically a low-energy construction material when compared to other major building materials. The GER for cement may range between 5 and 9 GJ/tonne depending on the country of production, whilst the GER of aggregate is approximately 1.5 GJ/tonne [4]. Since the volumetric fraction of cement in concrete is about 10%, the GER for plain concrete is approximately 2 GJ/tonne, whilst that of reinforced concrete is approximately 2.5 GJ/tonne due to the steel content. These GER values for concrete compare well with timber, which has a GER of 2 GJ/tonne, and is significantly lower than that of steel which is between 35 and 50 GJ/tonne. Reinforced concrete also has comparatively favourable cost/unit strength of 0.15 compared to either steel or timber, which is approximately 0.4 [5].

In spite of its comparatively low embodied energy, between 15 and 30% of the operating costs in the concrete construction industry is related to energy and materials. It has also been estimated that up to 65% of the total energy requirements of the construction industry is related to the manufacture of materials [4]. Therefore, there is significant scope for reducing costs and improving sustainability in the construction industry. Two of the areas in which the materials production and processing aspects of concrete construction can be improved are through the use of low-energy materials and energy-efficient processing techniques.

3.2 Low-energy materials systems

3.2.1 The embodied energy of recycled materials

The issue of low-energy materials is not a simple one since concrete as a *construction* material has many constituents. The binder content of concrete is approximately 10–15% by volume, whilst over 70% of the volumetric fraction is composed of aggregates. It may be thought that the use of a high recycled material content may lead to direct ecological benefits. In terms of embodied energy content, the use of recycled materials do not necessarily reduce the embodied energy of concrete. It has been estimated that the process energy of RCA is marginally lower than that of virgin natural aggregates. However, the energy required for transportation of RCA is much higher than that of natural aggregates. Consequently, the embodied energy of RCA is between 20 and 40% higher than that of natural aggregates, depending on whether the analysis is carried out by volume or mass respectively [unpublished CSIRO report, 1994]. This does not even take into account the embodied energy content of concrete prior to demolition. However, the increased use of RCA will ease the pressure on landfill, and possibly reduce disposal costs, whilst reducing the consumption of natural aggregate resources. Therefore, using RCA as an example, the economic and ecological benefits of using a high volume of recycled material need to be considered in terms of:

- the intrinsic embodied energy prior to demolition
- the energy requirements in recycling and reprocessing
- any increase in cement content required to compensate for marginal quality RCA
- any savings in reduction of disposal costs
- reduction in use of landfill
- reduced consumption of virgin material stock
- community acceptance of clean processes

3.2.2 Blended cements

Whilst the binder fraction of concrete is comparatively small, the estimated 5 billion tonne global consumption of concrete still means that approximately 1 billion tonnes of Portland cement is produced annually worldwide. This also means 1 billion tonnes of CO_2 emissions and more than 5 billion GJ of energy consumed.

The use of blended cements containing supplementary cementitious materials (SCM) will take on an increasingly important environmental context in future years with respect to reducing pressure on landfills and the consumption of raw materials and energy. This is particularly the case in blended cement systems, which use a high fraction of SCM. Examples of these include high volume fly ash (HVFA) with 40–50% of fly ash, high blast furnace slag (BFS) binders with 70% of BFS, and other ternary or quartenary binder systems with 20–30% Portland cement. Australia consumes some 6 million tonnes of cement annually but only about 20% is blended cement.

The use of increasingly large volumes of supplementary materials to replace Portland cement may seem to make good ecological sense. However, the overall sustainability equation outlined earlier also requires that structures serve their intended design life with minimised or acceptable maintenance costs. The extent to which SCMs can be used to replace Portland cement is limited by technical issues related to not only the properties of hardened concrete, but also governed by the placeability of the fresh concrete. In binary cements containing fly ash, the upper limit of cement replacement seems to be 40–50%. In the case of silica fume, the optimal cement replacement level seems to be approximately 10%. Some of the beneficial properties of BFS for reducing temperature development or increasing sulfate resistance become evident only at relatively high cement replacement levels of between 50 and 70% by weight of binder. However, very high BFS contents in binders result in relatively slow strength development at normal or low temperatures.

3.2.3 Binders with no Portland cement

Current strategies for enhancing sustainability of concrete construction centres on the issue of the energy consumption of Portland cement production and associated greenhouse gas emissions. Therefore, the prevailing philosophy is that of reducing the amount of Portland cement usage. However, there are technologies that are aimed at producing binders with virtually no Portland cement content. Advances in alkali activation of slags, typically using sodium silicate or sodium sulfate, may eventually provide mainstream acceptance of binders with very high proportions, i.e. up to 90%, of BFS and virtually no Portland cement content. There have also been recent advances in the development of geopolymeric binders produced from geosynthesis reactions involving naturally occurring silico-aluminates.

3.2.4 Low energy accelerated processing (LEAP)

An holistic approach to materials selection needs to consider the construction activity as a whole, with an understanding of the effects of prevailing in situ or processing environments. Clean construction calls for compatibility between processing technique and the material being processed. When heat accelerated curing is used in the precast industry to speed up production, it is equally important that thermally reactive binder systems are used. It is well known that cementitious binders are thermally reactive and that early age strength development is significantly affected by temperature. For instance, the thermal reactivity of slag cement binders increases systematically with increasing BFS content [6]. It has been found that the apparent activation energy of a mortar produced with a binder containing 30% Portland cement and 70% BFS was 49 kJ/mol as compared to 30 kJ/mol for a plain Portland cement mortar. Therefore, when compared to curing at standard temperature, the increase in early age strength due to heat curing is much more significant in concrete containing the high BFS binder [6,7]. This may be exploited for speeding up either in situ construction or precast production subject to durability requirements being met.

Even the most thermally reactive binders would not be beneficial if the method of heating is highly inefficient. Conventional methods of heating concrete, e.g. using steam or hot water, are thermal conduction/diffusion processes, which are quite slow and often wasteful. Depending on the type of element or product being processed, a high proportion of heat is used in heating up the formwork system and then the concrete. When boiler/generator efficiencies, transmission losses and transfer losses are taken into account, the overall efficiency of conventional heating techniques may not exceed 20%. Recently, the feasibility of applying microwaves to cure concrete has been demonstrated [8,9]. The potential for this method of heating lies with the fact the dielectric heating is characterised by a relatively deep penetration of microwaves into a material which converts electromagnetic energy into heat. Therefore, depending on the frequency of microwaves used, a relatively deeper penetration of heat may be achieved using microwaves when compared to conventional methods. The conversion of electricity to microwaves in magnetrons is 70% efficient and there are no primary sources of waste since all the energy in transferred into the element being heated. Current research at CSIRO indicates that the energy consumption required to achieve the same heat curing regime using microwaves may be 2–5 times less than that achievable even in a low-loss steam-curing system. It has also been shown that extremely high rapid heating rates are possible using microwaves without excessive drying [9]. Therefore, the combined benefits of energy efficiency, speed and low waste provide a potential 'clean production' technology for the precast concrete industry.

3.3 Eco-design in clean construction

Eco-design is the principle whereby environmental criteria are given equal weight to other performance criteria in design [10]. In the past, aesthetics and structural performance were the main design criteria. Increasingly, structures are being designed for durability where performance criteria are used to specify materials [11]. In addition, construction processes are also being modified to ensure durability, e.g. extended curing practices. The next stage is to design for ecological sustainability by considering the environmental impact of not only the actual construction itself, but also the activities leading up to the construction of a building or facility. For example, the design process needs to take a step backward to examine the environmental friendliness of the processes required to manufacture the building materials. In this respect, eco-design can be approached the same way as durability design, where environmental impact is managed through environmentally-based performance specifications aimed at encouraging sustainability.

4 Conclusions

Concrete will remain the dominant construction material in the early part of the next millennium. Whilst the economic benefits and performance of concrete is not in question, the environmental impact of concrete construction needs to be appraised in light of the global awareness of sustainable development. To ensure long-term sustainability, clean construction principles that encompass the life cycle of a constructed facility need to be considered. Beyond simply managing wastes by more efficient recycling or reprocessing, clean construction calls for an holistic approach that takes into consideration the environmental impact of materials production and service life performance. The use of low-energy materials and low-energy processing technologies provide examples where advanced concrete technology can play a key role in aiding the construction industry to achieve long-term ecologically sustainable development.

5 References

1. Environment Protection Agency, Australia (1996) *Getting Ahead of the Game – An Anticipatory Approach to Environmental Management – A Cleaner Production Handbook for Local Government*, EPA, Canberra.
2. Curwell, S. and Cooper, I. (1998) The implications of urban sustainability, *Building Research and Information*, Vol. 26, No. 1, pp. 17–28.
3. Storey, J.B. and Baird, G. (1993) Sustainable building design in the city, *Transactions, The Institution of Professional Engineers, New Zealand*, Vol. 20, No. 1/GEN, pp. 7–14.
4. Trankler, J.O.V. and Walker, I. (1994) Potential in quality improvement of processed building rubble by demolition and treatment technics', in *Environmental Aspects of Construction with Waste Materials* (eds Gourmans, Van del Sloot & Aalbers), Elsevier Science, pp. 621–632.
5. Kreijger, P.C. (1987) 'Ecological properties of building materials', *Materials and Structures*, Vol. 20, No. 118, pp. 248–254.
6. Mak, S.L. (in press) Thermal reactivity of slag cement binders and the response of high strength concrete to in-situ curing conditions, submitted to *Materials and Structures*.
7. Mak, S.L., Foster, G., Chirgwin, G. and Ho, D.W.S. (1997) Engineering properties of heat cured slag cement concretes for bridge infrastructure, Proc. Concrete 97, CAI 18th Biennial Conf., Adelaide, 14–16 May 1997, pp. 45–52.
8. Mak, S.L. (1997) Microwave accelerated processing for precast concrete production, Proc. 4th CANMET/ACI Int. Symp. on Durability of Concrete, Sydney, 17–22 August 1997, Supplementary Papers, pp. 709–720.
9. Mak, S.L., Shapiro, G. and Son, T. (1998) Accelerated heating of concrete with microwave curing, submitted to 4th CANMET/ACI Int. Conf. on Recent Advances in Concrete Technology, Tokushima, Japan, June 1998.
10. Cramer, J. (1994) Life cycle analyses: results of some case studies, in *Environmental Aspects of Construction with Waste Materials* (eds Gourmans, Van del Sloot & Aalbers), Elsevier Science, pp. 17–28.
11. Ho, D.W.S., Chirgwin, G. and Mak, S. L. (1997) Water sorptivity of heat cured concrete for bridge structures, Proc. 3rd CANMET/ACI Int. Symp. on Advances on Concrete Technology, Auckland, August 1997.

ENVIRONMENTAL PROPERTIES OF BUILDING MATERIALS

K. TUUTTI
Skanska Teknik AB, Danderyd, Sweden

Abstract
The concrete industry uses different types of binders in production processes. Local opportunities in various geographical regions are then a primary parameter. A major compound in all binder types is Ordinary Portland cement. Production of cement clinker, in the production of OPC, depends on the properties of raw materials and investments made for reduction of emissions and energy consumption. Modern cement plants with electrostatic filters, catalytic processes for reduction of NO_x and scrubbers for reduction of SO_x emissions have much better environmental properties than old plants with low investment characteristics. Additions used in the cement industry or in the ready mixed industry normally reduce the environmental impact due to minimization of the use of OPC and due to reduction of the global production of waste. A high priority task is the evaluation and comparison of the environmental impact for the use of different building materials. A general recommendation is that exchange of building systems of concrete, steel or wood will produce small environmental effects. However, it is very important to choose types of material to optimize the functional properties. Especially a long service life will minimize the environmental impact.
Keywords: Emissions, environmental properties, concrete, steel, wood, life cycle, service life, carbondioxide.

1 Introduction

The environmental impact from production and utilization of different products can be evaluated by the use of a life cycle analysis. In such assessments, it has been generally believed that the service life is of major importance. Bridges, off-shore structures, frameworks in buildings hold an exceptional position in comparison with installations,

Concrete Technology for a Sustainable Development in the 21st Century edited by O.E. Gjørv and K. Sakai.
Published in 2000 by E & FN Spon, 11 New Fetter Lane, London EC4P 4EE, UK. ISBN: 0 419 25060 3

cars or other consumer products designed for a shorter service life. Life cycle analysis, LCA, quantifies different environmental factors such as, use of resources and emissions from the stage of production to the ultimate stage in the landfill. The basis of valuation of all these factors is also the interaction between themselves.

A calculated service life for buildings is normally 50 years and for bridges above 75 years. In Sweden the national application document ratified by the Swedish Road Authority, demands a service life of about 100 years for foundations and other parts in the structure that will be difficult to replace in a bridge.

A long service life has consequences in a LCA assessment. Properties in the service phase such as maintenance together with energy consumption for heating and cooling of buildings has a major impact on the environment. Consequently, it is important to bring forward an accurate evaluation of the whole service period. The possibility of predicting beyond 100 years is not possible due to technical conditions in the future.

2 Life cycle of concrete

Concrete is a composite material consisting of gravel divided into different fractions joint together by a cement paste. During the past ten years, organic admixtures are used more frequently to improve the fresh or hardened properties of concrete. A schematic illustration of the life of concrete is shown in figure 1.

The difference between normal concrete with a compressive strength of 30 MPa and high performance concrete with a compressive strength of 80 MPa is very small. A typical difference in composition for a high performance concrete is approximately 5 % less aggregate and 5 % more cement expressed as percentage of the total weight. A high performance concrete contain also 5 to 10 % micro silica and 1 to 2 % super plasticiser expressed as a percentage of the cement content.

Standardized cement types can vary in composition due to utilization of resources such as flyash, slag and limestone fillers. Recycling of waste materials from other processes provides an opportunity to reduce the environmental impact in the cement production process. However, the essential environmental impacts depend on the cement clinker production. A raw material with a high concentration of sulfur can give serious SO_x emissions. Consequently, filters, scrubbers and catalytic processes are able to improve the environmental influence.

Simple and energy efficient processes in a gravel pit or in a mine produce sand and aggregate. Recycled concrete has been tested but only used in the concrete production in a very small scale.

Additives, such as flyash and slag, have been used for decades as filler and as a cement reducer. Utilization of waste materials optimizes the environmental impact as the amount of cement clinker is minimized. Many research results confirm that use of additives give different positive effects on both the fresh and the hardened concrete properties. However, the reactivity of a specific product is an important factor why large fluctuations decrease the opportunities to use the product. A new glass filler additive has been tested and classified in Sweden. That material is inert nevertheless, addition of about 50 kg/m^3 has an unexpected positive effect on workability.

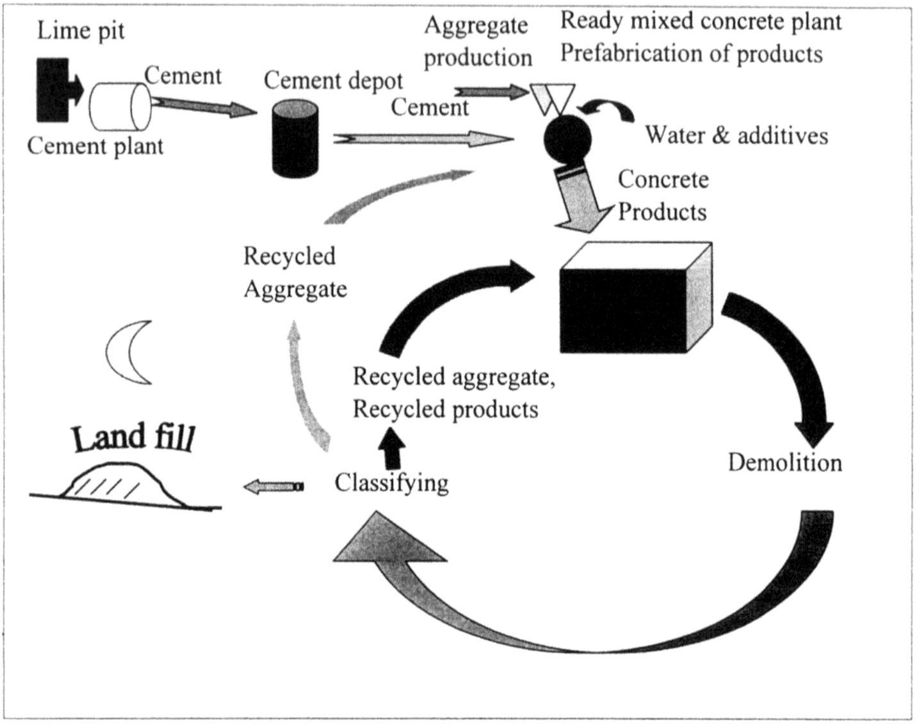

Fig. 1. Schematic sketch of the life cycle for concrete

Organic admixtures are used more frequently now than before. The amount varies with the type and does not normally exceed 2% by the weight of cement content. Chemicals are artificial products and attract public attention. The common apprehension is that chemicals are harmful to the environment and should be minimized in products. This is a misconception due to both the types of chemicals that are used and the way these products are encapsulated and bound in the concrete matrix. Utilization of chemicals improves several concrete properties tremendously. Above all the workability on fresh concrete has improved in such a way that concrete with W/C ratios as low as 0.25 could be cast on site with normal equipment. The service life of structures in a cold climate depend also on the use of air entrain agents. Many admixtures are used in concrete production for achieving acceptable durability properties.

3. Environmental profile of one cubic meter concrete.

Stiftelsen Östfoldsforskningen (1) has studied the environmental impact for the use of concrete. All components in concrete were analyzed from the cradle to the grave. The purpose was to present a list of quantitative figures on emissions into the atmosphere, discharge of substances in groundwater and the use of resources such as energy and raw materials. Fundamental studies like that enables engineers to summarize the environmental impact for different concrete products and to make comparisons between materials. This report has used a fictive concrete composition as an example for calculations and comparisons between materials. The composition represents normal concrete for buildings, see table 1.

Table 1. Normal concrete composition and average transport distance in Sweden.

Material	Amount, kg	Transport distance, km
Cement	300	50
Aggregate	1900	5
Water	190	0
SiO_2	10	400
Admixture, plasticiser	4	300
Distribution of concrete	2400	20

Analyze of the different compounds in plain concrete indicate that the major effect depends on the cement, see figure 2. The carbondioxide emission from use of concrete according to table 1 is 255 kg per m^3. That is comparable with the carbondioxide emissions for a car trip from Copenhagen to Paris.

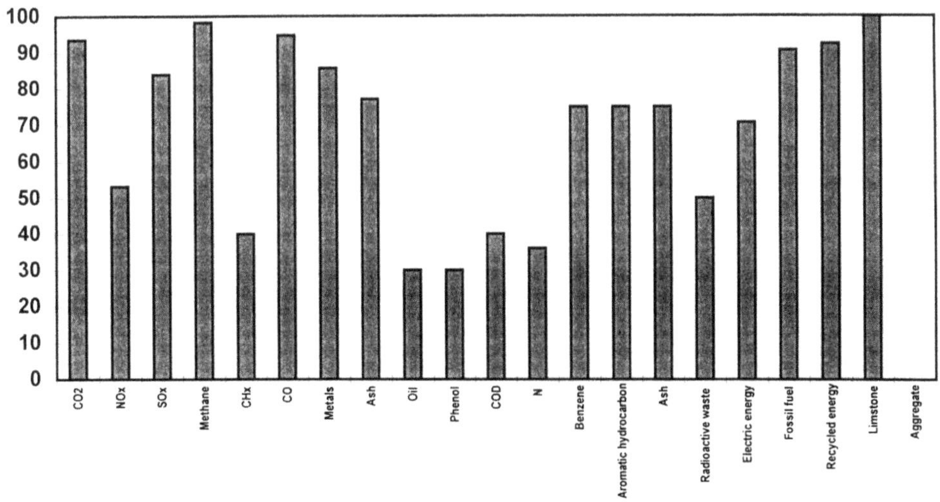

Fig. 2. Environmental impact from Ordinary Portland cement expressed as a percentage of the effect from the concrete in table 1.

Different concrete products demand specific concrete compositions. A low cement content can be seen as more favorable than high strength concrete with a high cement content. However, calculations indicate that a proper function, a long service life and the total energy consumption during the lifetime of a product is much more important than the effect of the material itself. An interesting aspect is that an extremely positive effect occurs if a recycled binder replaces OPC. That type of changed composition increases the need of knowledge of the change on the durability properties of the concrete structure.

4. Environmental profiles of specific products

4.1 General

Chalmers Institute of Technology (2) and (3) has analyzed different products in a comparable study of their environmental influence. Products that are analyzed are divided in three stages, construction, a service period of 50 years, demolition and recycling if possible. These analyses consider the impact on the ecological system, the health of human beings and use of natural resources. Social and economic effects are not included in the evaluations.

4.2 Building frame

The study includes both office buildings and residential properties. Seven different structures were analyzed including steel, precast concrete elements and on site cast concrete. The results proves that the major environmental impact of a building arises from the energy consumption during the service life, see figure 3. A heavy and a compact structure have such an energy storage capacity that the energy consumption is minimized compared with a light steel structure. A comparison between concrete, steel and wooden structures as different materials indicate insignificant differences. An energy storage capacity and a long service life are the most important criteria for an environmental impact.

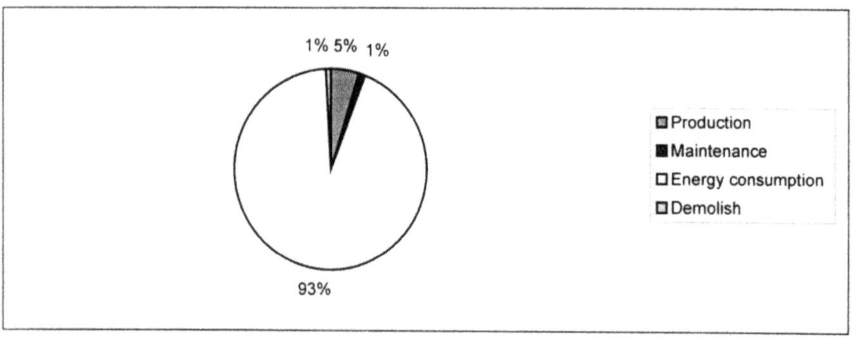

Fig. 3. Energy consumption of a building frame spliced into production, maintenance, heating during service life and the final demolishing procedure.

4.3 Beams and columns

A direct comparison of the environmental impact from different products is a complex question that raises several issues, such as:

- functional equality of different materials
- optimization of the use of specific products
- change of material properties during the service life
- uncertainties of demolishing procedures and recycling possibilities in the future
- evaluation of environmental properties and

However, products can be perspicuously compared by use of the relevant environmental parameters SO_2, CO_2, NO_x and energy consumption in production processes and the final demolishing procedure. A concrete, steel and wood beam and column, conditions described in figure 4, are designed and analyzed by this simple method (4). The results are presented in table 2.

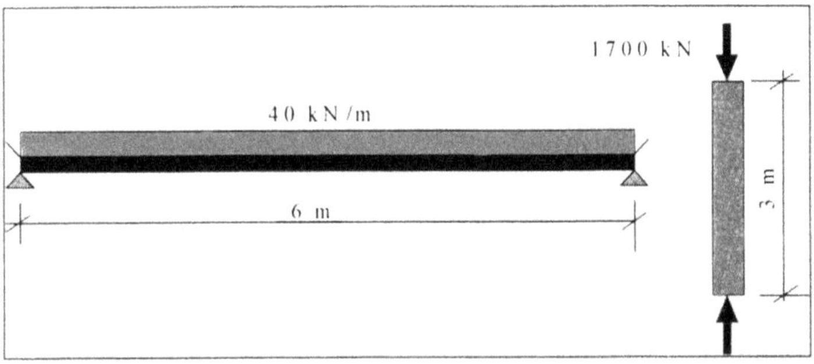

Fig. 4. Design assumptions in the evaluation procedure of comparing concrete, steel and wooden products.

Steel products such as beams and reinforcement in the concrete are calculated as three times recycled. Wooden materials are calculated as 75% used as fuel in the final stage. Maintenance works is not included. After demolishing, all materials are assumed to be transported 30 km to a recycling industry.

Table 2. Calculated environmental influences from a beam and a column produced by reinforced concrete, steel and glued wooden structure.

	Beam			**Column**		
	Concrete	Steel	Wood	Concrete	Steel	Wood
SO_2 to air, g	280	330	130	90	120	80
CO_2 to air, g	300	230	60	90	80	30
NO_x to air, g	940	660	960	300	240	570
Energy, MJ	2250	3770	2550	740	1360	1510

The results in table 2 indicate that products produced by reinforced concrete, especially columns, are more favourable with regard to the environmental impact. A general conclusion is that the comparison point out that it is more important to chose materials for functional requirements than to exchange one material with another.

4.4 Sewage pipes

A thousand-meter sewage system with a diameter of 225 mm was analyzed by Chalmers Institute of Technology (3). All digging and execution processes were defined according to standardized construction methods. Surprisingly, the main factor was identified as take away the surplus earth. An important factor is minimization of the soil transported from the site, which could be very positive for concrete pipes. An investigation was made on several pipe materials where comparisons were based on the whole sewage system, se figure 5. The results indicated concrete as a favorable material compared with other used materials.

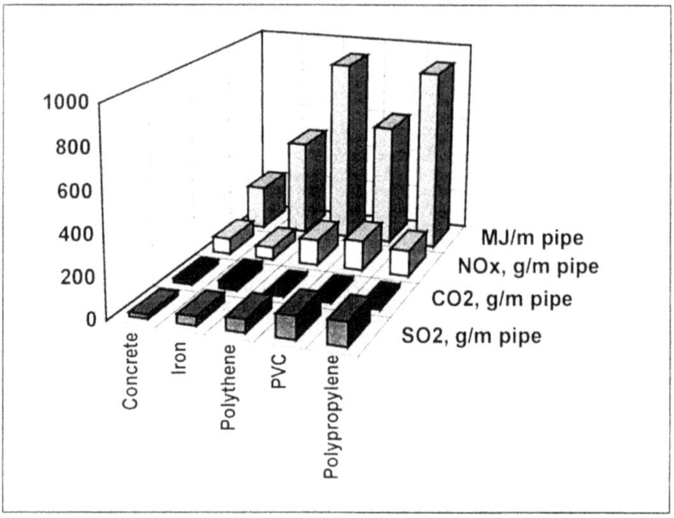

Fig. 5. An environmental comparison of different sewage pipes.

5. References

1. LCA of Cement and Concrete - Main report. Stiftelsen Østfoldsforskning, Fredrikstad, 1995.

2. LCA of Building Frame Structures. Chalmers Institute of Technology. Miljöplanering, report 1997:2.

3. LCA of Concrete Sewage Pipe. Chalmers Institute of Technology. Industriteknik, report 1996.

4. Environmental impact of utilisation of various concrete products. Miljöenheten, Skanska Teknik AB, report 1997.

HIGH PERFORMANCE GROUTS FOR SUSTAINABLE DEVELOPMENT OF INFRASTRUCTURE

Microsilica grout for ground injection

O. JØRGENSEN and P. FIDJESTØL
Elkem ASA Materials, Kristiansand, Norway

1 Abstract

Environmentally friendly methods for ground and rock injection will be an important part of a future sustainable construction industry. Future development of the infrastructure will often come into conflict with environmental concerns unless steps are taken. One solution is increasing use of underground space in infrastructure development.

Injection is frequently an important part of construction under ground. It is primarily used to prevent water leakages, but injection is increasingly used for ground improvement as well. Most injection work use cement-based grouts, while for very fine cracks or impermeable soil, chemical grouts have traditionally been required. This means that while prevention of water leakage is often necessary to protect groundwater levels, the result may be pollution of ground water, since some chemicals traditionally used for difficult ground conditions have the potential for polluting the groundwater.

In this situation, sustainable infrastructure development needs advances in the material technology of injection. One route is the development of particulate mineral grouts that can penetrate very fine cracks or pores so the need for potentially harmful chemicals is reduced. Quite another aspect is that chemical solutions tend to be very expensive and complicated to handle, while particulate grouts based on cement or lime are easy to handle.

Tunnelling, injection, silica fume, groundwater

Concrete Technology for a Sustainable Development in the 21st Century edited by O.E. Gjørv and K. Sakai.
Published in 2000 by E & FN Spon, 11 New Fetter Lane, London EC4P 4EE, UK. ISBN: 0 419 25060 3

2 Introduction

High-speed trains need to go in a straight, such a right of way would demolish existing towns, destroy fields and in general disrupt existing society. Building space for secondary use, such a storage, is already a scarce commodity in many cities and other challenges are part of the construction of highways in mountainous or hilly country, where surface roads would be costly expensive to build, a blot on the landscape and expensive to operate in terms of fuel consumption.

Underground construction can easily conflict with the ambient environment. Suffice it to mention ground settlement under existing buildings and possible drainage or contamination of groundwater. It is therefore important to provide the groundwater protection and underground strengthening without introducing new environmental hazards.

Injection is frequently an important part of construction in rock and soil, primarily it is used to prevent water leakages, but over the last few years, injection is also used for ground improvement. It is attempted to improve the quality of the rock to prevent falling rock, or to strengthen the soil in order to reduce the amount of necessary support. Most injection work use cement-based grouts, while for very fine cracks or impermeable soil, chemical grouts have traditionally been required. This means that while prevention of water leakage is often necessary to protect groundwater levels, the result may be pollution of ground water, since some chemicals traditionally used for difficult ground conditions have the potential for polluting the groundwater.

3 Basics of injection technology

Various principles of injection are in use. The system that is addressed here is called permeation grouting. The principle is that injection grout replaces water in the pore or crack system in a geological structure. The injection pressure is adjusted so that the formation is not disturbed. In permeation grouting, it is an axiom that particulate grouts can penetrate pores or cracks with a diameter or width larger than three times the maximum particle size of the grout. An additional experience is that reduced dry content in the grout improves penetrability.

The factor which works against reducing the dry content is that the grout at a water/powder ratio of 1. 5 -2 gets unstable, and the dry material settles out. This instability prevents the grout from entering the cracks or pores and is even more obvious when pressure is applied to the grout. This instability is one of the reasons that chemical solutions have been necessary for very fine cracks. One route to the development of an environmentally friendly injection material is therefore to develop a particulate grout with very fine particles that is stable at a low dry content.

4 Microsilica in injection grout

The use of selected quality microsilica slurry for injection is known from the "Micron-S" system [1, 2]. In this system, developed by the French contractor Bachy, microsilica is used in combination with slaked lime. The present paper, however, only considers cement-based injection materials.

Microsilica (silica fume) is known to stabilize concrete. The advantages of using microsilica in cement-based systems are well known from concrete technology, where typical benefits are reduced tendencies to bleeding and segregation, stronger and less permeable concrete, increased durability and resistance toward chemical attack. Several of these parameters are also considered important for injection grouts. Therefore the possibility of a microsilica-based product for use in combination with cement-based injection grouts was expected and it has been developed.

This knowledge has been taken further in a quality microsilica developed for use in combination with microcement- and cement-based grouts. While stability would be obtained with most qualities of microsilica, the problem is that the coarse fraction of the microsilica will be coarser than the microcement, and therefore can block the openings of the pores or cracks. Hence the challenge was to produce a material with maximum particle size smaller than in microcement, < 12-20 μm. The end product has been in field trials and sale for about a year.

Because both fresh and hardened properties of the cement-based grouts are improved, better cement-based injection is possible in finer soils and cracks than what has been possible until now. The main benefits are improved penetrability, which lead to reduced construction costs and less need for potentially harmful chemical injection materials, and improved quality of treated soil or rock, giving less water flow and stronger ground than previous material combinations[1].

The development was complicated, since standard test methods are scarce in this industry. Methods for comparative testing had to be developed and verified along with the development of the material itself.

4.1 Microsilica For Injection

The additive is based on special qualities of microsilica, selected and processed so that the content of coarse particles (>3μm) content is very small. As seen in figure 1, the microsilica[2] is very fine, with more than 90% of the volume of particles less than 1 μm. The base product is a 50/50 water/microsilica slurry.

As indicated above, several important properties of injection grouts are substantially improved by this additive:

- There is a strong stabilizing effect on the grout.
- The penetration depth is substantially increased.
- The permeability of the injected ground is reduced.
- Increased strength, improved microstructure and improved interface between the injection grout and the injected object (soil, rock).

---

1 Norwegian Patent # 178887 concerns the use of microsilica for ground injection.

2 Microsilica consists of glassy spheres with average diameter about 0. 15 μm. Varying numbers of spheres form agglomerates during production that are fused at contact points. The parameter determined in the particle counter and reflected in the diagram is in reality the circumscribed diameter of these primary microsilica agglomerates.

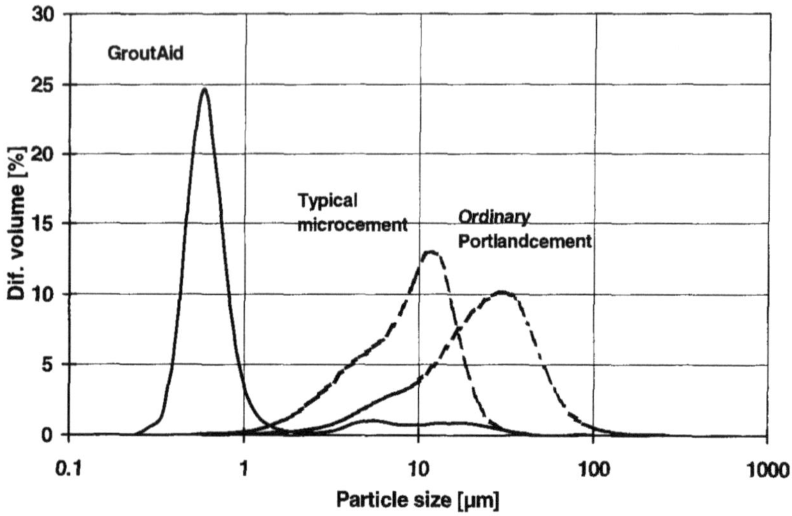

Figure 1 *Particle size distribution of GroutAid additive compared to typical microcement and ordinary Portland cement.*

4.2 LABORATORY TESTING

Though cement has been the grouting material of first choice for decades, surprisingly few standardized methods are available to characterize the properties of injection grouts. Apart from control of the actual grout volume injected and monitoring the injection pressure, little effort is published on methods for follow-up and control of the actual injection work, and little has been known about the final state of the material pumped into the ground. This lack of (standard) methods to characterize and qualify an injection grout also hinders laboratory testing and the ability to predict the performance of a grout in the field.

The test methods used were picked from other application areas, such as oil-well cementing and prepacked aggregate concrete. The following are results of laboratory testing according to these procedures on cement-based grouts with and without microsilica.

4.2.1 Viscosity

Viscosity was measured with a Marsh funnel. The Marsh-viscosity of tap-water at room temperature is approximately 26 seconds and injection grouts with a Marsh-viscosity of 27-30 seconds are said to have a very low viscosity. According to [3], injection grouts should have a Marsh-viscosity between 28 and 40 seconds. All injection grouts that were tested had Marsh-viscosities between 29 and 32 seconds. Since dispersing agents were used, it was not possible to see any effect of microsilica on the Marsh-viscosity.

4.2.2 Stability

To obtain a successful injection, a stable, non-segregating grout is necessary. The stability parameter should be separated into two different effects:

- Separation of water and solids due to gravity forces - bleeding
- Removal of water from the grout due to applied pressure - fluid loss

4.2.2.1 Bleeding

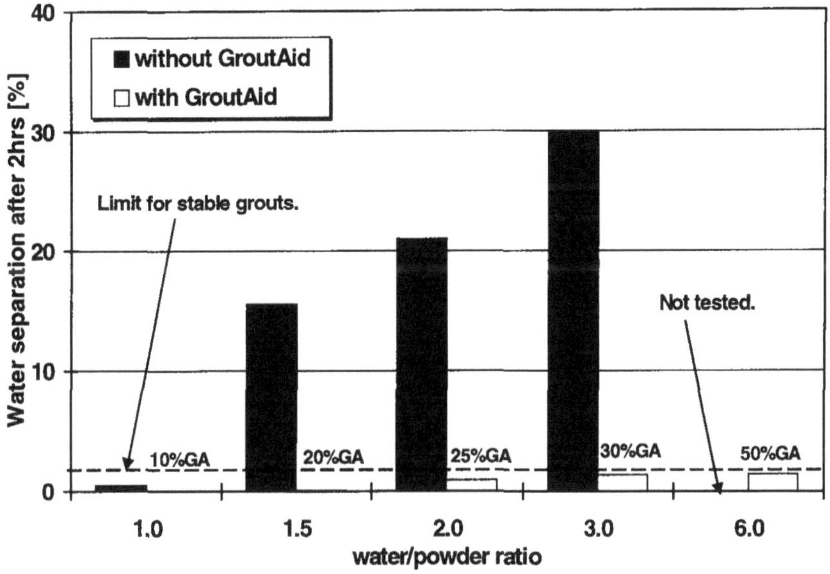

Figure 3 *Water separation after two hours, injection grouts with and without GroutAid additive.*

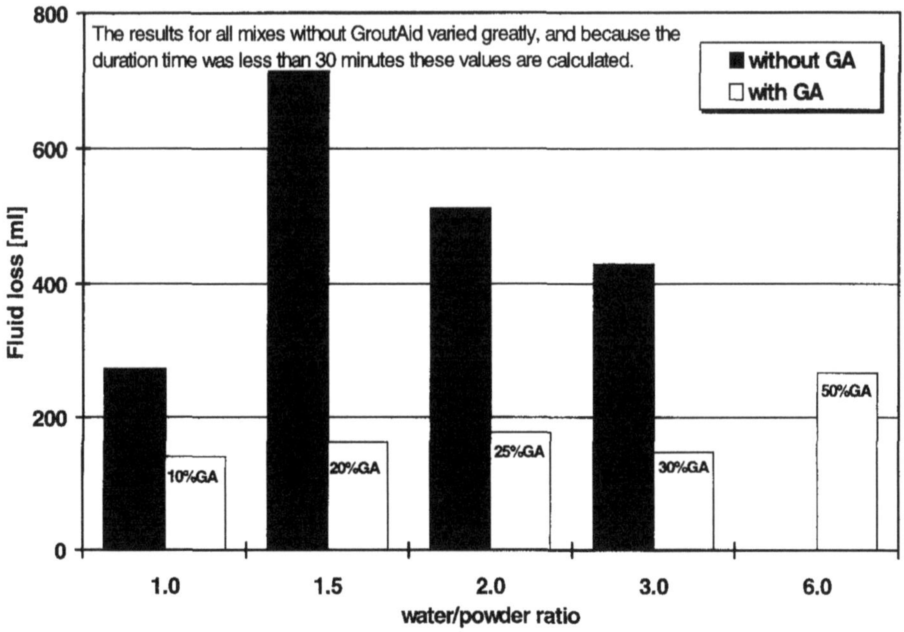

Figure 2 *Fluid loss after 30 minutes at 7 bars. Injection grout with and without GroutAid additive.*

According to [3], an injection grout is stable if the bleeding is less than 2%, while in [4] the limit 5% after two hours. Stable grouts will give cracks and voids a more complete filling and in turn reduce the need for re-injection. Important parameters to describe the stability of an injection grout are bleeding and fluid loss.

The addition of microsilica to a grout means that a large number of extremely small particles with very high specific surface area are introduced. The large surface area and the correspondingly high internal attraction forces keep the cement and microsilica particles in suspension with little settling in water, also at high water/powder ratios. Figure 2 shows water separation for injection grouts with and without the additive. A stable injection grout can be achieved at water/powder ratios as high as 6. 0.

4.2.2.2 Fluid loss

When an injection grout is placed in a permeable formation under pressure, filtration occurs. The water in the grout is squeezed out and escapes into the formation, leaving the particles behind. This process is called "fluid loss," and a test method common in the oil industry was used to determine the ability of the injection grouts to retain the water.

Even if bleeding is low, water can be expelled from the grout when a pressure gradient is applied. The amount and rate of water removal indicate the stability of a grout at a certain pressure gradient. Reference [3] suggests that:

- The fluid-loss of a stable injection grout is less than 100 cm^3 after 30 minutes
- For an unstable grout 100 cm^3 requires less than 30 minutes and/or the fluid-loss is well above 100 cm^3.

Grouts with microsilica often have very high water content, so it was felt that the time to 100 cm^3 is probably more important than the actual volume of expelled water before the test stops.

The internal stability caused by the microsilica furthers improved fluid-loss control by the attraction between water and solids and possibly by reduction in the permeability of the initial cement filter cake. Figure 3 shows that the water in neat cement grouts is expelled in a relatively short time, while grouts based on the additive are stable and retain water for the full duration of 30 minutes, even at water/powder ratios as high as 3. 0. These results confirm that a microsilica-based grout will resist separation due to pressure much better than a conventional grout. Thus, high injection pressures can be used in practice.

4.2.3 Penetration

In the laboratory, the penetration ability of the grouts has been tested by injection in a sand column. While the results obviously cannot be directly transformed into penetration in soil or rock, the method serves to distinguish between and characterize the grouts. As figure 4 show, addition of the additive provides a large improvement of the penetration of injection grouts, at similar dry content (same water/powder ratios). The most likely explanation is that the modified grout is much more stable, resulting in minimal bleeding and sedimentation, and thus, no blocking. Since water is not expelled during the injection, the solids do not settle out and deposit in cracks or pores close to the borehole, so that a homogenous grout flows into the formation.

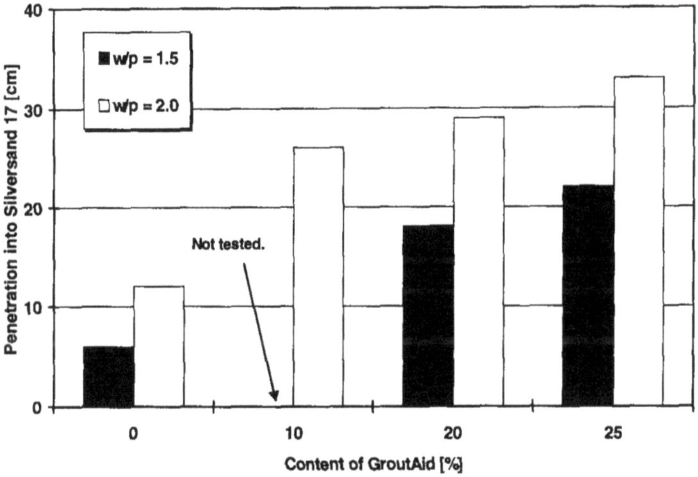

Figure 5 *Penetration depth in Silversand 17 (D_{50}=. 17 mm of injection grouts with water/powder ratios 1. 5 and 2. 0 and varying content of GroutAid additive.*

As mentioned above, one possible opportunity with microsilica additive, is the use of stable grouts with very high water/powder ratios. The viscosity of a grout decreases as the water/powder ratio increases, and the stable microsilica-based grouts with high water/powder ratios and low viscosity show much improved penetration compared to ordinary stable cement-based grouts with water/powder ratios around 0. 8 (figure 5).

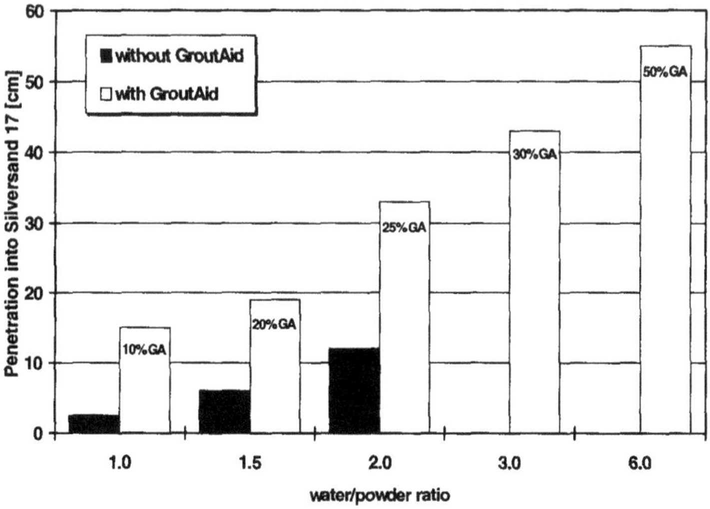

Figure 4 *Penetration depth in Silversand 17 for injection grouts with varying water/powder ratios, with and without GroutAid additive.*

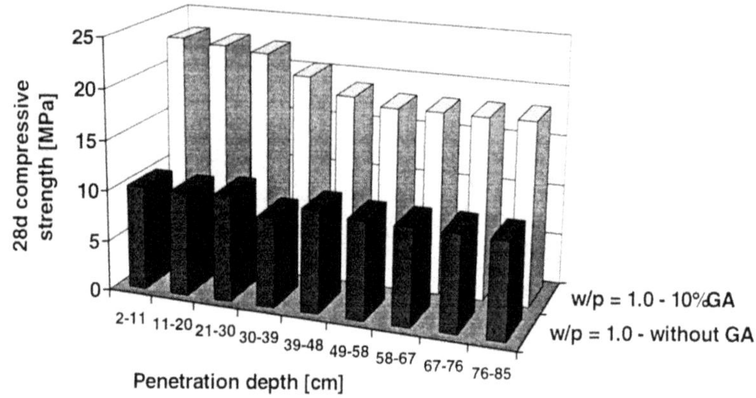

Figure 6 *Compressive strength of injected sand. Injection grouts with water/powder ratios 1. 0, with and without microsilica based additive.*

4.3 Compressive strength

From its use in concrete, it is established that microsilica reacts with the calcium hydroxide that forms during the hydration of cement. This creates an increased amount of binder, which provides strength and durability. In addition, the small microsilica particles provide an improved packing of the particles in an injection grout, and improve the interface between grout and sand particles or rock. As can be seen in figure 6, the compressive strength of a grout based on cement and microsilica additive is 1. 5 to two times higher than that obtained with a neat cement grout.

5 Other Benefits of Using a Microsilica-based Additive in a Grout

The benefit of microsilica in cement-based systems is well known from "ordinary" concrete technology, and over the years extensive research work has been carried out on fresh and hardened properties of concrete with microsilica, including long-term properties.

Cement-based injection grouts will function similar to the paste fraction in concrete, except that the water/powder ratio is much higher than usual in concrete. However, the mechanisms are the same, and therefore, most of the experiences from the use of microsilica in concrete can be applied to cement-based injection systems.

5.1 Durability

When cement-based injection grouts are exposed to chemical attack, it is normally in the form of leaching of calcium hydroxide. Since microsilica reacts with, and binds, the easily soluble calcium hydroxide, the use of GroutAid additive will reduce this leaching. The resistance to leaching ensures the durability of the injection.

5.2 Economy

Because of the enhancements provided by GroutAid additive, the need for very expensive chemical grouts is likely to be reduced. GroutAid additive makes cement-based injection possible in finer cracks and more fine-grained soil than what has been possible until now. This can make injection more cost-effective.

The improved penetration of grouts containing GroutAid additive can be used to reduce the total costs in several ways, for example:

- by increasing the distance between the injection holes, thus making it possible to reduce the number of holes to be drilled.
- by drilling longer holes, making it possible to increase the number of drill and blast sequences between each injection.

In addition, materials' costs can be reduced because grouts with higher water/powder ratios, but still stable, can be used.

5.3 Environmental aspects of the grout

As the use of microsilica-based grouts with enhanced penetrability reduces the need for chemical grouts, some of them potentially harmful to the environment, the introduction of GroutAid additive will have a positive effect on the environment. Since even complicated injections can be made using mineral grouts, organic materials will not be introduced into the ground.

6 Grouting as part of sustainable and environmentally safe development

6.1 Tunneling

The development of new grouting materials allows safer tunneling in complicated rock. As recent experiences in Scandinavia, good materials are not sufficient, proper execution of the work is also required. The particulate grouts can give environmentally safe injections, even for very fine pores and cracks, and problems related to potentially toxic chemicals in ground water can be greatly reduced.

6.2 Waste disposal sites

A serious global problem is the handling of existing deposits of waste, and the procedures to choose for establishing new storage areas. The number of class A hazardous waste sites probably number in six digits in Western Europe alone, and the capital and technology to finance immediate renovation does not exist. Thus a solution combining horizontal and vertical

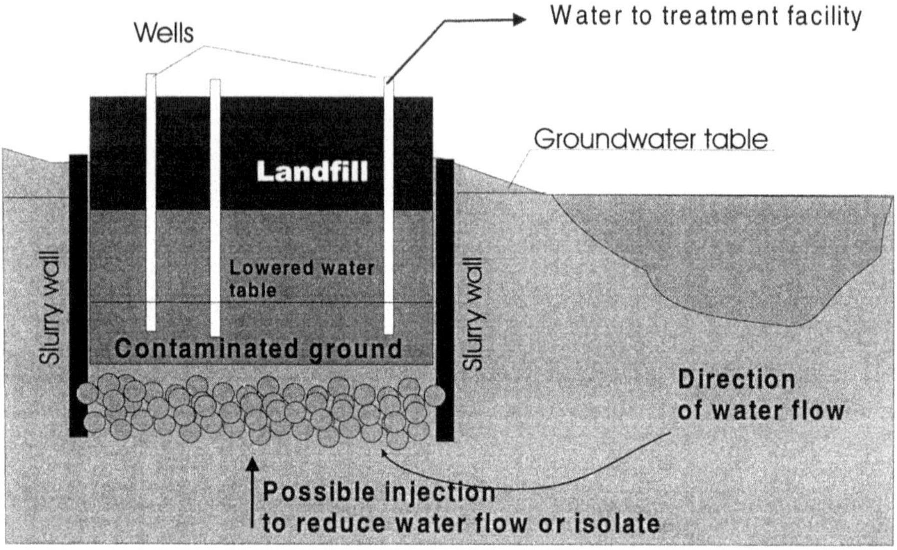

Figure 7. Groundwater pollution is often prevented by control of groundwater level in the landfill. Depending on the permeability of the ground, this procedure can require large amounts of water to be removed and treated. Injection will reduce the amount of water that need to be pumped and cleaned, ensuring better safety against contamination

injection combined with slurry walls may give the breathing space necessary to finance and perform the cleaning of existing waste pits, while the same procedures can give safe storage of future waste. Figure 7 compares current and possible future configurations for safeguarding groundwater around a disposal site.

In the same fashion as in the figure, ground injection for prevention of seepage to the environment can be actively used to prevent contamination in simi

7 Summary

A special quality of microsilica for use in combination with cement- and microcement-based grouting has been developed. The additive is environmentally safe and it has a strong stabilising effect, which improves injection in general and also enables stable grouts with very high water/powder ratios. Summing up, it provides improved penetration of particulate mineral grouts, reduced permeability of the injected formation and improved strength and durability.

These enhancements lead to a more cost-effective grouting process with better quality than what has been possible up to now.

8 References

1. Fidjestøl, P. , Dingsøyr, E. , Gandais, M. , Marchini, H. "Microsilica-based grouts for ground injection". Presented at CANMET/ACI International Workshop on Silica Fume in Concrete. Washington, USA, 1991. Out of print, copy available from the author.
2. Sherwood, D. E. , Gandais, M. "Neutral mineral micro particle grout and its application in recovery of a collapsed tunnel in Switzerland". Grouting in the Ground. Thomas Telford, London, 1992. A.L.Bell ed. pp. 631-648
3. Håndbok nr. 1 - Fjellinjeksjon , (Handbook no.1 - Rock Injection), Norwegian Society for Rock Mechanics, Oslo, 1995.
4. International Society for Rock Mechanics, Commission on Rock Grouting. Final Report, May 1995.

PRESENT AND FUTURE VIEWS OF ENVIRONMENTALLY-FRIENDLY CONCRETE

M. TAMAI and T. OKINAKA
Kinki University, Higashi-Osaka City, Japan

Abstract
The present and future views of environmentally-friendly concrete, such as environment mitigatable concrete and organisms adaptable concrete, are briefly discussed in this item. Recently, Eco-cement has been developed to utilize the demand to use municipal waste and incineration ash of sewage as a raw material of cement. A wide variety of functions can be rendered to "Porous Concrete(POC)"through which water and air can pass freely, in contrast to ordinary concrete, although mechanical properties of POC such as strength and durability are inferior to those of ordinary concrete in general. Various attempts have been made for the wide usage of POC. For example as the mitigatable usage, POC boards function to absorb noise at particular frequencies which can be controlled by the thickness and porous structures of the POC boards. It is noteworthy that NO_x-gas absorbing property is added to the noise-absorbing POC board recently. Also this as organisms adaptable usage, water can be purified by proper microorganisms living on the interior and exterior surfaces of the POC. In this paper, fixing and growing lawns, weeds, and ferns in POC are showed as the application of POC to a planting concrete.
As a result, Eco-cement and POC are thought to be useful in the establishment of the sustainable development of mankind.
Keywords; Continuous Voids, Eco-cement, Environmentally-Friendly Concrete, Environmentally Mitigatable Concrete, Organisms, Planting, Porous Concrete, Water Purification,

Concrete Technology for a Sustainable Development in the 21st Century edited by O.E. Gjørv and K. Sakai.
Published in 2000 by E & FN Spon, 11 New Fetter Lane, London EC4P 4EE, UK. ISBN: 0 419 25060 3

1 Introduction

Due to the increase in world population, the depletion of fossil fuels, the desertification of tropical rain forests, and the increase in wastes, a significant turning point in the global environment might occur between the years of 2030 and 2050 [1] .

Under these circumstances, the nations participating in the Global Environment Summit held in Rio de Janeiro, Brazil in June, 1992, and it's engaged in this summit to make efforts suitable for the Global Environment Age. The developed countries agreed to give the appropriate support to the environmentally developing countries. The government of Japan established the Environment Basic Law in November of 1993. The basic concepts of this law are to "influence present and future generations such that an environmentally sound global environment, which is jointly owned by and exits for all of mankind, can be maintained in the future" and to aim the sustainable development and the ability of mankind to live together in harmony with animals and plants without polluting the earth and restoring natural purification activities.

Up to the present, concrete has been used for the construction of civil engineering structures, such as roads, harbors, and water supply and drainage facilities in many countries, and it has contributed to the development of economy and culture in these countries. However, since concrete has been used to thoroughly enhance the function of structure, those structures have disrupted the natural ground and the water edges on which rich green grows . This has resulted in the distraction of the creatures that inhibit the earth. Moreover, a large number of concrete building producing heat from air-conditioners has caused a heat-island phenomenon in cities. Such problems has been addressed for a long time, and the idea of being environmentally conscious is spreading throughout the world these days.

In this paper, utilizations of environmentally-friendly cement and porous concrete (POC), manufactured as an environmental load-reducing raw material from wastes for environmental load reduction and its forefront are briefly reviewed. Then, the present state and future views of the environmentally-friendly utilizations for ecosystems which can accommodate a biological environment are described [1][2].

2 Load-reducing environmentally-friendly concrete

2.1 Utilization of industrial by-products and wastes

Blast-furnace slag and fly ash have already been used for manufacturing cement and concrete [1]. The application of municipal refuse and drainage sludge to raw materials for producing cement is being promoted as an "Eco-Cement Plan" under the auspices of the Ministry of Health and Welfare and the International Trade and Industry in Japan.

It is said that the total cost of collecting, transportating, and incinerating municipal refuse and drainage sludge for slag making is approximately 100 to 200 yen/kg, respectively. It should not be surprising that the steel and cement industries may be converted into waste treatment industries in the future.

Considered to be an environmental problem, the estimated amount of limestone deposits available as the main raw material for normal portland cement in Japan is 10 billion tons. Assuming the annual consumption of limestone for cement, fertilizers, chemicals, and others is 0.2 billion tons, the supply will be depleted in about fifty years. Although it is said that Japan has inexhaustible deposits of clay, it is predicted that the reserves will be sharply reduced in regards to the conservation of nature. It is of urgent necessity to use energy efficiently and to develop technology which can control the emission of CO_2 in order to solve the problem of global warming. For the lack of terminal waste treatment plants it is expected that municipal refuse will be used as the main raw material and that various types of slag and limestone powder will be blended with the raw material. Although powdery wastes among municipal refuse have a tendency to melt at 1200 to 1400 ℃ when used as a slag aggregate for volume reduction and the containment of harmful heavy metals, the melted material is worth as little as crushed stone. Therefore it is a problem to try to enhance the value of the powdery wastes. According to the chemical analyses of powdery wastes among municipal refuse, these wastes contain CaO, SiO_2, Al_2O_3, and Fe_2O_3, which are regarded as the main ingredients of cement [3] . It is, rational therefore, to manufacture cement step by step, raising the temperature to 1500 ℃ instead of melting. In a comprehensive program by the Ministry of Construction in Japan entitled, "Development of Energy- and Resources-Saving Technologies for National Land Construction"[4], the consumption of resources and energy and the emission of CO_2 have been investigated for the purpose of developing an energy-saving cement in order to establish a process for manufacturing low-earth environment load cement. The raw materials used in the experiment to create such a cement include type A (municipal refuse, incineration ash from refuse, drainage sludge, etc.), type B (industrial wastes including blast-furnace slag and waste quartz sand), and type C (substituted by a mixture of powdered limestone and blast-furnace slag). type A has high premature strength, while types B and C are suitable for preparing mass concrete because the rise in adiabatic temperature is small, although the premature strength is even a little lower than that. Although the incineration ash from the municipal refuse is unsuitable for cement for RC, because it contains approximately 3% of salt, it is considered suitable for use as a soil conditioner in soft ground.

2.2 Reduction in environmental load with POC

POC is an open-cell cellular material. The following applications are being studied and same practical applications are being executed as well:

1) Heat insulating properties, structure, and floating, etc.

Since closed-cell type autoclaved lightweight concrete (ALC) is not permeable to air and water and its specific strength to density ratio is higher than that of plain concrete, although it has excellent heat insulating properties, it is used as a construction material. It is an ultra light-weight aggregate without water- absorbing properties, is synthesized from waste glass and POC, and is manufactured using expanded polystyrene foam as the aggregate float on water. Accordingly, these materials are applicable to floating water-weeds and pontoon [5].

2) Water-permeable paving material

POC, using No.7 or No. 6 crushed stone, has recently been used for footpaths and light roadways in cities for the stability and comfort of cars as well as for its water permeability and the heat-insulating properties of the pavement.(Fig.1)

Fig.1 Situation of water permeable pavement in side walk

A 27.7-km-long section of Kobe's Hanshin Expressway, damaged by the Hanshin Earthquake, was restored with the water permeable asphalt pavement. Since POC absorbs less solar radiation than black asphalt, it is hoped that a paving technique for water-permeable material using POC will be established [5]. Although conventional POC has a shortcoming in that it peels off with drying shrinkage, elasticity can be imparted to the POC by injecting high-quality polymer emulsion into the pavement. This study is being made in Japan, but a paving technique has already been established and applied to the pavement of expressways in Germany and Belgium [6]. It is, expected therefore, that the technique will also be practically applied to the pavement of expressways in Japan in the near future.

3) Sound absorbing material

Sound energy transmitted into a porous material is absorbed by it as frictional heat and vibrational energy. When sound energy is transmitted into a porous material with pores of large diameter or space, the sound disappears by the interference of sound waves. This study has been put to practical use in Germany. The sound absorbing characteristics can be further improved by properly configurating the appropriate aggregate and components with the appropriate thickness, according to the frequency of the specified sound [7]. The speed of a superexpress Shinkansen train in Japan will be increased from 250 to 300 km/hr, and the nation lost a lawsuit for noise pollution in the Nishi-yodogawa area of Osaka. It is hoped, however, that measures to counter the vibration and noise generated from roads and railways will be improved by the nation, public corporations, and the Japan National Railway Co.(JR).

4) Harmful gas absorbing material

The total surface area of the outside and the inside of 10 cm-thick POC, prepared using a 1 mm-diameter aggregate is approximately 500 times the surface area of one side of plain

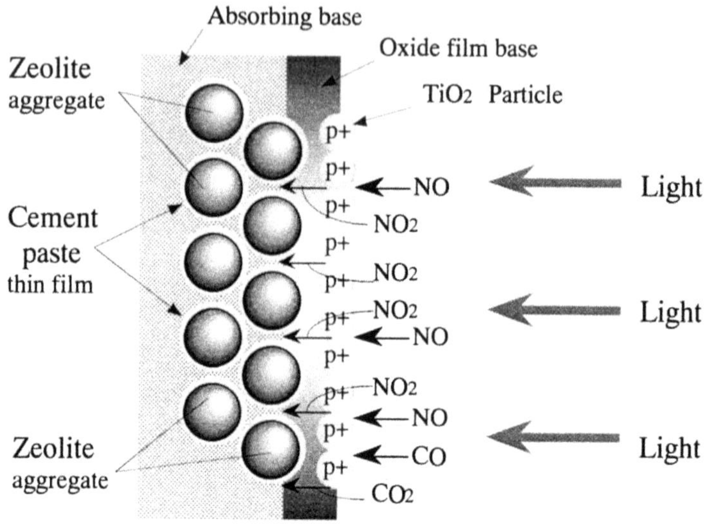

Fig.2 Schematic representation of NO removal: Combind mechanism of NO-oxidation to NO2 on TiO2 photocatalyst and NO2-absorption on zeolite,

concrete. Specified gases in the atmosphere, including CO_2, SO_x, and NO_x, can be absorbed and made harmless by a harmful gas-absorbing material as a substitute for the aggregate and binder in the POC. Gas-adsorbing materials mainly composed of inorganic materials such as artificial zeolite, with a large ion-exchange capacity, can adsorb SO_x and NO_x. Those harmful gases can be further oxidized, fixed, or made harmless by applying a anatase type of titanium dioxide as the photocatalyst to the surface of the POC. Although this technique can be applied by spraying it on the surface of plain concrete, sound-absorbing concrete for absorbing and making the above-mentioned harmful gases harmless can be developed by combining it with sound-absorbing concrete [8].

5) Humidity-controlling material

Since Japan is located in a monsoon area, it is humid in summer but dry in winter. Earthen walls were used in ancient Japanesque architecture to control the humidity naturally. The surfaces of the walls of concrete buildings built in recent years gather mold because they are condensed during the humid summer season. In a room waited with special POC-typed material, the humidity in the air is absorbed during the humid season and then the air in the dry season is humidified. The Agency of Industrial Science and Technology of the Ministry of International Trade and Industry in Japan is promoting the development of a ceramic humidity-controlling material this year (1998) as a measure to counter the energy- and resources-saving. The material is characterized by a large internal surface area, a pore diameter on the surface as small as 10 to 20 μm, and by being producible even from the POC of a specified material [9].

6) Heat energy storage material
POC is available to a material for temporarily storing solar heat and incineration heat and releasing them at any time. High-specific gravity magnetite is used as the aggregate for increasing the amount of heat to be stored. The waste heat generated in a plant and solar heat are temporarily stored in the POC through a liquid or gaseous heat medium and used as occasions demand. When the temperature difference with the heat medium is 10℃ or higher, it is applicable to the generation of electric power or to air-conditioners through a heat pump.

3 Creatures including animals and plant-conscious material

The Japanese government announced a policy of "National Strategy for Biological Diversity" in October of 1995. The policy is mainly promoted by the Environmental Agency to restore and to create a biological environment by incorporating concerned Ministries and Agencies into one body. The Ministry of Construction set forth a policy whereby conventional concrete would not be placed on the surface of river banks for repairing the rivers or for river improvement or used in forestry conservation, but it would turn the policy in the opposite direction to concentrate on a more creature-conscious multi-nature type of river improvement [2]. The administration of roads will also be conducted by paying attention to the future protection of green plants, symbiosis, and the construction of comfortable roads. This means that the nation has finally begun to seriously tackle problems related to the protection of living things and the conservation of endangered species. Tens of thousands of species, including microorganisms, are dying out every year, and 80 out of 174 species of mammals presently living in Japan are thought to be close to the point of extinction.

1) Planting concrete
It is often found that fine cracks in paved roads and joints of footpaths are grown over with weeds. Grasping a clue from that phenomenon, planting concrete has been developed in such a way that either open-cell type POC or that which is covered with earth is sown with grass or weed seeds or seedlings are planted in any type of POC to grow plants [10].

Since the root stalks spread deeply throughout the pieces of aggregate and fasten securely into the soil, either No. 5 (13-20/mm) or No. 4 (20-30/mm) crushed stone is suitable for the application. The works are executed by placing the concrete on-site or by using its secondary product. After the execution of this work, it is necessary to leave the place as it is for approximately one month in order to neutralize the surface of the pores. Liquefied ferric sulfate may be sprayed on the product for neutralization in order to cause rapid vegetation. Then, the plantation bed can be made by filling in the pores of the POC with water-retaining earth. The planting concrete can be applied to side-slopes for creating rivers, roads, and housing sites. The side-slopes placed with planting concrete are vegetated with grass, weeds, and shrubs.(Fig.3,Fig.4)

In addition to vegetating concrete, shotcrete is available for protecting the side-slopes. After arranging blocks made of organic material and fertilizer on the specified side-slopes, grass-planting shotcrete prepared by blending a swelling, water-retaining,

Fig.3 Practical example of river banks for creating natural rivers placed with planting concrete(after 5 months)

Fig.4 Practical example of side-slopes for roads placed with planting concrete (after 3 months)

grass-planting aggregate with POC for spreading the root stalks of plants deep into the pores of the POC is sprayed upon the side-slopes which are protected by fastening the roots of various plants to the pores of the POC. The strength of shotcrete is generally as low as 3 to 6 MPa.

The above-mentioned planting POC has already been experimentally applied to 43 sites throughout Japan to investigate the grass-planting technique for river wall protection under the auspices of the Ministry of Construction and the Hokkaido

Development Agency, and the investigation Committee of the Execution Technique of Planting Concrete has been established. It is estimated, therefore, that the execution standards of strength, durability, voids ratio, pore diameter, etc. will be prepared within the 1998 fiscal year and practical works will be rapidly executed in the near future.

2) Habitable space for creatures

Small land and aquatic animals adhere to and inhabit the rough surfaces and open pores of POC placed at the edge of rivers, lakes, marshes, and the sea. A food chain is formed by their interaction and symbiosis. POC can be used as the basis of this chain. Algae, shellfish, and small aquatic animals adhere to and inhabit the surfaces of the POC placed at the edge of the water, while protozoans, polychaeta, and the larvae of various types of insects inhabit the open pores of the POC. Thus, the basis for biological diversity is created, and POC can indeed be regarded as a material for restoring and recreating the devastated global environment.(Fig.5)

3) Water purifying concrete

Since various microorganisms adhere to the large internal and external surfaces of POC placed in the water areas of rivers, lakes, and marshes, and on the coast, the quality of water is also indirectly purified by the biota. Bacteria, including aerobic heterotrophic bacteria, ammonia-oxidizing bacterium and nitric acid-oxidizing bacterium adhering to those surfaces of POC decompose organic matter and ammonia and convert them to inorganic matter. Flora existing nearby is the cause of the primary production of organic matter from inorganic matter and CO_2 by photosynthesis, and is followed by the multidimensional production through the secondary production, Thus, a food chain is formed [11].

Since algae and grasses grow naturally on the floating weeds of a floating structures of POC in eutrophicated lakes or marshes, as mentioned above, not only is the quality of the water raised by purification, but also a balanced, diversified flora is restored or created.(Fig.6)

Table 1 Void ratio and interior surface area of POC with glass ball (30,40,50% of the voids filled with paste)

Packing model	Ball size (mm)	Void ratio (%)			Interior surface area(cm²/ ℓ)				
		30%*	40%*	50%*	0%*	F-2**	30%	40%	50%
Cubic lattice	10	33.4	28.6	23.8	3141	1642	2373	2116	1861
	5				6282	3284	4746	4232	3722
Rhombic lattice	10	27.7	23.7	19.8	3627	1011	2615	2277	1940
	5				7254	2022	5230	4554	3880

*Voids filled with paste of F-1 range, **0% filling

*** At the critical point of F-2 range

Fig.5 Situation of attached macro-algae and many kinds of animals in POC after one year immersion into natural seawater

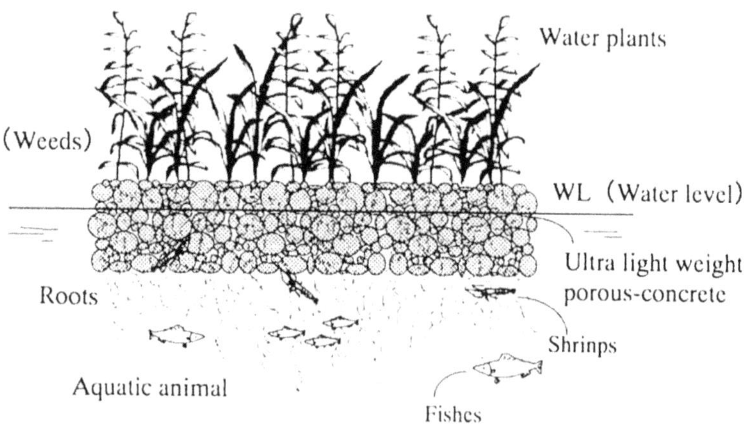

Fig.6 Floating structure using ultra light weight porous-concrete

4 Concluding remarks

Researchers and engineers in the field of concrete are making efforts to spend the by-products of cement to manufacture ultra-durable concrete which do not disrupt the global environment. Although such an attempt has just been started this possibilities to convert municipal powdery wastes into raw materials for cement, and to make an environmental load-reducing type of open-cell cellular concrete, and a biologically-accommodating concrete to the materials for protecting our environment. Research works on most of this type of technology are just beginning. It is expected, however, that such research will progress much further, with the aim of restoring and recreating the devastated global environment as a technique for contributing to the sustainable development of mankind.

5 Acknowledgment

A part of the research was done by a scientific grant No.09480137 of Japanese Ministry of Education. Authors are grateful for this financial support.

6 References

1. JIC (Japan Concrete Institute)(1995): Proc. of JCI Symposium on the Present and Future Views of Environmentally-Friendly Concrete, pp.1-136.
2. Tamai, M. Mizuguchi,H. Okamoto,T. (1997); Organism Adaptable concrete, JGEE, Vol.3, pp.65-75
3. Tamai, M.(1987); Treatment of High Water-Content Sludge Using Powderery Wastes, Prediction and Performance in Geotechnical Engineering, A.A.BALKEMA /ROTTERDAM,pp.355-360.
4. Public Works Research Institute of Japan (1997); Report on the Development of Using Technology of Energy Reducing Cement, pp.1-74.
5. Tamai, M.(1994): Permeable Concrete, Concrete Journal, JCI, Vol.32, No.7, pp.134-138.
6. Beeldens, A. and Gemert, D. V.(1997); Development of Porous Polymer Cement for Highway Pavement in Belgium, Proc. of the Second East Asia Symp. on Polymer in Concrete, pp.121-130.
7. Sakata, N.(1997); Porous Concrete Made Precast Sound Absorption Panel, Cement & Concrete, (JCA), No. 607, pp. 42-47.
8. Tamai, M., et al.(1997); Sound Absorbing Concrete with Characteristics of NO_x Absorption, JCA, Proc. of Cement & Concrete, No.51,pp.870-875.
9. Tanaka, M., et al.(1997); Void and Moisture Absorptivity of Porous Concrete Used Shirasu Pumice, JCA, Proc. of Cement & Concrete, No.47, pp.888-893.
10. Tamai, M.(1994); Concrete Materials (Planting & Concrete), Concrete Journal, JCI, Vol.32, No.11, pp.64-69.
11. Tamai, M., et al.(1995); Epilithic Organisms to Porous Concrete Immersed into Natural Sea Water and Water Purifying, JCI, Proc. of JCI Symp. on the Present and Future Views of Environmentally-Friendly Concrete, pp.83-90.

NEW TECHNOLOGY FOR THE RECYCLING OF CONCRETE - JAPANESE EXPERIENCE

F. TOMOSAWA and T.NOGUCHI
The University of Tokyo, Tokyo, Japan

Abstract
Demolished concrete accounts for the largest part of construction waste. If all demolished concrete is disposed of, final disposal areas, which are even now running short, will soon be filled up. On the other hand, aggregate resources for concrete are nearly being exhausted. The development of new sources for aggregate is becoming also difficult. Reproducing aggregate from demolished concrete and recycling for concrete are paramount technologies for resolving the problems of shortage of waste disposal areas and exhausted aggregate resources, while preventing environmental pollution. However, conventional recycled aggregate has been of low quality for structural concrete and has not been accepted by owners of such structures as building and bridges requiring high performance. Problems also exist in the concrete production and supply systems, whereby normal aggregate and low-quality recycled aggregate cannot be easily changed according to uses. To solve these problems, it is necessary to improve the quality of recycled aggregate, i.e., to minimize the cement portions adhering to recycled aggregate to attain the quality approximating to original aggregate. It is more advantageous in the long run to use completely recyclable concrete, which can incorporate byproducts of other industries. Such technologies have been being developed recently, realizing the production of high quality recycled aggregate.
Keywords: concrete, construction byproducts, demolished concrete, recycled aggregate, recycle, waste.

1 Introduction

Since the period of high economic growth that began in the 1960s, the construction investment in Japan has been maintained at a high level, leaving an enormous stock of concrete structures. The early portion of this stock as well as the building stock of the

Concrete Technology for a Sustainable Development in the 21st Century edited by O.E. Gjørv and K. Sakai.
Published in 2000 by E & FN Spon, 11 New Fetter Lane, London EC4P 4EE, UK. ISBN: 0 419 25060 3

postwar reconstruction period are beginning to enter the age of renewal, when large amounts of substances are discharged as waste from demolishing sites. It is said that whether a new construction project can be started will in the near future depend on whether the demolition waste can be successfully disposed of.

Concrete is a durable material. Reinforced concrete buildings are said to last for several decades to more than a century, but they are actually demolished and removed much earlier. Since the high economic growth period, the annual use of concrete in Japan has been approximately 200 millions m^3, or 440 million tons on average. In other words, 6 billion m^3, or 13 billion tons of concrete have been accumulated in Japan in the last 30 years. About half of this amount has been used for buildings.

Among construction waste which is abundantly generated and will certainly continue to increase, demolished concrete is the largest in amount and most difficult to reduce and recycle, requiring immediate measures.

While the largest part of recycled concrete waste is applied to road subbases, a large percentage of demolished concrete is still disposed of, i.e., dumped. The use of demolished concrete for concrete aggregate must now be pursued, while natural aggregate resources are in the process of being exhausted.

This paper reports on the state of demolished concrete generation in Japan, as well as the problems and newly developed technologies.

2 Legislative stipulation on demolished concrete

In order to understand the current state of demolished concrete in Japan, it is necessary to know how waste is classified, defined, and dealt with by law.

Domestic and industrial waste has been disposed of in accordance with the Waste Treatment Law. In 1991, the Recycling Law was enforced to curb the volumetric increase of waste and promote recycling of useful resources included in it. According to the Waste Treatment Law, waste is classified into domestic waste (e.g. from households) and industrial waste. Waste from construction works also include both. Industrial waste is categorized into sludge, construction timber waste, scrap metal, glass and ceramic waste, and construction waste, to which concrete rubble is assigned. Transportation (conveyance), storage, intermediate treatment, and final disposal of industrial waste are required to be carried out under strict administrative control by the Waste Treatment Law.

On the other hand, the newly enforced Recycling Law defines subsidiary matters generated from construction works as "construction byproducts," which are classified into two categories: one is "recyclable resources," which can be used as raw materials, and the other is "waste," which cannot be recycled. In other words, what have been designated as "waste" in the Waste Treatment Law (construction waste) are divided into useful matters (recyclable resources) and unnecessary matters (proper waste) to promote the effective use of recyclable resources. These relations are shown in Fig. 1.

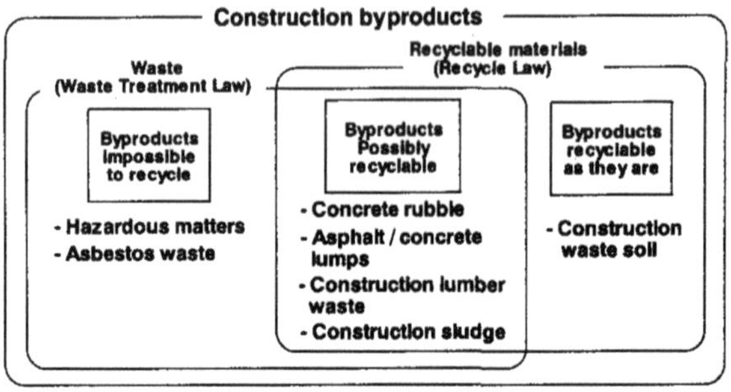

Fig.1. Construction byproducts controlled
by Waste Treatment Law and Recycling Law

According to a survey by the Ministry of Construction (MOC), the state of waste generated from construction works and their recycling and disposal in 1995 are as shown in Fig. 2[1]. Concrete rubble is the largest in quantity, and naturally assigned to "recyclable resources" while final disposal sites are being filled up (Table 1), and its recycling is strongly required. The Recycling Law designates the construction industry as a "special industry," which is required to promote recycling of concrete lumps as a measure to stimulate the use of recycled resources. It also classifies concrete lumps as a "designated byproduct" to stimulate recycling.

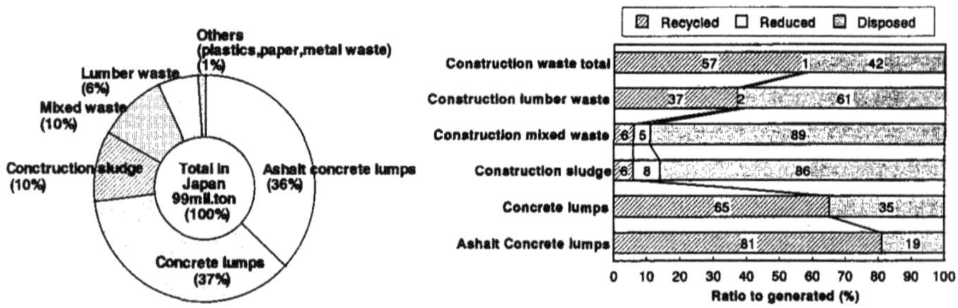

Fig.2 Investigation into production and reuse
of construction byproducts in 1995 (Ministry of Construction)

Table 1 - Remaining capacity and remaining number of years
for final disposal of industrial waste (1994)

Area	Required annual amount of final disposal (mil.ton)	Remaining capacity of final disposal sites (mil.ton)	Remaining number of years (year)
Metropolitan area	23.60 (25.72)	17.72 (14.71)	0.8 (0.6)
Great Osaka area	12.60 (14.24)	44.09 (48.01)	3.5 (3.4)
All Japan	84.00 (89.00)	210.72 (200.65)	2.5 (2.3)

( ) : Numbers for the previous year (Ministry of Health and Welfare)

The MOC formulated the "Action Program on Management of Construction Waste/Byproducts - Recycling Plan 21" in 1994 to promote recycling of construction byproducts and reduce waste to be dumped. With the aim of reducing construction waste to nearly zero in the future, so-called "zero-emission," this action plan intends for the construction waste to be reduced by 10% from the estimated value and the final disposal amount to be half of the 1990 value by the year 2000. In this plan, the target recycle ratio of asphalt/concrete lumps and concrete rubble is 90% for the year 2000, which was 50% in 1990. Whereas the recycling ratio of asphalt/concrete lumps is increasing, that of concrete rubble still remains low.

3 Recent production and recycling of demolished concrete

According to the MOC, the amount of demolished concrete was approximately 36 million tons in Japan in 1995 as shown in Fig. 2. However, the actual amount is estimated to be much greater. Calculating from the original units of concrete used (amount used per square meter) for buildings and the statistical data and estimations of demolished buildings, waste concrete from reinforced concrete, steel framed reinforced concrete, and steel structures amounted to 56 million tons already in 1990 and is considered to have exceeded 70 million tons in 1995[3] (Fig. 3). This will further increase when waste concrete from civil structures are added.

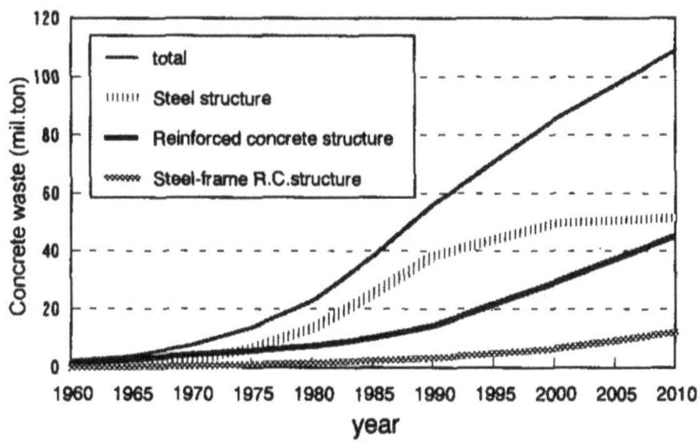

Fig.3. Concrete waste from removed buildings - Estimation from the statistics of removed buildings and original units of concrete used for buildings

Measures including the Recycling Plan 21 have radically increased the recycle ratio of demolished concrete from 48% in 1990 to 65% in 1995. In other words, approximately 23 million tons out of demolished concrete of 36 million tons in 1995 were used as recycled resources, and the rest, 13 million tons, were disposed of at final disposal areas. The destination of the majority of recycled concrete is still road subbases. The use of demolished concrete as recycled aggregate remains in the experimental stage. The total demand for subbase materials is said to be 300 million tons annually, suggesting much room for accepting demolished concrete. However, subbase materials are generally required at remote areas from where concrete is demolished. It is therefore difficult and unrealistic to apply all demolished concrete to subbases in consideration of the transportation costs.

Where has the large difference between the survey and estimated values of demolished concrete gone? In fact, the largest part of construction byproducts is "construction waste soil," which amounted to 446 million m^3 in 1995. Being out of the scope of the industrial waste covered by the Waste Treatment Law, such soil has been used for land-filling, back-filling, and ground raising. Part of the demolished concrete may be included in such soil when structures, particularly underground portions, are demolished. However, if good quality aggregate can be reclaimed from demolished concrete, dumping it with soil should not be recommended from the standpoint of effective utilization of resources.

4 Researches on recycling of concrete

Among the various uses for demolished concrete, the most effective use with a large potential demand is to crush it into aggregate and reuse as recycled aggregate for concrete.

Active research has been conducted on recycled aggregate concrete by a number of researchers, with quality specifications for recycled aggregate and recommended practice for recycled aggregate concrete being proposed. It is used for actual construction works, though experimentally. However, such aggregate still presents unresolved problems and has not been available on the market except as a commercialized subbase material, since it has not been approved as aggregate for ready-mixed concrete for general use.

Research projects carried out so far and their results are as follows:

(1) Studies on the Treatment and Recycle of Construction Waste (1974-1977)

Committee for Treatment and Recycle of Construction Waste, Building Contractors' Society : "Proposed Standard for Recycled Aggregate and Recycled Aggregate Concrete", BCS, 1977.

This was the first research project with an objective of using waste concrete for recycled aggregate. However, being made simply by crushing and classifying demolished concrete, the aggregate was of low quality and usable for only limited applications. This project's importance lies in its being the first research conducted by the construction industry with the aim of recycling construction waste.

(2) MOC comprehensive R & D project : Technology for using waste for construction Work (1981-1985)

Sub-committee for Public Works : "Proposed Recommended Practice for Design and Construction of Concrete Structures Made Using Recycled Aggregate "(1986)

Sub-committee for Building : "Proposed Specification for Concrete Containing Recycled Coarse Aggregate" (1986)

Such qualities as density, absorption, grading, sulfate soundness, and percentage passing a 75 μm sieve were specified as the requirements for recycled aggregate. The quality types of aggregate were also established. The design strength range and uses for recycled aggregate concrete containing aggregate of each type were specified.

(3) MOC : a comprehensive R & D project, "Technology for suppression of generation and promotion of reuse/recycle of construction waste and byproducts" (fiscal 1992-1996)

To meet the strong demand for recycling of construction waste, the MOC organized again a research project to stimulate recycling of construction waste soil, construction sludge, and concrete rubble, which account for a large part of construction waste. As for concrete rubble, a large number of experiments, tests, and experimental applications to actual structures were conducted to enhance and commercialize the results of the project mentioned in (2) above.

Tables 2 to 5 give the specifications for recycled aggregate and recycled aggregate concrete proposed as a result of researches under this project. Different specifications were proposed for civil structures and buildings, but the quality requirements for recycled aggregate will be unified.

In any event, the quality of aggregate is classified according to the absorption. A high absorption indicates a high attached cement content, which generally leads to concrete with an inferior strength, durability, and deformation and shrinkage properties. Accordingly, the maximum allowable design strength and the members and portions to which such concrete may be applied are limited by the quality of recycled aggregate.

Table 2 - Proposed quality standard of recycled aggregate for concrete for public works (1994)

Type	Recycled coarse aggregate				Recycled fine aggregate	
	Type 1	Type 2		Type 3	Type 1	Type 2
Absorption (%)	3 or less	3 or less	5 or less	7 or less	5 or less	10 or less
Sulfate soundness(%)	12 or less	40 or less	12 or less	---	10 or less	---
		40 or less (1)				

(1) Where freezing and thawing resistance is not required
(2) Grading and content of injurious impurities are also stipulated.

Table 3 - Proposed types of recycled aggregate concrete and suggested uses in public works (1994)

Type	Coarse aggregate	Fine aggregate	Suggested design strength (MPa)	Suggested use of recycled aggregate concrete
I	Recld. coarse aggregate Type 1	Normal fine aggregate	18 to 24	Reinforced or plain concrete; lower structure of bridges, tunnel lining, retaining walls, etc.
II	Recld. coarse aggregate Type 2	Normal or recld. fine Type 1	16 to 18	Plain concrete; masonry units, bases for road attachment, gutters, gravity type retaining walls, etc.
III	Recld. coarse aggregate Type 3	Recld. fine aggregate Type 2	Less than 16	Subslab concrete, back filling concrete, leveling concrete, etc.

Table 4 - Proposed quality standard of recycled aggregate for concrete for building (1997)

Type	Recycled coarse aggregate			Recycled fine aggregate	
	Type 1	Type 2	Type 3	Type 1	Type 2
Absorption (%)	3 or less	3 to 5	5 to 7	5 or less	5 to 10

(1) Grading, under 75 μ particles and content of injurious impurities are also stipulated.

Table 5 - Proposed types of recycled aggregate concrete and suggested uses in building (1997)

Type	Coarse aggregate	Fine aggregate	Suggested design strength (MPa)	Suggested use of recycled aggregate concrete
A	Recld. coarse Type 1	Normal fine	18 or more	Ordinary reinforced concrete buildings
B	Recld. coarse Type 2	Normal fine	18 or more	Concrete attached to ground; foundation, cast-in-place concrete piles, concrete slabs on steel decks, etc.
C	Recld. coarse Type 2	Recld. fine Type 1	18 or more	Foundation slabs, earthen floor slabs, subslab concrete, back filling concrete, leveling concrete, etc.
D	Recld. coarse Type 3	Recld. Fine Type 2	18 or more	Subslab concrete, back filling concrete, leveling concrete, etc.

5 Problems in promoting utilization of recycled aggregate concrete

The results of the research projects (1) and (2) above have not been put into practice, and recycling has not been in operation. Though research project (3) formulated technical specifications for recycled aggregate and recycled aggregate concrete, it is still ambiguous whether the systems of recycling would be successfully disseminated.

The problems of recycled aggregate concrete when compared with normal concrete include low strength gains, wide scatter of qualities, high drying shrinkage and creep, and low modulus of elasticity. These can be resolved by applying it to limited uses and adopting suitable design methods. Such concrete can be sufficiently usable now, if it is applied to specific uses. The technical specifications obtained from the above-mentioned research project suggest members and portions to which each type of such concrete can be applied.

The problems to be tackled now are related to systems rather than technology. There are problems remaining untouched in the distribution systems for recycled aggregate, production and supply systems for recycled aggregate concrete, and quality assurance systems for aggregate and concrete. Problems also exist regarding cost and general tendency towards avoidance of recycled materials. Solution of systematic problems is a prerequisite to address these problems. The supply system of recycled concrete should particularly be established. In other words, plants for producing recycled aggregate concrete should be adequately located, readily providing such concrete of a desired type in the form of ready-mixed concrete. Such a system is difficult to coexist with the present systems of production and distribution of ready-mixed concrete. This hampers the commercialization and dissemination of recycled aggregate concrete.[4]

Whereas it is vital to resolve these problems, it should not be forgotten that these problems arise from the fact that the quality of recycled aggregate is lower than that of virgin aggregate. This is the reason why users of concrete do not dare to use recycled aggregate concrete and fear that virgin aggregate might be mixed with recycled aggregate. Such a feeling of distrust is aggravated by lower cost due to lower quality of recycled aggregate.

It is therefore of paramount importance to improve the quality of recycled aggregate and develop technologies to produce high quality recycled aggregate. The concept of conventional technologies is to minimize cement paste adhering to aggregate to improve the aggregate quality by repeating crushing of concrete. This tends to crush aggregate as well, reducing the yield and producing a high percentage of fine powder, which is not useful. The inclusion of both hardened cement and powdered aggregate in the powder hampers the use of such powder.

New technologies needed to solve such a dilemma are being developed. The following sections report on these technologies.

6 New technologies of recycling concrete

When recycling concrete, it is most efficient to separate aggregate from cement paste as much as possible. Aggregate can then be used as good quality recycled aggregate, and cement containing little crushed aggregate can be recycled as a material for cement. This forms a closed loop between concrete and its materials, bringing the system towards zero-emission. New technologies to realize this are being developed. Some of them are introduced as follows:

6.1 New techniques for producing high quality recycled aggregate [5]

A study is being conducted as a research project assigned from the Ministry of International Trade and Industry by Nuclear Power Engineering Corporation to recycle waste concrete generated in abundance when a nuclear power plant is demolished. The objective of this development is to reclaim a high percentage of aggregate of a quality level usable for the construction of a new nuclear power plant. This technology is also applicable to the case of producing recycled aggregate from general demolished concrete. When commercialized, this technology will realize high quality recycled aggregate that is usable for structural concrete in the same manner as virgin aggregate.

The targets of this development are significantly high as follows:

- Recycled coarse aggregate: a density in the absolutely dry state of not less than 2.5 and absorption of not more than 2%
- Recycled fine aggregate: a density in the absolutely dry state of not less than 2.5 and absorption of not more than 3%
- Aggregate reclamation ratio: 70% of concrete

(1) Improved Mechanical Grinding Method (Being developed by Takenaka Corporation)

This is a method in which concrete is crushed into rubble of a size of around 50 mm and then ground with an improved horizontal grinding mill into recycled coarse aggregate 25 to 5 mm in diameter, rather than being simply crushed down to this size. Pieces less than 10 mm in size are ground into fine aggregate using a mill having a similar mechanism. The cylindrical mill is equipped with an internal cylindrical rotor with special tooth-shaped projections. Pieces of concrete rubble are rubbed against one another in the space between them, efficiently removing attached mortar and cement paste. An improved jaw crusher, ball mill-type crusher, impact crusher, or pneumatic impact crusher may be combined where required to provide optimum operation conditions. Test operations produced qualities conforming to the requirements of the Japanese Architectural Standard Specification for Concrete Works (JASS 5) for normal aggregate. However, production of fine aggregate was found slightly difficult.

Figures 4 and 5 show the relationship between the grinding time and the qualities of the resulting recycled aggregate.

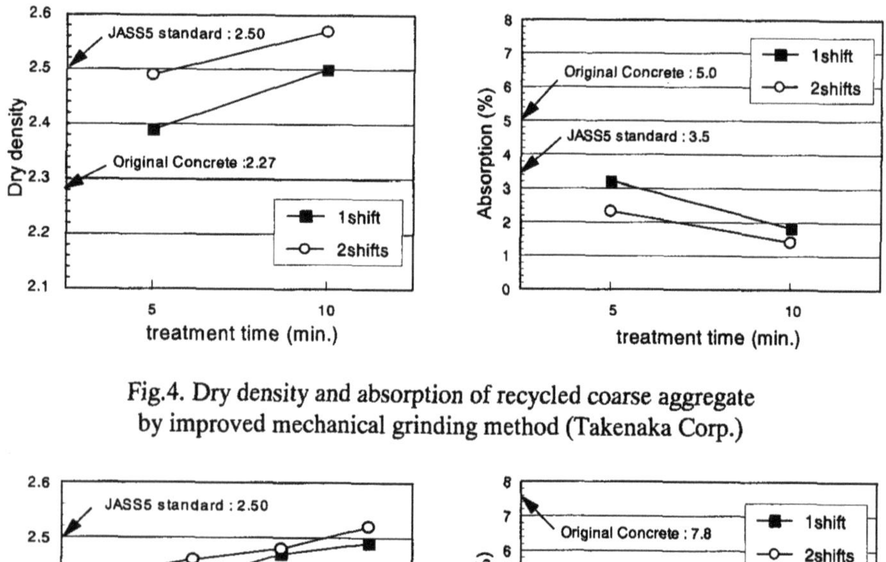

Fig.4. Dry density and absorption of recycled coarse aggregate by improved mechanical grinding method (Takenaka Corp.)

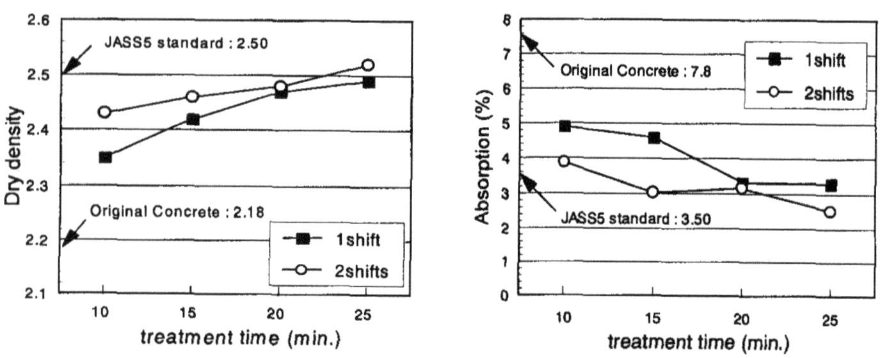

Fig.5 Dry density and absorption of recycled fine aggregate by improved mechanical grinding method (Takenaka Corp.)

(2) Selective Heating and Grinding Method (Being developed by Mitsubishi Heavy Industries Ltd.)

This method selectively heats the cement paste phase of demolished concrete rubble by irradiating microwaves to make it brittle, and then grinds the rubble with a horizontal grinder to produce good quality recycled aggregate. Experiments were conducted first on a small scale to verify the effectiveness of this method, and then on a larger scale of 0.2 t/hr. The optimum heating and crushing conditions were determined, and the targets of the development were attained. The qualities of the resulting coarse and fine aggregates exceeded the targets, with the recycled percentage being over 70% of concrete and over 85% of aggregate.

Figures 6 and 7 show the relationship between the heating time and the qualities of the resulting aggregate.

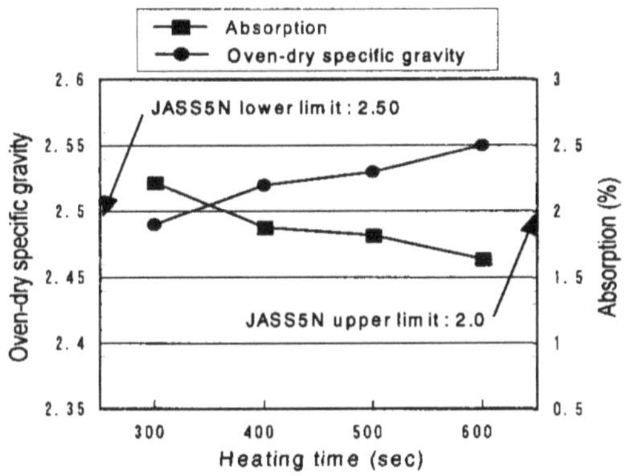

Fig.6 Heating time and quality of recycled coarse aggregate (Mitsubishi Heavy Industries, Ltd.)

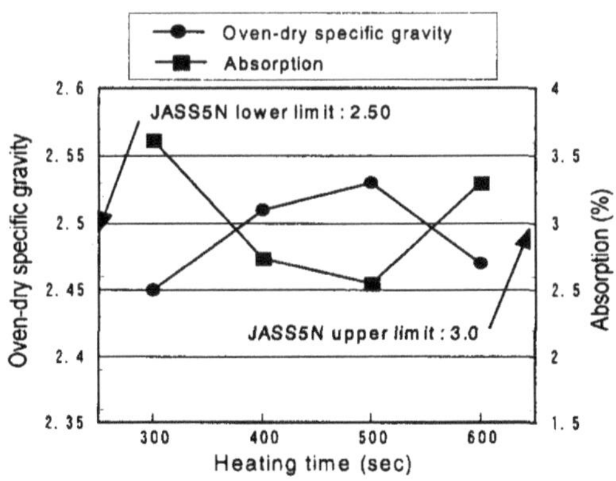

Fig.7 Heating time and quality of recycled fine aggregate (Mitsubishi Heavy Industries, Ltd.)

(3) Air Heating and Grinding Method (Being developed by Mitsubishi Material Corporation)

This is a method whereby concrete rubble crushed to 50 mm is charged into a vertical furnace, and heated with hot blast from below, which makes the cement paste phase brittle. The rubble is then ground with a tube mill to separate cement paste from the aggregate. Various heat sources can be applied to heat the concrete at around 300℃. The qualities of the resulting aggregate met the targets, and the recycled percentage was sufficiently high. The qualities of the concrete made using the resulting aggregate were nearly the same as the original concrete.

Figure 8 shows the relationship between the heating temperature and the qualities of the resulting aggregate. Table 6 gives the qualities and recycled percentage of the resulting aggregate.

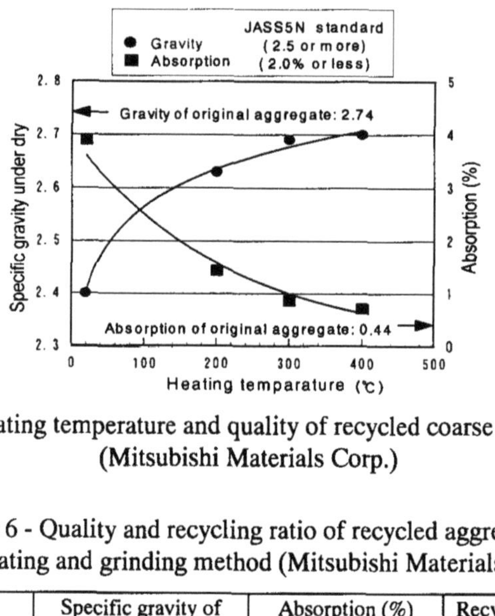

Fig.8 Heating temperature and quality of recycled coarse aggregate (Mitsubishi Materials Corp.)

Table 6 - Quality and recycling ratio of recycled aggregate by air-heating and grinding method (Mitsubishi Materials Corp.)

Aggregate	Specific gravity of oven dry condition	Absorption (%)	Recycled ratio (wt %) (to original concrete)
Coarse	2.69	0.84	41.1
Fine	2.53	2.28	32.8
Recycled ratio (Coarse and Fine)	73.9%(to original concrete)		

6.2 Completely recyclable concrete

Concrete is generally regarded as a material difficult to recycle completely. However, the author has proposed "completely recyclable concrete," which allows repeated recycling similarly to steel and aluminum[6]. This concrete is composed only of cementitious materials, and is reprocessed as cement, allowing repeated recycling. Part of the concrete is used as recycled aggregate, which is made into completely recyclable concrete again. This enables concrete to be recycled in a fully closed system excepting the reduced carbon dioxide emitted during calcining cement.

Powders generated during producing recycled aggregate can be used as a cement material without any further processing, facilitating the production of high quality recycled aggregate.

Moreover, completely recyclable concrete can include a large amount of byproducts of other industries, such as blast-furnace slag, fly ash, and products made from these. These materials can also be recycled repeatedly. Investigation is now being conducted to commercialize such concrete.

WHAT IS COMPLETELY RECYCLABLE CONCRETE ?

Completely recyclable concrete is a concrete or mortar whose...

- **binders,additives and aggregates**
  are all made of cement or materials of cement
- **all of these materials**
  can be used as raw materials of cement or recyclable aggregate after hardening

▼

A concrete containing recycled aggregate from a completely recyclable concrete is **also a completely recyclable concrete**

...In this way, such a concrete **can be recycled endlessly.**

Fig.9. Definition of completely recyclable concrete

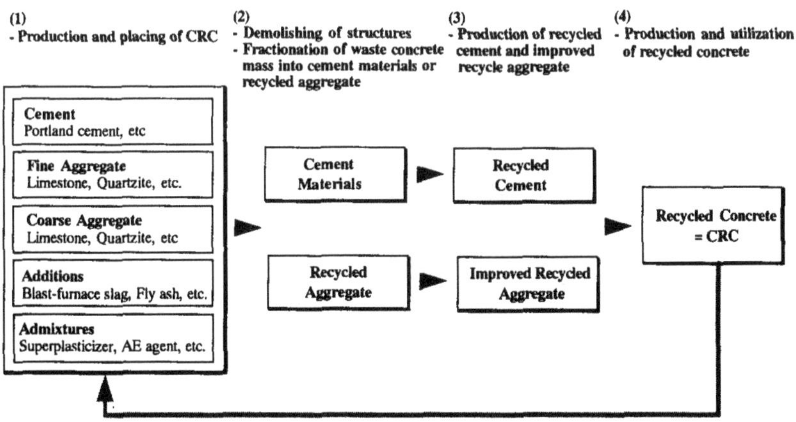

Fig.10. The concept and basic pattern of completely recyclable concrete

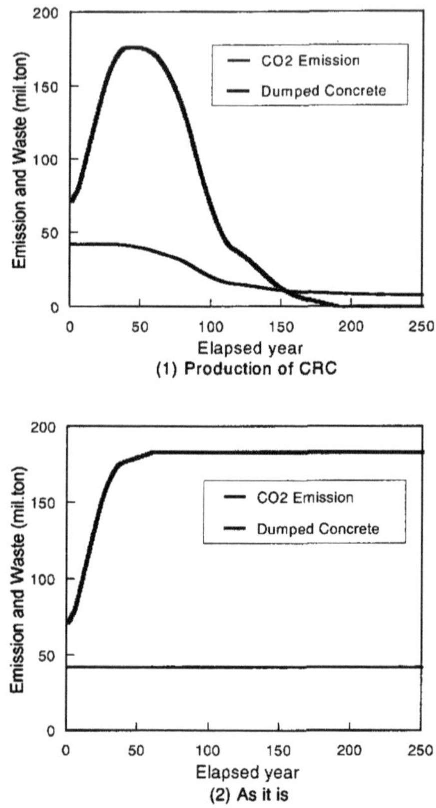

Fig.11 Changes in CO_2 emission and quantity of dumped concrete (in the case of 5% yearly replacement of conventional concrete by completely recyclable concrete)

7 Concluding remarks

In Japan, houses, urban facilities, and infrastructures for society and industry have up to now been actively constructed for nearly 40 years, accumulating enormous amounts of concrete. However, structures made of concrete include those with a low quality and those outdated and not meeting the demand of the present-day society and its people, due to the rapidly changing social structure. These are demolished and removed to be replaced with new structures, generating abundant construction waste and concrete rubble.

On the other hand, the development of new aggregate resources is becoming increasingly difficult for the sake of environmental protection, increasing the significance of the problems of recycling of concrete, particularly recycling as recycled aggregate. Recycled aggregate produced by conventional technologies is of low quality and hardly accepted by users for general structural concrete. Technologies to produce recycled aggregate having the same quality as virgin aggregate are in demand and being developed.

Formulation of a liberated distribution system for concrete conforming to the quality requirements is essential for disseminating recycled aggregate concrete, whatever its quality class may be. To this end, it is necessary to promote the specialization of ready-mixed concrete plants or enlargement of or cooperation between such plants so that a wide variety of concretes can be made available.

Completely recyclable concrete, which can entirely be recycled as a cement material, is also being developed. This will prove to be an important technology from the standpoint of lime resource protection as well.

The civilization of the twentieth century is characterized by mass production and mass consumption of everything based on the development of scientific and industrial technologies. Another characteristic, I dare say, is the absence of consciousness of the method of disposal after use during the designing, producing, and using of such mass-products. Though this may have been beyond human wisdom, the debt is steadily and rapidly making its presence felt towards the turn of the century. This also applies to construction.

Construction waste is losing its destination, while potential waste, larger in amount, awaits to claim more spaces. Nevertheless, demolishing and reconstruction cannot be ceased. How to recycle the waste to be generated has now become a critical problem. What we have to do now is to establish new design and production paradigms for buildings and civil structures aiming for zero emission, so as never to go into such blind alleys again.

8 References

1. Investigation on the present situation of generation, recycle and final disposal of construction waste/byproducts (Fiscal 1995), Bureau of Construction Economy, Ministry of Construction, March 1997.
2. Action Program on Management of Construction Waste/Byproducts - Recycling Plan 21, Ministry of Construction, 1994.
3. MOC Comprehensive R & D project: Technology for Suppression of Generation and Reuse/Recycle of Construction Waste and Byproducts, Report for Fiscal 1993, March 1994.
4. F. Tomosawa: How to recycle concrete - Architecture as huge garbage's, Journal of Architectural Institute of Japan, Vol. 112, No. 1401, pp.21-22, February 1997.
5. Report for 1997 (tentative): Techniques for Recycling Demolished Concrete from Decommissioning Nuclear Power Plant, Nuclear Power Engineering Corporation, March 1998.
6. F. Tomosawa, T. Noguchi: Towards Completely Recyclable Concrete, Integrated Design and Environmental Issues in Concrete Technology, E & FN SPON, pp. 263-272, 1996.

Barriers for Sustainable Use of Concrete Materials

H.Kawano
Concrete Division, Public Works Research Institute, Ministry of Construction, Japan
Tsukuba, Japan

Abstract

From the viewpoint of the sustainable use of resources and environmental conservation, the use of low quality aggregate, construction by-products and other industrial by-products is an important issue in the concrete field. Relevant research and development has been active on use of, for instance, recycled aggregate, low quality aggregate, slag from various metal refining, fly ash, GGBFS and others. But only limited utilization has come into practice so far in Japan. This paper shows the examples of technical developments on the sustainable use of concrete resources and examples of practical application. Then the paper analyzes the reasons, technical as well as social and economical, which are obstructing the progress of their utilization and considers the way out.

Keywords: By-products, concrete, demolition, recycling, reuse, waste management

1. Introduction

December 1997, the International Congress on Global Warming was held in Kyoto. The major topic was reduction of carbon dioxide.

Many concrete engineers may be discouraged when they talk about CO_2 reduction. Because two major materials for reinforced concrete structures, i.e. cement and steel, are produced through the process where much CO_2 is a by-product. These materials have been fundamental for infrastructures supporting industries and daily life, but discussion on environmental conservation raises questions as their use.

As for aggregate resources, we also have a difficulty. Production of aggregate does not generate as much CO_2 per weight. However, the amount of aggregate is so large that it raises other environmental problems.

Many concrete engineers and researchers recognize this problem. In fact, there

Concrete Technology for a Sustainable Development in the 21st Century edited by O.E. Gjørv and K. Sakai.
Published in 2000 by E & FN Spon, 11 New Fetter Lane, London EC4P 4EE, UK. ISBN: 0 419 25060 3

has been extensive research and development on saving resources, recycling and reuse of concrete materials to date. And some of these have been on a practical stage. But this is not the case for many research and development projects. Considering the reasons that prevent the results of this research and development from being practical may point to a solution.

2. Relevant research and development so far

Many research and development projects have been conducted in order to extend the effective use of concrete materials. Those could be divided into three categories as following.

Category 1 : Using by-products from other industries in concrete

Category 2 : Using concrete by-products, including demolished concrete, in other structures.

Category 3 : Using concrete by-products in concrete again

The typical examples for Category 1 are slag and fly-ash. Blast furnace slag has been used as aggregate from the first stage and as GGBFS recently. Fly-ash has a long history as a mineral admixture for mass and structural concrete. Other than those materials, many other materials are there expecting to be used in concrete, i.e. bricks, tiles, crushed glass, ash from several industries, and so on. However, very little has been used practically.

As for Category 2, much crushed demolished concrete has been used as sub-base materials for road pavement and foundation materials.

Category 3 includes recycled aggregate, sludge from ready-mixed concrete plants, reuse of the mortar adhered to the inner surface of the drums of agitator cars. We have tried to extend the use of recycled aggregate, but the speed of extension has been much slower than expected so far, in spite of much research over the past 25 years.

3. Actions enhancing reuse

The Ministry of Construction, Japan, implemented several action programs in order to extend reuse of construction by-products, concrete, asphalt, wood and soil. Through these programs, in April 1994, "The Draft for quality specifications for recycled materials from demolished concrete" was issued. The MOC also presented the "Recycle 21" program in which numerical targets were shown for recycling of several kinds of construction by-products. The target for the recycling ratio of demolished concrete in the year 2000 is 90 percent, and the actual results for 1990, 1993 and 1995 were 48 percent, 67 percent and 65 percent, respectively.

Some Japanese Industrial Standards (JIS) have been or are being revised including standards for recycled materials. GGBFS for concrete was standardized in JIS in 1995. Copper slag aggregate was added to "slag aggregate for concrete" which formerly consisted of blast furnace slag and FeNi slag. The standard for "Ready-mixed concrete" was revised as mortar adhered to the agitator drum could be used in newly mixed concrete with a special chemical agent. The standard for fly-ash for concrete is under revision to extend usage.

4. Barriers against reuse

As shown above, numerous research and development projects on the reuse and recycling of demolished concrete and adding by-products to concrete have been conducted. However, the well-going cases are limited to some exceptional materials, i.e. GGBFS, Fly-ash and so on.

We have to consider what prevents extending the sustainable use of concrete materials. The reasons could be listed as following:

a. lack of suitable laws
b. lack of codes, specifications, standards and guidelines
c. cost
d. poor image
e. lack of experience
f. low quality
g. variations in quality
h. too many kinds and too large amounts of by-products
i. inefficient supply system
j. shortage of proper information
k. others

a. Lack of suitable laws

The Japanese Government has made several laws and rules to enhance recycling and the reuse of many kinds of materials and by-products, for example low interest official loans, low rate of taxation. As for the construction by-products, these laws do not seem to have had much of an effect so far.

On the other hand, we have a law which controls the transportation and treatment of wastes. This law was made in order to control wastes harmful to human health, so it is very strict. Many kinds of construction by-products are not harmful to human health, but their transportation and treatments are controlled by this law. When construction by-products are purchased and then reused, no regulation is raised. But once they are treated whether free or with a fee, they are labeled "waste" and their transportation and treatment are controlled strictly by the law. We can not even store demolished concrete without a permit until a site is found where it can be effectively used. This is the case for fly-ash, slag and similar harmless by-products which have potential use as concrete materials. Appropriate laws and regulations are needed for these mass harmless materials.

b. Lack of codes, specifications, standards and guidelines

Almost all concrete codes, specifications, standards and guidelines deal with concrete using ordinary cement and aggregate. If concrete using a slightly extraordinary material has the same characteristics as ordinary concrete, the material will be accepted smoothly. Small changes may be covered by modifying details of the design, quality control and/or concreting in-situ. For example, GGBFS can be used as a mineral admixture with a slight change in mix proportioning and curing. However, in many cases, using recycled materials or by-products brings about an innegligible change of concrete characteristics which cannot be covered by a small change in details of design. In these cases, we must change the design of the structure and/or the concreting method. These procedures cannot be accepted easily in general construction projects. Therefore, we must prepare specifications or manuals.

c. Cost

The price of ordinary Portland cement has not risen for past 25 years in Japan. And the quality has been kept at a high level. The increase in aggregate cost has been smaller than increase in general consumer prices. This fact is thanks to efficient mass-production.

The transport and treatment costs of demolished concrete are not low, because lots of demolished concrete are small and the sites are separated. Quality varies considerably, so much energy and significant costs are required to treat demolished concrete.

However, we can observe remarkable improvement in the treating machines.

d. Poor image

People tend to think that by-products are not main stream and are not of high quality. Of course, if the by-product had higher quality, it would already be in the market. We have to recognize that they have some disadvantages.

However, we often use over-specified materials in applications where by-products may satisfy the requirements. We must put the right materials in the right place.

e. Lack of experience

In construction areas, when we introduce new materials or new construction methods, experience is required in order to ensure safety. This may be a contradicting requirement. This fact sometimes becomes a barrier in the way of introducing new technology. Toward this end, the MOC has conducted demonstration projects and/or test projects in order to introduce new technologies to construction sites. Recently, it appears that it is taking a shorter period to create standards for new construction materials than in the past.

f. Low quality

Generally, recycled materials are of lower quality than virgin materials. However, for concrete materials, there are many techniques to apply those material to structures without degrading the quality of structures. Performance-based-design methods may be favored for using recycled materials.

We have to recognize that trials to raise the quality of recycled materials advance technologies. For example, treating machines for recycled aggregate from demolished concrete were substituted by a crushing machine for new aggregate, but the requirement to obtain more recycled aggregate with low water absorption improved the machine. And, recently, new concept treating machines have been developed which enable higher quality recycled aggregate with lower energy consumption.

g. Variations in quality

The quality of demolished concrete varies from site to site and the amount from sites is small. Almost all recycling plants are of small size without large stock yards. These facts bring about wide variation in the quality of recycled materials from the plants. Wide variation requires frequent quality control testing and may trigger concern about the quality of structures using such concrete.

However, do we have to apply the same quality control for recycled materials? For example, aggregates in demolished concrete have already once passed the standards for concrete aggregate, so we may be able to eliminate some testing. Additionally, easy and frequent quality control test methods could be more effective when using recycled materials.

h. Too many kinds and too large amounts of by-products

Today's mass consuming society brings many kinds of and much waste and by-products. Repeated recycling use for the same purpose is ideal, but we have economical wall to consider. The easiest way is disposal, but this leads to environmental problems and, consequently, exhausts resources. People like to think that the second easiest way is to use these materials as constructing materials and that almost all kinds of waste can be used as either concrete materials or asphalt concrete materials, and if this is not possible then as a soil substitution. I often hear reports that someone is trying to use crushed glass as a concrete aggregate, or that someone is trying to use shredded plastics, and so on.

Are construction sites litter boxes? We should not choose the easy way out. We must consider the feasibility of re-reusing these materials as well as the performance of the structures. Crushed glass aggregate will not only create an alkali-aggregate reaction problem but will also create difficulty when we reuse glass aggregate concrete. The same case applies for plastics aggregate.

i. Inefficient supply system

The supply system for concrete materials is overly efficient and steady. It is, therefore, very difficult for new materials to enter into the market. A rise in supplying cost is vital for low price concrete materials.

The key to suppressing the cost of recycled materials is information. An information network connecting demolishing sites, treatment plants and re-usable sites is essential. Such a network will minimize the cost of transportation and storage.

j. Shortage of proper information

There are two aspects for information. One concerns "b. lack of codes, specifications, standards and guidelines." Have we been announcing the correct information on recycled materials? For example, what kinds of recycled material are available, and what kind of usage is recommended, etc.

The other aspect is information on supply mentioned earlier.

k. Others

5. Future actions

We concrete engineers and researchers have been concentrating on the study of hardware for the reuse of concrete materials. From an economical viewpoint only, their reuse will not be extended without support. We must study software in order to proceed with reuse. At least, we must evaluate the life cycle cost. Also, it is important to introduce an evaluating system of environmental efficiency relative to reusing concrete materials.

DEMOUNTABILITY AND RE-USE OF CONCRETE STRUCTURES

H.W. REINHARDT
Institute of Construction Materials, University of Stuttgart and
FMPA BW (Otto-Graf-Institute), Stuttgart, Germany

Abstract
Recycling of concrete is current practice. However, crushing of concrete is not environment friendly and concrete made of crushed concrete needs at least as much cement as other concrete. Therefore, it would be preferable to design concrete structures demountable such that they can be easily demounted or adjusted to new needs and maybe reused. The paper deals with requirements on demountable structures and the current situation in Central Europe.
Keywords: Demountable structures, concrete, re-use, life-cycle assessment.

1. Introduction

It is rather common to use a product for a certain time and to throw it away when it has become oldfashioned, is worn out, does not function any more or is damaged by misuse or an accident. The same may happen to a structure or structural element. To throw something away means to waste energy, material, and ressources, furtheron it needs dumping space. This custom is opposite to sustainability.

What can be done to improve the situation? The material, i. e. concrete, should be used and not consumed, the longer the better. Concrete should be made of constituents which do not need so much energy and raw material for their production, i. e. furnace slag and flyash are by-products with a high puzzolanic or latent-hydraulic potential. Partial replacement of portland cement which needs most energy for production and causes most emissions during production is rather effective. Fig. 1 shows that the production of portland cement has the largest impact on the environment.

Concrete Technology for a Sustainable Development in the 21st Century edited by O.E. Gjørv and K. Sakai.
Published in 2000 by E & FN Spon, 11 New Fetter Lane, London EC4P 4EE, UK. ISBN: 0 419 25060 3

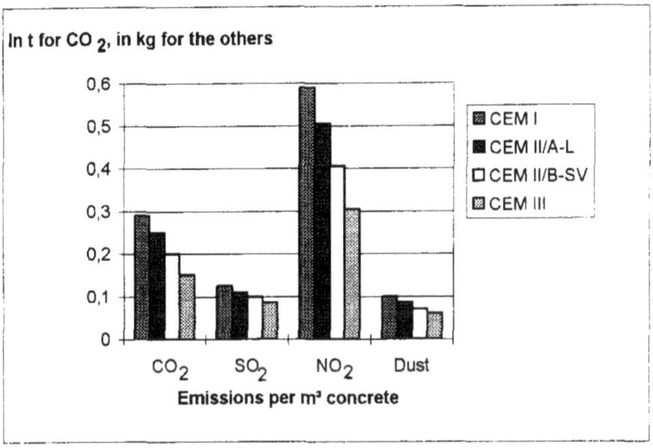

Fig. 1 Emission due to cement production related to one m^3 concrete with 330 kg of cement [1].

The energy consumption for producing portland cement (CEM I) amounts to 4050 MJ/t whereas slag cement (CEM III/B) needs only 3080 MJ/t. There are many parts of buildings which do not need a high grade concrete, such as foundations and walls.

Today, it is often cheaper to use the same concrete for all parts of a building rather than to distinguish between the specific requirements. By-products can also be used as aggregate, i. e. some types of slag. So, one answer to the question is to make a concrete with low input and to use it as long as possible.

However, every structure comes to its end. Today, it is good practice to tear a structure down, to separate different materials and reuse them as much as possible. As concrete is concerned it will be crushed and can be used as aggregates. The environmental profit is rather limited because crushing of concrete needs much energy and at least the same amount of new cement is necessary to make a new concrete. Table 1 shows that the use of land, water and sand and gravel is less, but the energy consumption is even more than in case of concrete with original materials. This means that the savings are not large. On the other hand, recycling of concrete is certainly preferred compared to dumping.

Another solution could be to design structures demountable and to aim at a (partial) re-use of elements.

Table. 1 Comparison between a primary and secondary concrete C35 [2]

Item	Unit	Quantity		
		Primary	Secondary	Difference
Gypsum	kg	11.6	11.6	0
Limestone	kg	385.7	385.7	0
Clay	kg	55.1	55.1	0
REA-Gypsum [1)]	kg	2.9	2.9	0
Sand/gravel	kg	1795.2	718.3	1076.9
Energy	MJ	1740.0	1750.0	-10.0
Landuse	m^2	0.157	0.069	0.088
Water	m^2	2.643	1.188	1.455
SO_2	g	200.3	200.9	-0.6
NO_x	g	743.9	749.8	-5.9
Dust	g	116.4	117.3	-0.9
CO_2	kg	299.0	299.5	-0.5
CO	g	149.6	150.7	-1.1
CH_4	g	271.0	271.2	-0.2
NMVOC [2)]	g	37.9	38.4	-0.5
N_2O	g	2.4	2.4	0
HCl	g	2.4	2.4	0
HF	g	0.1	0.1	0
Overburden	kg	200.7	102.8	97.9

[1)] From fue-gas desulphurization
[2)] Ozone forming potential

2. Requirements on demountable structures

2.1 General

The ideal situation would be if one could design a structure such that all parts could be demounted like with a mechano toy, i. e. load bearing parts, partitioning walls, screeds, curtain walls, energy and water supply, etc. would be connected by disconnectable couplings. After a certain time the structure could be demounted. It could also be adapted to new needs and services if the owner changes or a production process changes if the building were an industrial one. After demounting all parts which are still technically adequate could be reused. The savings would concern energy of dismantling, immission of dust and noise during these activities and material and energy necessary for producing new parts. Technically speaking, used concrete parts could even be superior because creep and shrinkage are less. Demounted elements could also be checked by non-destructive testing before re-use. However, this is an ideal picture which does probably not work in the real world. It is questionable whether used parts would fit into a new design after many years and whether people would accept such parts which could be regarded as inferior to new parts.

In the real world, one has to accept that the re-use of elements will be very limited. So what is the advantage? Energy consumption and dust and noise generation during demounting would be less than during dismantling. The different materials could be separated easily which means that recycling would concern non-contaminated materials. If a structure would be partially demountable it could be adjusted to new services. This could mean that the main structure would remain in service for a longer time which makes the life-cycle assessment better.

2.2 Requirements on the design

A demountable structure uses prefabricated elements. The number of different elements should be limited to some standard elements like columns, slabs, walls. However, these should also be special elements which allow to realise a layout which is not rectangular and has not a regular shape. Since many similar elements are necessary the production could be automated. The size of a single element is mainly limited by transport facilities at one end and by fire resistance at the other. The elements can be reinforced and/or prestressed. The geometrical tolerancies have to be very small (less than a millimeter) in order to make an easy and exact mounting possible. Shrinkage and creep has to be anticipated.

The joints have to be disconnectable which may require steel parts. These can be screwed or sometimes welded. A welded joint can be cut without great effort. One can also use post-tensioning from element to element. Fig. 2 shows two beam to column connections.

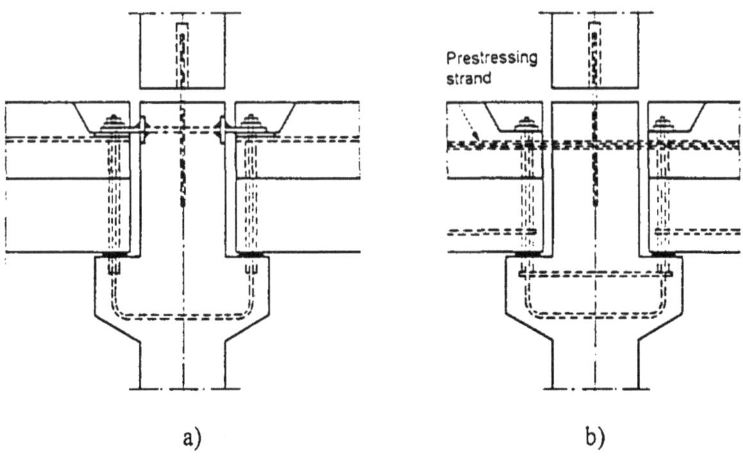

Fig. 2 Beam to column connections [3]
a) Screwed steel parts, b)Prestressing without bond.

Fig. 3 shows a connection between a hollow core slab to a bearing beam. Coupling bars are situated in the joint between two slabs and connected to a steel plate which is integrated in the beam.

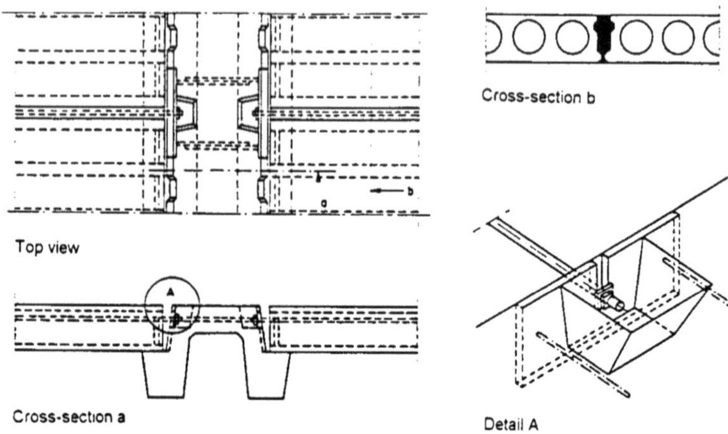

Fig. 3 Connection between hollow core slab and beam [3].

The force is transmitted via steel bars from plate to plate of the beam. It has to be admitted that the joints between the slabs and between slab and beam have to be grouted which makes them not dismountable in a rigorous sense.

Demountable structures are composed of elements which are connected to each other. Most elements will have a single span. Both aspects make a structure less stiff than a monolithic one. Therefore, the structure has to be thoroughly analysed and designed such that shear walls or cores are available which can take horizontal forces. Joints between slabs have to transfer shear forces as schematically shown in Fig. 4.

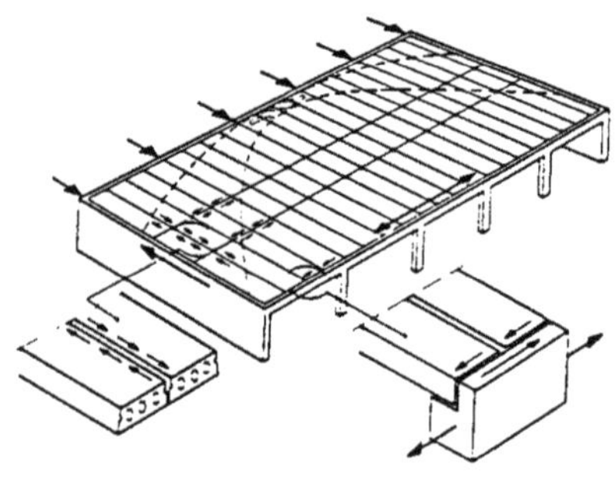

Fig. 4 Floor system with compressive arch, tension chord and shear forces in joints [3].

A second loading path is necessary for the case that a primary element fails due to some unforeseen loading. Fig. 5 gives some examples of an alternative load path. The safety factor for such loadings can be one. Some destruction of elements is acceptable as long as the equilibrium of forces is maintained.

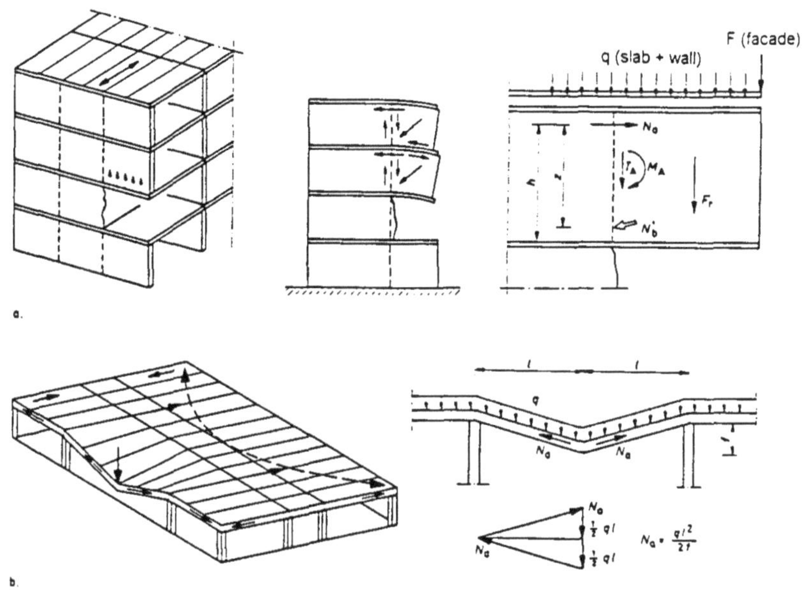

Fig. 5 Examples of an alternative load path [3].

The load bearing structure and the interior finish has to be separate. This would enable the separation at the time of complete dismounting or allow the modernization of the finish at an earlier time.

3. Examples

There is a publication edited by the Netherlands Building Office (Rijksgebouwendienst) [4] which contains a catalogue of five demountable building systems. All systems consist of mainly three types of elements, i. e. columns, slabs and walls. The slabs are either reinforced or prestressed and are coffered slabs or double-T slabs. The columns have the smallest size with respect to fire resistance. The slabs are always supported at the corner which is connected to the column by steel plates and bolts or pins.

The publication [4] contains many details about the structures, the mounting speed, the dismountability, the cost and realized structures.

During the workshop, examples will be presented. It will be shown that a demountable structure can be erected at a great speed, almost not hampered by the weather. This makes the works on the building site precisely systematic and on schedule.

4. Conclusions

Demountable concrete structures have a great potential with respect to energy and material saving due to easy demounting and re-use of structural elements. If a demountable concrete structure will not be re-used it is still beneficial for the environment because it can be adapted to new services and keeps the main structure longer in service.

The current situation of demountable concrete structures is characterized as follows: there are closed systems available which are limited with respect to their layout. In the future, systems should be developed with a greater variability of design and use of different types of concrete. Furtheron, demountable structures should be more promoted in the education of architects and engineers.

5. References

1. Lünser, H. (1998) Ökobilanzen im Brückenbau. Zur Berücksichtigung von Umweltbelastungen bei der Bewertung von Brücken. Birkhäuser Verlag, Basel
2. Breitenbücher, R. et al. (1996) Umweltgerechter Rückbau und Wiederverwendung mineralischer Baustoffe. DAfStb, Bulletin No. 462, Berlin
3. CUR. Demontabel bouwen (1990). CUR rapport 135, Gouda
4. Demontabele bouwsystemen in beton (1996). Rijksgebouwendienst, Den Haag

RECYCLING OF OFFSHORE CONCRETE PLATFORMS

K.HØYLAND and T.O.OLSEN
Dr.techn.Olav Olsen a.s, Oslo, Norway

Abstract
Decommissioning and recycling of offshore concrete platforms have not yet been performed, but it will be in the near future. There is a large focus on this, not only from concrete engineers. The special effects of difficult accessibility and gross dimensions are discussed. Also how the need for recycling can affect the design and choice of type of structure is considered, and how a robust concrete lifting vessel can be used for decommissioning other offshore structures.
Keywords: Concrete platforms, recycling.

1 Recycling in general

Recycling of materials and products have been accepted in many countries and many industries for many years.

The construction industry is project oriented, and has generally a shorter perspective than the production industry. However, the concept of recycling must be understood and respected in all phases of the process, including the design stage. Concrete structures are no exception.

Norway is in a somewhat peculiar situation here, with its huge supplies of aggregates and also space available for disposal without reuse. But considerable focus is now directed towards the recycling of concrete structures by the industry, research institutes and construction owners.

2 Recycling - decommissioning

All structures have functions; the function of the offshore platform is to support a hydrocarbon production process. Recycling of offshore concrete platforms is therefore only a part of a much larger topic, namely the decommissioning of the platform.

Concrete Technology for a Sustainable Development in the 21st Century edited by O.E. Gjørv and K. Sakai.
Published in 2000 by E & FN Spon, 11 New Fetter Lane, London EC4P 4EE, UK. ISBN: 0 419 25060 3

Decommissioning of offshore platforms is a complex of political, economical and environmental considerations, as well as the technical ones.

In general there are several ways of decommissioning an offshore platform:

- leave in place
- demolish in place
- lift off and dump at a deep water site
- lift off and tow inshore for recycling

Combinations are also possible.

In any case considerable cleaning is involved. The production at and in the platform may have left unwanted materials (such as oil).

Theoretically the platform may be reused at a different location.

The high cost of removing a platform may make it interesting to leave it in place. As the offshore concrete platforms are durable [1,2], the life of the structure may be considerably longer than the production life of the offshore oil or gas field. If left in place provisions for safety of ship traffic and the fishing industry is required, and maintenance of the structure must be performed.

3 Offshore concrete platforms - background

Fig. 1 shows the largest gravity based offshore concrete platform ever built, the Troll Phase I Condeep platform. It is one of many offshore concrete platforms in the hostile North Sea. It is located west of Bergen in 303 m waterdepth. When the deck was mounted, the concrete structure was lowered in a Norwegian fjord, such that the maximum hydrostatic pressure was 325m. Not many structures are designed for such loads, and the shape of the structure, the high quality and the high strength concrete were prime tools to make it possible.

Reaching for these types of achievements is one of the major reasons for Norway being so actively engaged in the research and development of high performance concrete over the later years. This R&D focus are now also directed towards recycling.

During tow-out the structure displaced more than 1 mill.m^3 of water, making it the heaviest object moved by man ever. It is, including the deck structure, 472m tall, built of 245 000 m^3 of high strength concrete with 100 000 t of ordinary and 11 000 t of prestressed reinforcement.

Due to soft soil conditions the gravity based Troll structure is equipped with long, cylindrical skirts. The purpose is partly to enclose and confine the soil, and partly to reach deeper and stronger layers. The length of the skirts are 36m, and the total area of the base is more than 16.000 square meters. This is required to resist the environmental loading of the North Sea.

The design life of the Troll Phase I concrete structure is 70 years. Its operational life is planned to be 50 years.

A structure like Troll Phase I has a function of a vessel in some stages, and a function of a support structure subjected to extreme loadings in other phases of its life.

Detailed information about the Troll structure is found in [3], some general information on marine concrete structures, including references, are given in [4].

Important here is that the platform is designed to be removed. The removal process is the reverse of the installation process.

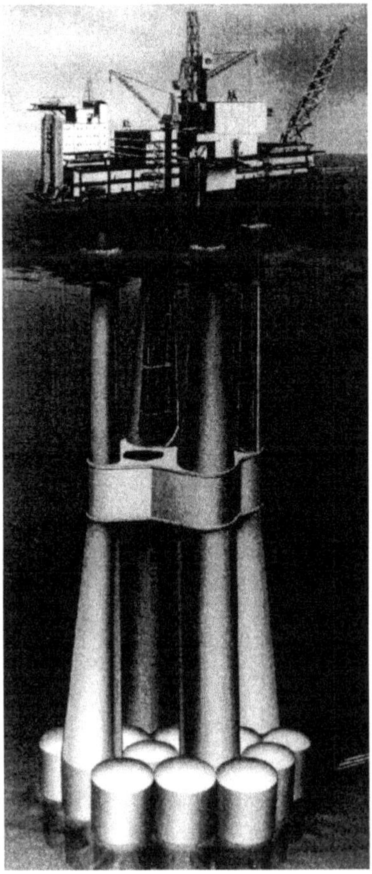

Fig. 1. The Troll Phase I Condeep Platform.

Fig. 2 shows the Heidrun TLP, a tension leg platform. This is a floating platform connected to the seabed by tethers. Consequently it may be removed relatively simply. The suction anchors may possibly be allowed left in place. Fig. 3 shows an example of suction anchors for the Snorre field.

Including the full service life of the structure, also decommissioning and reuse, makes structures that are easy to remove from its operational location favourable. This line of thought is used not only for offshore structures, but also for bridges, as a Submerged Floating Tunnel (SFT) in Fig. 4 may be considered favourable as compared to a traditional bridge. Fig. 5 shows a semi submersible monohull, named Semo, recently developed by Dr.techn.Olav Olsen a.s for offshore hydrocarbon production. Again the ease of removal and reuse at other locations or other purposes may be important.

Fig. 2. The Heidrun TLP Platform.

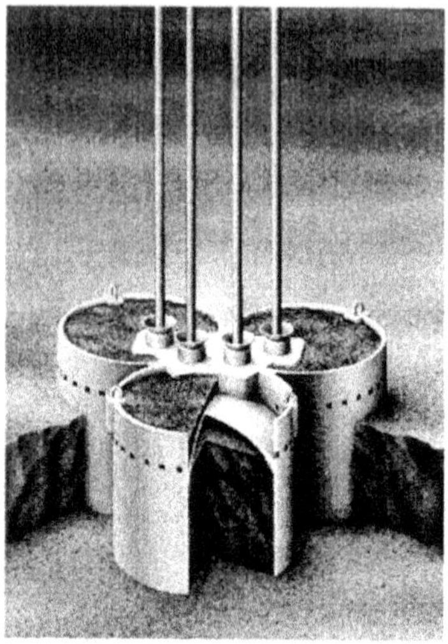

Fig. 3. Suction anchor

Fig. 4. One of the concepts for the Høgsfjord SFT crossing

Fig. 5. Semo-Semi submersed monohull

4 Recycling of concrete structures in general

It is possible to recycle concrete structures. Modern mechanical tools can handle cutting of relatively large chunks of reinforced concrete pieces. Explosives primarily developed for military use are made available to the civil market. It is today possible to design demolitions such that the reinforcement is physically separated from the concrete debris, for easy removal and transportation to the steel plant. The concrete debris may be used as fill material as is, or in new concrete.

5 Decommissioning of offshore concrete structures

The offshore structure requires some special consideration. Firstly it is very heavily reinforced, both with ordinary and prestressed reinforcement. It is believed that also these sections can be handled, but this is not done in practice in a large scale.

The location and accessibility of the structure needs planning. The hostility at the offshore location makes operations very expensive and difficult to plan with respect to schedule.

Additionally most of the structure is below the water. Cutting tools are available, but again cost and ability to handle large pieces may be prohibitive.

An obvious method of decommissioning is to refloat the platform and tow it inshore. The deck can then be part of the platform, and sheltered inshore operations are easily handled.

This is the reverse of the platform installation process, and therefore proven feasible in many respects. However, the structure is not the same any more after many years of operation. The deck structure may be altered, debris may be located on the seafloor around the platform, the mechanical equipment for ballasting may not be in good shape. Very important are the holes that are drilled through the platforms. In many cases these must be closed, after removing protruding pipes, to assure sufficient buoyancy. Some platforms are installed in clay, and it is important to be able to pull the platform out of the seabed. Piping is provided for under base pressure build-up.

Since 1978, there has been a requirement on the Norwegian continental shelf that offshore concrete platforms shall be designed in such a way that they can be re-floated. For all structures installed after Statfjord A, the theory exists and the platforms should have been equipped with required outfitting and systems to be able to remove the concrete substructures - though this has yet to be demonstrated.

There are concerns among several parties about the ability and feasibility to carry out this operation in a safe and controlled manner, and reliable cost estimates for the refloating operation hardly exist for the older concrete platforms.

This is the background for a joint industry project now being performed in Norway.

The main objectives of the first phase of the study are to:

- improve the basis for decisions with respect to final disposal of offshore concrete substructures and possible inshore decommissioning of topside structures
- present a summary of results from previous work on the subject of refloating
- perform an in-depth investigation of the challenges related to refloating, identify possible expected/unexpected problems and present practical methods/solutions where the safety aspects are duly considered
- give a technical documentation of the possibilities for refloating typical older concrete platforms on the Norwegian continental shelf

Aspects related to eventual initial removal of topside and possible subsequent consequences for the remaining substructure have been dealt with. This comprises evaluation of structural integrity of the substructure and geotechnical stability. Possible impact on eventual later refloating is also discussed.

A survey has been made on relevant information on refloating of concrete GBS's from previous studies, reports, papers, articles etc.

In addition to general evaluations applicable to most fixed concrete structures the Gullfaks C gravity based structure, which is almost as large as the Troll Phase I platform shown in Fig. 1, has been selected as a model for developing specific refloating scenarios and exemplifying technical and operational challenges to be handled.

The first phase of the project is completed, and the project continues in a phase 2. This phase will focus on recycling of structures/structural parts/materials, demolishing techniques, risk and safety assessments and environmental impact from defined operations. A model evaluating energy consumption/cost for retrieval of concrete offshore structures will be developed with emphasis on objective accounts for environmental costs.

6 The use of concrete structures to decommission other structures

Realising the need for recycling offshore structures, it is only natural to consider that concrete structures possess potential in terms of robustness and price.

The heavy lift Lettkon barge [5] is shown in Fig. 6. The purpose of the barge is to lift offshore installations at the offshore site. This can be either in an installation situation, or in a field closing situation. Here the vessel, rather than laying on the side of the installation, lifts straight up and down utilising ballast water. The barge is developed by Dr.techn.Olav Olsen a.s in the Norwegian joint industry project Lettkon, financially supported by the Norwegian Research Council and a large number of Norwegian companies in the concrete industry.

Fig. 6. The Lettkon Liftbarge

7 Conclusions

Reinforced concrete structures are recycled. For offshore concrete platforms the high reinforcement density requires special attention. So does the removal process. However, if carefully planned, it is quite feasible.

8 References

1. Fjeld, Hall, Hoff, Michel, Robberstad, Vegge, Warland (1994) The North Sea concrete platforms - 20 years of experience. OTC, Houston
2. Fjeld, Røland (1982) In - service experience with eleven offshore concrete structures, OTC, Houston
3. Skogseid, H., Bidne, K. (1995) Challenges in Design of the World's tallest Offshore Concrete Platform - Troll Gravity Base Structure. Concrete 95, Towards Better Concrete Structures, Brisbane, Australia
4. Olsen, T.O. (1998) Marine Concrete Structures. Consec `98, Tromsø
5. Olsen, T.O., Mathisen, L.L. (1998) Development of a special purpose lifting vessel. XIII FIP Congress 98, Amsterdam

USE OF RECYCLED AGGREGATE IN SHOTCRETE

C. CHAN, N. BANTHIA
Department of Civil Engineering, The University of British Columbia,
Vancouver, Canada
and
K. SAKAI
Department of Safety Systems Construction Engineering, Kagawa University,
Takamatsu, Kagawa, Japan

1 Abstract

As societies progressively become more environmentally conscious, many practices that pose a potential threat to our environment become matters of serious consideration. Handling of waste concrete is no exception. Concrete consumption has risen in most countries of the world with a growth in population and urbanization; and, naturally, along with an increased consumption, there is also an increase in the amount of wastes concrete societies have to deal with.

The paper discusses the potential application of recycled aggregate from waste concrete in wet and dry-process shotcrete, and argues that in many instances the use of recycled aggregate in shotcrete is a more viable proposition than in cast concrete. The paper also presents data to demonstrate the feasibility of using recycled aggregate in shotcrete and identifies areas where continued research will benefit.
Keywords: Environment, rebound, recycled aggregate, shotcrete, waste management.

2 Introduction

Waste concrete can be produced from a number of different sources. The most common one is the demolition projects. Many concrete structures like buildings, bridges, sidewalks and roads are razed after a period of time into their service life for purposes of replacement or landscape changes. Other sources of wastes include: natural disasters like earthquakes, avalanches and tornadoes; human-made causes like war and bombings; and structural failures. All these contribute to vast quantities of waste concrete which must be managed in some way. Clearly, the actual amount produced will vary from place to place; but, the following will provide a feel for the quantities [1] [2] [3] [4]:

Concrete Technology for a Sustainable Development in the 21st Century edited by O.E. Gjørv and K. Sakai.
Published in 2000 by E & FN Spon, 11 New Fetter Lane, London EC4P 4EE, UK. ISBN: 0 419 25060 3

- In Japan, about 25 million tons were produced in 1990. This amount rose to 71 million tons by 1995 and is expected to be as high as 110 million tons per year by 2001.
- In Western Europe, about 0.7-1 ton per inhabitant is produced yearly.
- In the United States, the amount of waste concrete produced is roughly 40-50 million tons per year.

Traditionally, waste concrete has been managed mainly through their use in landfilling and recycling. The proportion of the total waste stream that is recycled also varies from place to place but lies in the range of 0-35% [1]. With the large number of other wastes competing for this stream and an assured increase of it in the future, landfilling has become a major problem, particularly in countries where land is scarce. Besides scarcity of land, other problems facing the landfill option include their siting, transportation costs, tipping fees, and public opposition. Thus, recycling has been gaining wider attention as a viable option for the handling of waste concrete.

The recycling of waste concrete, apart from reducing the volume of waste to be landfilled, also offers other benefits. One of the main environmental benefits is the conservation of natural resources, especially in regions where aggregates are scarce. If the recycling plants are centrally located and easily accessible, transport costs can also be minimized. Moreover, there is indication of unhydrated cement particles in the old concrete which can be triggered to further reacting. If attainable, it could lead to reductions in the use of cement whose manufacture bears large environmental hazards due to the emission of COx and NOx gases and dust. It is, therefore, clear that efforts must be placed by the research community and by the industry to increase the rates of waste concrete recycling as there are serious implications from an environmental standpoint that should be dealt with. This is particularly true in industrialized countries where waste volumes are high and resource availability is low. Examples of these are: Germany, the Netherlands, Belgium, and Japan, where the rates of recycling have reached up to 50% [5].

There are several incentives for the consumer to use recycled aggregate in concrete. These include:

- Cost: The cost of recycled aggregate compares quite favorably with virgin aggregates in most countries. In Canada, recycled aggregate costs $8 per ton as opposed to consumers paying up to $15 per ton for virgin aggregate.
- Municipal Approval: Most municipalities allow the use of recycled aggregate if it meets certain criteria.
- Waste Management: With the general need to reduce and manage waste, use of recycled aggregate in concrete is one of the most effective ways.
- Lower Densities: Lower densities of recycled aggregates generally mean lower concrete densities and higher yields.

3 The current practice

Presently, the major markets for recycled aggregates are in [1]:

1. General bulk fill
2. Base or fill in drainage projects
3. Sub-base or surface material in road construction
4. New concrete manufacture

In particular, recycled aggregates have seen their greatest use in the construction of pavements. This is because crushed concrete has a number of properties which suit those needed for such an application and the process is easy and economical. For instance, the pavement concrete to be crushed is usually in the exact place where the new pavement will be constructed. This saves on the quarrying and transport costs. Besides, concrete from pavements tends to be more or less consistent in quality and free from contaminants especially other building materials. Moreover, failure of a concrete pavement does not have the same potential consequence including loss of life as does the failure of concrete in a structure. Crushed concrete pavements have been used as sub-base or base course material, in bituminous mixes, as filter material and as low quality stabilization/filler material in the construction of new pavements.

There are also other minor applications for recycled aggregates such as fill for breakwaters and levees, rip rap, ballast, roofing granules, neutralizing beds, filtration beds, thermal reservoirs, sound barriers, masonry and cat litter boxes.

4 Properties of recycled aggregates

Depending on the purpose in mind, tests are generally carried out using existing standards (such as ASTM C33) to determine the suitability of a given recycled aggregate for use in new concrete. The properties so determined, can have an effect on the production of concrete mixes, but adjustments can often be made to achieve desirable properties. As well anticipated, recycled aggregates have a number of properties different than those of virgin aggregates. These include:

- A coarser and more angular shape as a result of the crushing operations; a varying particle size distribution depending on the crusher settings; and a lower density, which has sometimes led to their use as lightweight aggregates.
- Attached mortar and cement paste on the surface of the original aggregate. The extent of these depends on the degree of crushing and the properties of the original concrete. This attached mortar gives rise to other properties of the aggregates including a lighter weight, much higher water absorption, poorer bonding capacity and a lower abrasion resistance.
- The presence of foreign substances, known as contaminants [6]. These are present in recycled aggregates from the original demolition debris which gets passed on to them. They may encompass clay balls, bitumen chunks, gypsum from drywalls, bricks, organic material like wood, glass, steel pieces and other

metals. Contaminants generally pose adverse effects on the mechanical properties and the durability of concrete mixed with recycled aggregates. Therefore, much work has been devoted to find ways of reducing the contents of these in the stream of waste concrete, to mitigate their effects, and to find uses where these deleterious effects are acceptable.

5 Properties of concrete with recycled aggregates

Based on the foregoing, it is well expected that the mechanical properties of concretes made with recycled aggregates will be different than those made with virgin aggregates. Compressive strength values have been reported to be in the range of 0-40% lower than those of comparable concretes containing virgin aggregates [1] [7] [8]. For recycled aggregates, other properties of interest like the tensile strength, flexural strength, shear strength and modulus of elasticity are also inferior, and creep and shrinkage tend to be higher. The magnitude of the difference for each of the properties depends, among other things, on factors such as the fraction of total aggregates that is recycled, the characteristics of the original concrete, the nature and level of contaminants present, the amount of fines and the quantity of attached mortar, etc. Investigations have aimed at determining the "optimum" combination of these factors that will economically produce a recycled aggregate concrete that is suitable for a given application.

The durability of concretes containing recycled aggregates is also affected by the factors mentioned above. However, the most significant factor is the presence of contaminants. Many of these induce harmful reactions in the concrete that can shorten its life. For example, gypsum from drywalls, if at significant levels, causes expansive reactions which promote cracking in concrete. Another important factor affecting durability is the extent to which the original concrete has undergone harmful reactions and to what potential these have for continuing in the new concrete. Furthermore, the permeability and water absorption of recycled aggregate concrete are increased due to the attached mortar; so, durability may also be adversely affected.

6 The use of recycled aggregates in shotcrete

Extensive research is currently underway at the University of British Columbia to investigate the use of recycled aggregates in shotcrete—an entirely new application for recycled aggregates. Detailed results and analysis of the data may be found elsewhere [9]; only a brief summary is presented here. The premise of the study is simple: shotcrete is a pneumatically projected and compacted concrete with very different compaction dynamics from conventional concrete. The mode of interaction between the mix water, the cementitious part and the inert (filler) part of the mix are different, and a different rheology and spatial distribution of these components within the mix imply that the kinetics of strength gain and paste-aggregate interface development in shotcrete are decidedly different from cast concrete. It is entirely conceivable therefore that the use of recycled aggregates in shotcrete may produce and depict trends which are different from cast concrete. In fact, the use of recycled aggregate in shotcrete may prove to be potentially more beneficial than in cast concrete. This, however, remained to be demonstrated.

Both dry and wet processes of producing shotcrete were investigated (Fig. 1). The recycled aggregate used in the project was obtained from a local supplier that processes municipal concrete waste. At the mill itself, large pieces of waste concrete are sent through a series of jaw crushers, and the resulting finer products are separated and piled in various size-ranges. At the university, the recycled aggregates were further processed. The processing included: 1) dry-mixing in a pan type mixer for 25 min to remove the loose material from the surface, and 2) adjustment of gradation curves after dry-mixing through selective blending to match the ACI shotcrete gradation requirements. Panels were then shot using virgin and recycled aggregates, and identical panels were *cast* using the same mix as employed for the wet-mix shotcrete. Use of fibers was investigated at a volume fraction of 1% for selected mixes. Cores (50 mm diameter) were retrieved from the hardened panels for an assessment of compressive stress-strain response and splitting strengths as a function of age. For mixes containing fibers, beams (350 mm x 100 mm x 100 mm) were retrieved for flexural toughness measurements. The rebound data are given in Table 1.

The compressive strengths of the six mixes (Table 1) as a function of age are shown in Figure 2. The reduction in strength due to the use of recycled aggregate is clearly visible. It also appears that recycled aggregates are generally more effective in wet-mix shotcrete than in dry-mix shotcrete; and, in the case of recycled aggregates, cast concrete develops reasonably the same strengths as the wet-mix shotcrete. At later ages, however, with virgin aggregates, cast concrete develops

Fig. 1. Use of recycled aggregates in the wet-process shotcrete.

substantially higher strengths than both wet and dry-mix shotcretes. In Fig. 3, splitting tensile strengths are plotted as a function of age. Notice that, as in the case of the compressive strength, mixes with recycled aggregates developed lower splitting strengths than those with virgin aggregates, but the differences in the three processes (wet-mix, dry-mix and cast concrete) were not as significant.

Table 1. Rebound data

Process	Type of Aggregate	Mix Proportions*	Aggregate Rebound (%)	Fiber Rebound (%)**
Dry-Process	Virgin(DV)	1.0:0.42:3.01:1.26	25	-
Shotcrete	Recycled (DR)	1.0: -- :3.01:1.26	21	-
Wet-Process	Virgin (WV)	1.0:0.32:1.93:1.85	18	32
Shotcrete	Recycled (WR)	1.0:0.32:2.08:1.70	15	25
Cast	Virgin (CV)	1.0:0.32:2.12:1.67	-	-
Concrete	Recycled (CR)	1.0:0.32:2.12:1.67	-	-

*Cement : water : sand : coarse aggregate

**Hooked end fiber at design mix volume fraction of 1%

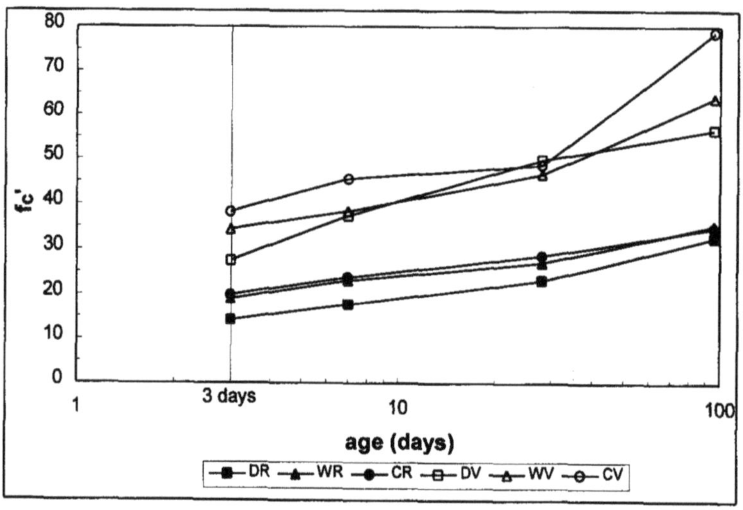

Fig. 2. Compressive strength vs. age relationship for different mixes.

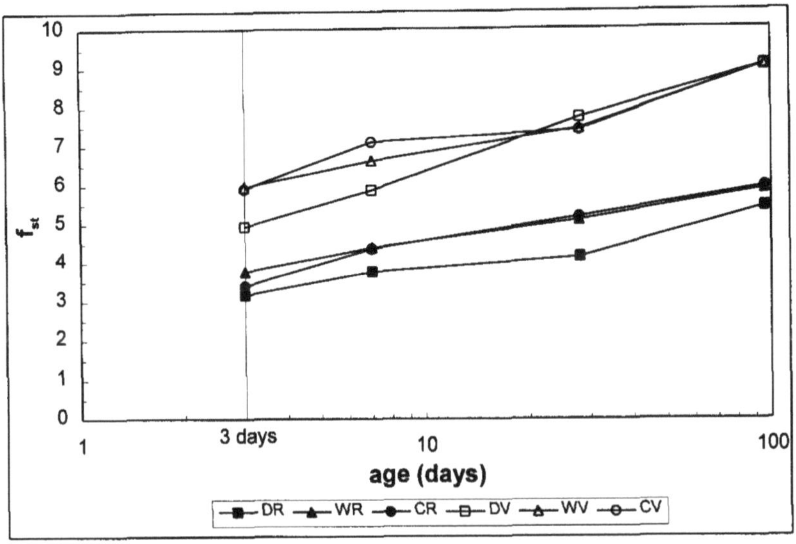

Fig. 3. Splitting tensile strength vs. age relationship for different mixes.

In Figs. 4, 5 and 6, stress-strain diagrams in compression are plotted, respectively, for the dry-process shotcrete, wet-process shotcrete, and cast concrete comparing virgin and recycled aggregates at an age of 28 days. The general increase in the strength (peak value of stress) due to the use of the virgin aggregate is clear and was previously mentioned. What is more interesting to note, however, is the increased strain at the peak stress in the mixes carrying recycled aggregates. Also evident is the catastrophic and sudden drop of load in the case of virgin aggregates as opposed to the slow and more stable drop of load in the case of recycled aggregates. The observations were quantified by calculating the failure strain at peak stress (Fig. 7) and the energy ratios *(see inset for definition)* at specified strains of 0.01, 0.02 and 0.03 (Fig. 8). Note that, except at early ages, mixes with recycled aggregates supported much higher peak strains, and the response in the post-peak part of the curve at larger strains of 0.01, 0.02 and 0.03 was also significantly tougher. It appears that the development of an inferior bond in mixes with recycled aggregate allowed the material to support higher strains and also absorb greater post-peak energies.

The foregoing observation may have interesting implications in the case of shotcrete, where deformability and toughness may be, in some instances, more important than the strength itself. In many applications, especially involving soft ground conditions, deformability of the shotcrete material is considered more important than its ability to carry higher stresses.

In Fig. 9, the load-displacement behavior of fiber reinforced mixes is shown. The strength and JCI flexural toughness factors (F_b) are also given in Fig. 9. Noting the steeper drop in load in the case of shotcrete with recycled aggregate (and the resulting lower flexural toughness factors, F_b), one can infer that fiber reinforcement is not as effective in mixes carrying recycled aggregates. In spite of the lower toughness values, the material may still be considered reasonably tough and may find uses in the underground and above-ground shotcrete applications.

The inferior toughness of fiber reinforced shotcrete with recycled aggregate is somewhat enigmatic. Generally, it is believed that the toughness of fiber reinforced concrete and shotcrete is governed by the bond-slip mechanisms between fibers and the surrounding mortar matrix, and aggregates play only a relatively minor role, if any. It is clear from the data given here that this belief is questionable, and that an excessively weak bond between the aggregate and the surrounding cement-paste may also adversely affect the fiber-matrix bond-slip performance and the overall composite toughness.

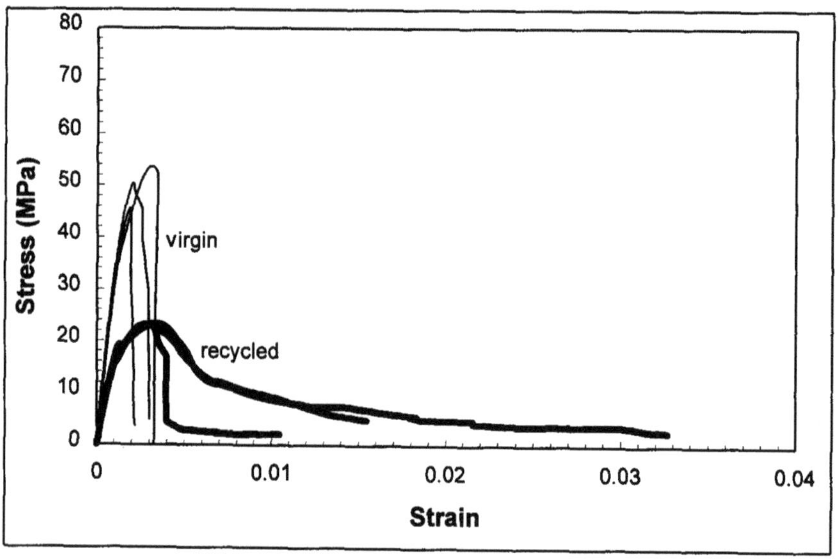

Fig. 4. Stress-strain diagram for recycled and virgin aggregates in dry-mix shotcrete.

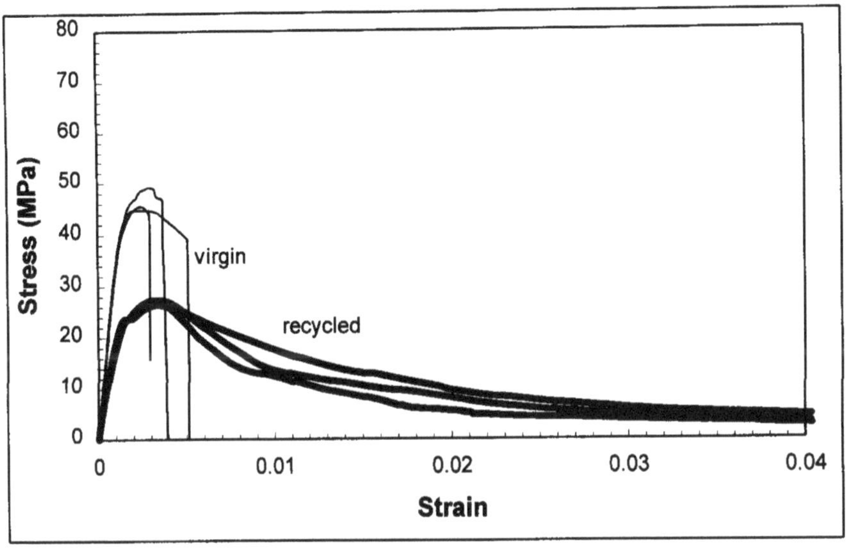

Fig. 5. Stress-strain diagram for recycled and virgin aggregates in wet-mix shotcrete.

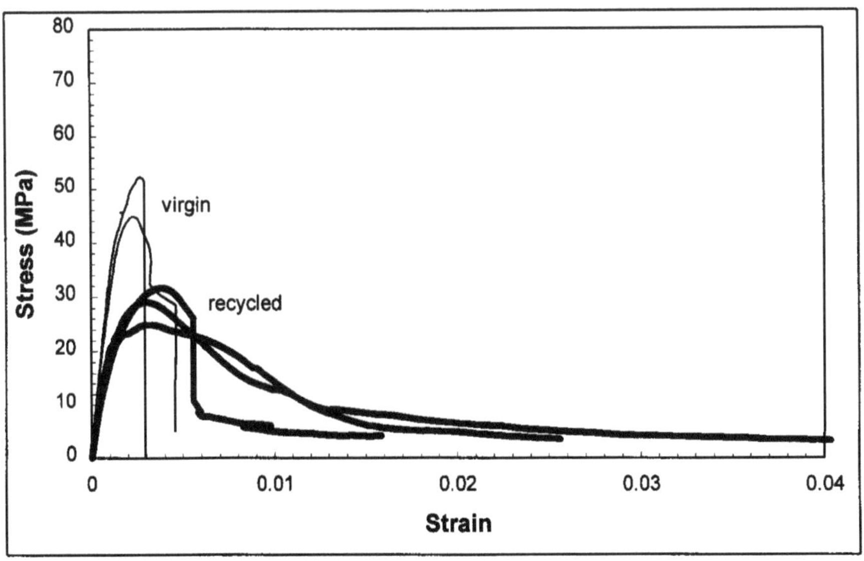

Fig. 6. Stress-strain diagram for recycled and virgin aggregates in cast concrete.

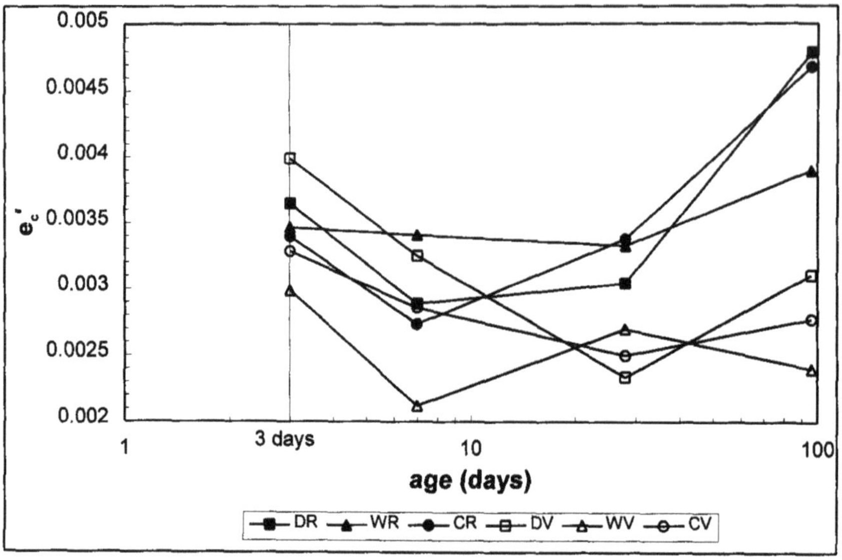

Fig. 7. Failure strain at peak stress vs. age.

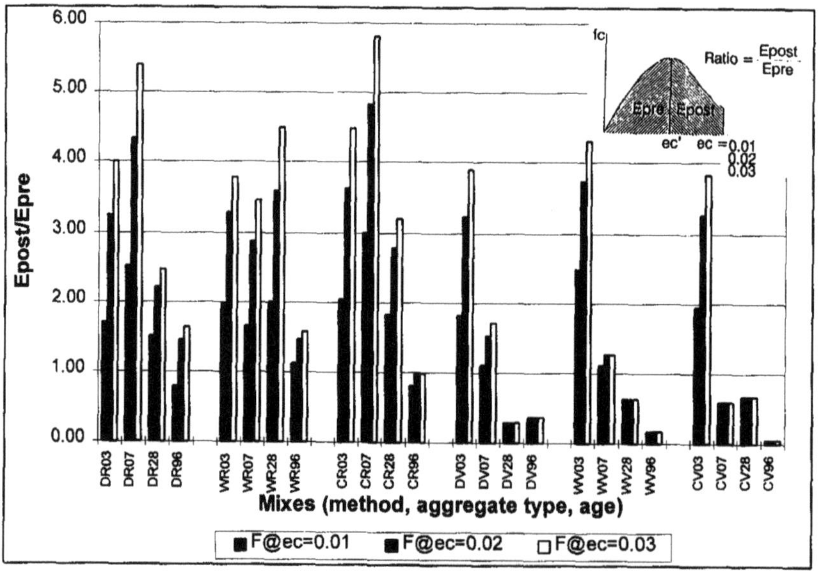

Fig. 8. Post-peak energy ratios at specified strain values.

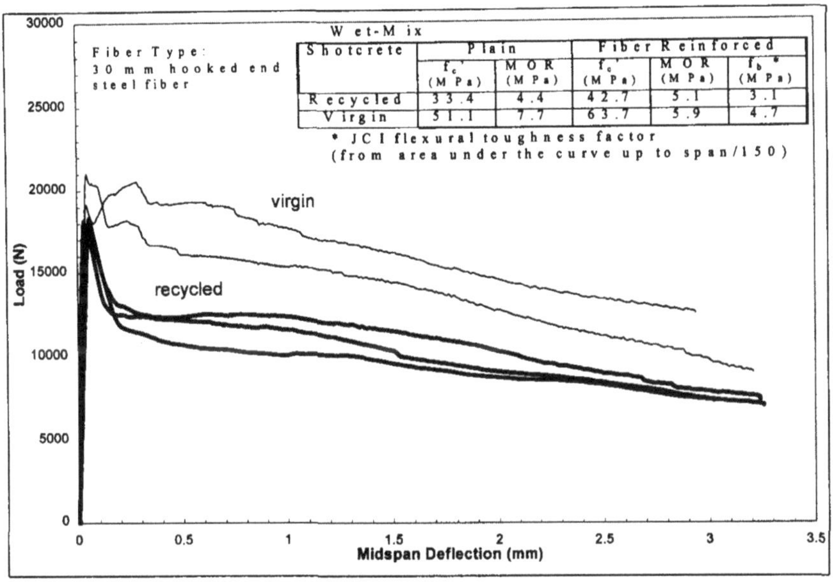

Fig. 9. Load-deflection diagrams for fiber-reinforced beams.

7 Concluding remarks

The acceptance of recycled aggregates for production of new concrete will depend on whether or not they satisfy certain specifications and criteria. While these may be different from locality to locality and from application to application, the paper identifies some such requirements and issues and argues that the use of recycled aggregate in concrete may be the only alternative from a sustainability point of view. Some countries have made considerable progress in the use of recycled aggregates back in concrete, and with time more and more will be required to head in this direction.

The paper demonstrates that the use of recycled aggregates in shotcrete is a viable proposition so long as some trade-offs are accepted. The only question is whether or not the economic market and the political environment of a place will allow for such technology to be implemented.

8 References

1. Hansen, T.C. (1992) *Recycling of Demolished Concrete and Masonry*, RILEM Report of Technical Committee 37-DRC, E&FN Spon, London.
2. Kibert, C.J. (1994) Concrete/masonry recycling progress in the USA, in *Proceedings, Third International RILEM Symposium on the Demolition and Reuse of Concrete and Masonry*, (ed. E.K. Lauritzen), E&FN Spon, London, pp. 83-92.

3. Simons, B.P. and Henderieckx, F. (1994) Guidelines for demolition with respect to the reuse of building materials: guidelines and experiences in Belgium, in *Proceedings, Third International RILEM Symposium on the Demolition and Reuse of Concrete and Masonry*, (ed. E.K. Lauritzen), E&FN Spon, London, pp. 25-34.
4. Kasai, Y. (1994) Guidelines and the present state of the reuse of demolished concrete in Japan, in *Proceedings, Third International RILEM Symposium on the Demolition and Reuse of Concrete and Masonry*, (ed. E.K. Lauritzen), E&FN Spon, London, pp. 93-104.
5. Schulz, R.R. (1994) The processing of building rubble as concrete aggregate in Germany, in *Proceedings, Third International RILEM Symposium on the Demolition and Reuse of Concrete and Masonry*, (ed. E.K. Lauritzen), E&FN Spon, London, pp. 105-116.
6. Young, J.F. (1981) Contamination Problems in the Recycling of Concrete, in *Adhesion Problems in the Recycling of Concrete*, (ed. P.C. Krejiger), Plenum Press, New York, pp.91.
7. Tavakoli, M. and Soroushan, P. (1996) Strengths of Recycled Aggregate Concrete Made Using Field-Demolished Concrete as Aggregate. *Journal of the American Concrete Institute*, Vol. 93, No. 2. pp. 182-190.
8. Topçu, I.B. and Güncan, N.F. (1995) Using Waste Concrete as Aggregate. *Cement and Concrete Research*, Vol. 25, No. 7. pp. 1385-1390.
9. Chan, C. (1998) Characterization and Use of Recycled Aggregates in Shotcrete, Master of Applied Science thesis (in preparation), The University of British Columbia, Vancouver.

MANAGEMENT OF CONCRETE DEMOLITION WASTE

M. TORRING
Norwegian University of Science and Technology,
Department of Geology and Mineral Resources Engineering, Trondheim, Norway

Abstract
Recycling of concrete demolition waste can provide opportunities for saving resources, energy, time, and money. Furthermore, recycling and controlled management of concrete demolition waste will save use of land and create better opportunities for handling of other kinds of waste.

There are a number of opportunities for utilization of concrete demolition waste apart from dumping. Recycling of concrete can be accomplished by: reuse of concrete products, processing into secondary raw materials for use as fill, road bases and sub bases, or aggregate for production of new concrete.

The prospect of using concrete demolition waste depends on a range of factors related to the building and construction industry, and to the consumption of resources and energy, where the main three factors are: population density, occurrence of and access to natural materials, and level of industrialization.

To optimize the use of natural resources and concrete demolition waste, to fulfill the requirements for materials for construction, and an appropriate operation of recycling plants, there is a need for long term action plans concerning use of materials and co-ordination between different interests among authorities and companies within the building and construction industry.

In the present paper the major constituents of an economic model for management of concrete demolition waste. The objective with the model is to assist decision makers to achieve the goals of economic efficiency and environmental protection simultaneously.

An analysis of prospects of successful recycling of concrete waste in various parts of the world are presented, and aspects of the implementation of waste management plans are discussed.
Keywords: Concrete, cost-benefit, demolition, recycling, waste management.

Concrete Technology for a Sustainable Development in the 21st Century edited by O.E. Gjørv and K. Sakai.
Published in 2000 by E & FN Spon, 11 New Fetter Lane, London EC4P 4EE, UK. ISBN: 0 419 25060 3

1 Introduction

In the European Union with a population of approximately 350 million, it is estimated that the annual generation of construction and demolition waste is between 175 and 370 million tons or equal to 0,5 to 1 ton per capita per year [1]. This makes up between 25 to 50% of all municipal waste. The figures are probably representative for most of the industrialized world. A global construction and demolition waste production of 2 billion tons per year may be a best guess. The major part of these quantities consist of concrete demolition waste or other rubble such as masonry.

In general, concrete demolition waste stripped for metals is considered chemically inert from an environmental point of view, for which reason there is normally no limitation for use when regarding the pollution issue. However, concrete waste consists of huge masses of materials, which are often inviting illegal deposit of other kinds of waste. Reinforced concrete may also be regarded as slightly harmful due to the risk of reinforcement bar corrosion.

Utilization of concrete demolition waste by recycling can provide opportunities for saving resources, energy, time, and money. Furthermore, recycling and controlled management of concrete demolition waste will save use of land and create better opportunities for handling other kinds of waste. Another great challenge is to reduce the impact of disasters, natural-, technical- or human made, especially wars. It is evident that there is a great need for management of concrete demolition waste after such events. In Fig.1, a model of total costs for traditional and selective demolition is presented. The top figure shows traditional construction and demolition where all natural resources are new and all demolition wastes are dumped. The figure below shows construction, selective demolition and recycling, where a part of the natural resources is substituted by recycled materials. This option often saves costs of transport, supply of natural materials and disposal of demolition waste.

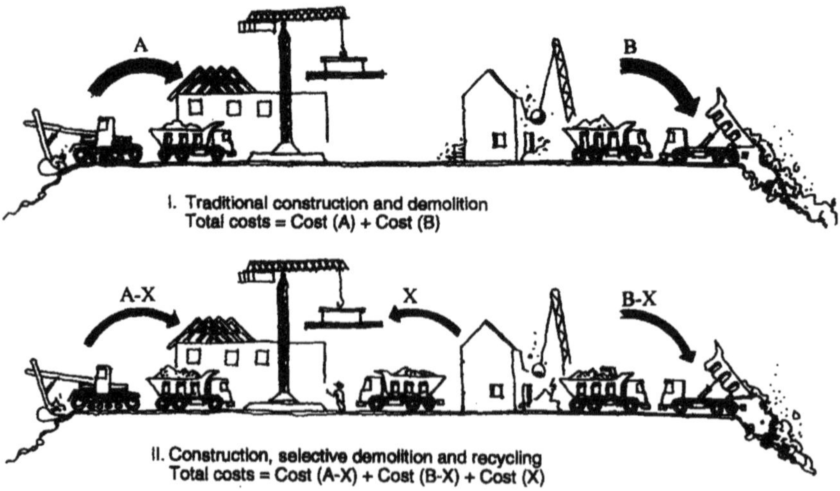

Fig 1. Model of total costs for traditional and selective demolition.

At present, the major part of the concrete demolition waste in the world is dumped. The future use of recycled materials in construction is, however, expected to increase due to the following reasons:

- A general development in the public opinion regarding environmental issues leading to a political pressure in the direction of minimizing the generation and transport of waste, and regulations for waste depositing that will aim at making the recycling option more competitive.
- The rapid depletion of the remaining natural aggregate resources (primarily sand/gravel), which will lead to a lack of such materials in many regions, followed by higher prices, a need for public regulations and for replacements like crushed and/or recycled aggregates. Moreover, new sites for production will be located further and further away from urban areas.
- The technical development of the production and use of recycled materials will lead to more cost efficient demolition methods and recycling plants, a better control of the quality, and more knowledge on application technology.

2 Technical properties

Recycling of concrete can be accomplished by:

- Reuse of intact concrete elements, (beams, columns and panels).
- Crushing of concrete into secondary raw materials for use as:
  - Fill
  - Unbound road bases and sub bases
  - Bound road bases
  - Aggregate for production of new concrete

2.1 Reuse of intact concrete elements

It is possible to move complete prefabricated concrete structures, or reuse parts from a demolished structure in a new structure. However products often become damaged during separation, and remaining bearing capacity and other technical properties can be difficult and expensive to determine. The design of most existing concrete structures is made with the aim of being crushed after use. The main resources in terms of energy for the concrete are bound in the cement. Even though concrete demolition waste can be recycled as crushed aggregate, design solutions should preferably be made where concrete elements to a larger extent can be reused.

2.2 Processing of concrete waste

The techniques for crushing and screening is similar to that for extraction of virgin aggregates. Jaw-, impact-, or cone crushers, in single or multiple steps modified with an overband magnet are used as mobile or fixed plants. Air or water separation units can be used to remove lightweight materials, such as wood, plastic etc.

2.3 International technical standards and recommendations

Valuable work has been done during the last decade within the RILEM network, leading to several state-of-the-art reports and congress proceedings [2], and finally to technical recommendations given by the RILEM Committee 121 [3]. These documents have served as a basis not only for some national standards, but also for the emerging European standards to cover this item. A separate working group has been set up within the European Standardization Committee (CEN) to develop recommendations with regard to the use of recycled aggregate.

2.4 General experience with recycled aggregates from concrete

Some main points characterizing the properties of recycled aggregates from concrete, in comparison with ordinary aggregates are [2]:

- high porosity due to high mortar/cement paste content.
- high fines content upon crushing.
- crushed concrete will have a bulk density somewhere in-between rock materials and light weight aggregate.
- risk for contamination of organic compounds, heavy metals and other environmental hazardously substances, for example from traffic and chemical industry.
- quality of recycled materials can be hard to control if the aggregate production is to be based on a general reception of urban building waste from a variety of sources, which is the case for some major commercial plants.
- adequate properties for use as a base course or fill material
- crushed concrete has good drainage properties compared to natural materials. On the other hand crushed concrete serves poorly as capillary break. Therefore, it should not be used as sub grade material.
- less resistance to impact and abrasion in road pavements
- water requirement and shrinkage in concrete is higher
- lower strength and E-modulus in concrete
- durability aspects should be controlled, (alkali, chlorides, sulfur etc.)
- adequate properties for part replacement as coarse aggregate for concrete in normal strength classes.

2.5 Use of the concrete fines

The fine fraction (<approximately 4mm) which consists of 30-60% hardened cement paste, normally constitutes 15 to 25% of the crushed concrete. This fraction may be hard to find applications for. In new concrete, the recycling fines will normally cause excessive water demand, and at the present time, its potential use in concrete is a question dealt with on research level. However, there are some examples of qualitative use of the concrete fines:

- as a filler replacement/addition to asphalt. However, this is a question which needs some more research, and which would call for fairly advanced separation equipment in the production. But if successful, this area of use can offer a relatively high price level.
- as a base course or fill material in a broad grading (e.g. 0-32 mm) the crushed

materials may serve satisfactorily without a need of sorting. Although sold at fairly low prices, this alternative often turns out to be the most economic solution for two reasons; 1) the option of getting all size fractions into the market at the same time (no problem with fines deposits), and 2) the need for a simpler plant (no need for size sorting) and less complicated logistics / transport.

- as neutralizer of acid soils or lakes.
- as raw material for cement production.

2.6 Mixed products in road bases

Practical experience from Denmark [4] has shown that a mixture of 50% crushed asphalt and 50% crushed concrete provides an excellent unbound road base material. The lubricating properties of asphalt reduces the effort required to compact the material by rolling. The mixed granular material is stronger than compacted crushed concrete and it supports static loads much better than pure crushed asphalt, and has become so popular among contractors that the sale of this mixed material exceeds both the sale of pure crushed concrete and pure crushed asphalt. This in spite that no certification exist for this product. Another frequently used product is a mixture of crushed concrete and masonry, since these materials are often mixed at the site of demolition.

2.7 Recycling of the reinforcing steel

There is a well established down stream solution for steel remains, at price levels varying with market prices, somewhat depending on the degree of cement paste contamination.

3 Model for management of concrete demolition waste

The decision of how to treat concrete demolition waste normally has both political and financial aspects. Problems to be taken into consideration can be: whether a specific waste product should be sorted and recycled, where a permanent recycling plant for a specific region should be established, and at what capacity, or if materials shall be processed on-site or off-site for a specific project. Other problems to analyze can be the effects of different regulations and fees.

An economic model for management of concrete demolition waste can help decision makers to achieve economic efficiency and environmental benefits simultaneously. This decision is often referred to as «Best Available Technique Not Entailing Excessive Costs» (BATNEEC).

Such a model can be achieved by optimizing the total costs and benefits of a waste management system taking in consideration the following main items:

- Optimizing the use of natural resources (by substitution with recycled products)
- Minimizing energy- and water consumption (processing, dust suppression etc.)
- Minimizing production of residual products (load on landfill, pollution)
- Minimizing the transport load
- Minimizing the risks in working environment (cost to keep a certain standard)
- Minimizing impact from dust-, vibrations-, CO_2-, noise emissions
- Minimizing capital costs (investments in recycling plants etc.)
- Minimizing of operational expenditures (salaries and maintenance etc.)
- Maximizing the value of recycled products (material, energy)

Input to the model should preferably be a monetary value for as many of the items as possible. Public preferences are often expressed by taxes (landfill, transport, water and energy). Moreover, monetary values should be placed on pollution as well as dust-, vibration- and noise emissions. The impact from these items will vary widely from place to place, and the definition of monetary values will require considerable efforts. Many items can alternatively be represented by a non monetary value (level of impact), when different alternatives are evaluated, by using the model mainly to structure a problem.

During optimization, it is important to consider all levels involved in the process, from the micro economy at the building site, to the global economy, in order to obtain a fair distribution of responsibilities and costs. When the inputs have been stipulated, the cost-benefit should be analyzed at the following levels:

C - for a contractor (e.g. demolition, transport, disposal, value of material)
L - for a local level (e.g. pollution, road maintenance, costs for landfills)
R - for a region (e.g. pollution, use of resources and land)
N - nationally (e.g. road maintenance, fees or subsidies)
G - globally (e.g. CO_2-emissions, use of resources)

Whether or not the goal at a certain level is maximum profit, or cost minimization, a general economical goal to minimize the total cost (T) can be presented as follows:

T-cost = C-cost + L-cost +R-cost +N-cost +G-cost, T-cost --> min

During cost minimization, the items mentioned above are often in conflict. As an example, it can be mentioned that the best solution in a transport perspective may be to have many small recycling plants at different sites, though the most feasible alternative for an operator would be to have one big centrally located plant.

In order to establish goals for a recycling effort in a region, it is desirable to assess the existing situation. Qualitative and quantitative evaluations must be made of the production and streams of concrete wastes in all phases of the life cycle of buildings and structures.

Attempts must also be made to forecast the probable development within a foreseeable future exemplified with different scenarios. The quantities and streams of concrete waste are often difficult to determine. Large quantities of waste produced never appear in official waste treatment systems, and are therefore not registered.

The inventories should include assessment of existing treatment facilities for concrete waste and the existing market for virgin aggregates which are to be substituted. Studies should also be made of the possibilities to co-ordinate recycling of concrete demolition waste with waste products from other industries, for instance power plants and domestic refuse incineration plants.

The objective for many countries has so far been the recycling of a certain percentage of the total amount of waste. Instead targets should preferably be set for the best qualitative use of the materials. There are often discussions about avoiding «downcycling» of materials and to produce «high quality» products. «High quality» reuse of concrete waste does not always correspond to production of the product with the highest value, but rather the most feasible product in a specific project or region. It is important to consider that resources are also used to produce recycled products.

As an example, it can be mentioned that crushing of concrete into concrete aggregates will use more energy, than if the material was used as a coarse filling material, and water used to reduce dust from crushing will create a residual product. Moreover the concrete made from recycled aggregates often requires use of more cement which will give both higher energy consumption and costs. Concrete demolition waste will in many cases contain impurities such as soil, bricks, plastics etc. The most practical use may therefore be as road building materials, where small amount of «impurities» does not affect the properties. The quality of the original concrete will affect the properties of the crushed aggregates. Test for fines content after crushing and other properties should be undertaken before any efforts to recycle the material, in order to get the most feasible use.

It is important that the model used to analyze the options for handling of the waste is presented in a form, which all parties involved in, and depending on a decision, can easily understand and accept. A detailed mathematical model might not be useful if the decisions based on the result are difficult to implement.

4 Prospects of recycling of concrete demolition waste world wide

The three main factors affecting the prospects of recycling of concrete demolition waste are: population size and density, availability of virgin aggregates, and level of industrialization. In Table 1, the prospects of systematic recycling of concrete demolition waste in different parts of the world have been analyzed according to these factors rating from "+" (doubtful) to "++++" (profitable) [3]. This table is followed by three examples highlighting the analysis.

Table 1. Prospects for successful recycling of concrete demolition waste [3].

Area code	Population density	Natural raw materials and deposits	Level of industri-alisation	Example	Prospects of successful recycling
I	High	Adequate	High	Many cities in EU, USA and the Far East, (Hong Kong), Copenhagen, (Denmark)	+++
II	High	Adequate	Low	Many megacities in the Third World: Mexico City, Jakarta, (Indonesia)	+++
III	High	Scarce	High	Amsterdam/Rotterdam (The Netherlands)	++++
IV	High	Scarce	Low	Dacca, (Bangladesh) Shanghai, (China)	++++
V	Low	Adequate	High	Rural Scandinavia	++
VI	Low	Adequate	Low	Rain forests and mountain regions in developing countries	+
VII	Low	Scarce	High	Kuwait, Abadan, (Iran)	+++
VIII	Low	Scarce	Low	Steppelands, Sandy deserts or tundra's in developing countries	+

4.1 Amsterdam/Rotterdam, the Netherlands

The region is densely populated. Natural raw materials are scarce and the level of industrialization is high. It is not surprising that the highest possible value recycling of concrete demolition waste is practiced in the Netherlands. The idea of deliberately mixing wastes which can be separated does not appeal to the Dutch. Larger volumes of concrete demolition waste is produced than what is needed for road building purposes, and therefore, calculations have been made that 15-25% of the aggregate in new concrete should be concrete waste. Local authorities in Rotterdam have passed regulations demanding that new concrete shall contain recycled aggregates.

4.2 Copenhagen, Denmark

The city is densely populated and the level of industrialization is high. However, because there is an ample supply of natural raw materials, economy and common sense dictate lower value recycling than in Amsterdam. Although approximately 80% of the construction and demolition waste is recycled, this can only substitute approximately 20% of the need for aggregates. Production of pure fractions of concrete waste has proved uneconomical and the commercial market has shown preference for blended products with bricks and asphalt. This is so in spite of the fact that public authorities for environmental reasons have discouraged the use of mixed materials. Recycled aggregates in new concrete comprise less than 0,01%. The major part of the concrete demolition waste is used as road building materials. An early prototype of the waste management model previously has been used for construction and demolition waste management in Copenhagen (including suburbs 1 million inhabitants) [4]. For treatment of the rubble fraction (concrete, masonry, asphalt), the optimal solution was found to be a recycling plant with a capacity of 500.000 tons per year, which is now established 4 km from the center of the town.

4.3 Dacca, Bangladesh

The city is densely populated. Access to natural raw materials is difficult, and like other Third World cities, Dacca struggles with the problem of reducing huge quantities of municipal solid waste that daily adds to already immense and unsanitary landfills. It would appear that the use of concrete rubble as road construction materials could substantially assist in solving both a waste disposal problem and a raw materials supply problem. This is true also for other large cities in the Third World, which, contrary to Dacca, are located close to abundant raw materials deposits, but where permanently congested traffic conditions make access to the deposits difficult. Mexico City is such an example.

5 Implementation

In order to reach the goals of concrete demolition waste management, it is necessary that all barriers and obstacles are detected and considered. A long-term action plan has to be made, combined with adequate development and research. Implementation of the necessary economical-, technical- and administrative instruments require that initiatives, involving legislation's and regulations are taken.

5.1 Economy

In a market economy, the choice between recycled and natural materials depends upon price and quality. The quality of concrete with recycled aggregates can be the same as that of concrete with natural aggregates, but recycled concrete aggregates are at present regarded with suspicion.

Hence, recycled aggregates will be preferred only when the price of such aggregates is considerably lower than that of the natural materials, even when the recycled aggregates meet given specifications. The economic barriers can be overcome by introducing economic instruments which encourage recycling and use of recycled materials. As examples, taxes demonstrating the true cost to society for waste accumulation, and fees in favor of recycling should be considered.

5.2 Policies and Strategies

Concrete demolition waste must be considered as a specific type of waste associated with the building and construction industry. It is important that the management and handling of waste is carried out by the industry itself. Generally, the building and construction industry is relatively conservative, and changes in normal procedures often take time and need long-term policies and strategies.

One of the most important barriers is the many different interests. It is usually the politicians, departments and public offices concerned with environmental issues that prepare the policies concerning waste recycling and reduction, whereas the building and construction industry is controlled by politicians, departments and offices concerned with housing, construction and public works. To co-ordinate the interests of all parties, it is necessary that long-term policies and strategies, with respect to achieving goals for recycling of concrete demolition waste should be adopted. These goals should then be continuously revised in accordance with the political and environmental situation, and followed up by adequate legislation and regulation at all levels, national, regional and local.

5.3 Certification of recycled materials

Even when recycled materials fulfill current standards for virgin materials, and the prices are competitive with the prices of virgin materials, certain obstacles still exist. Due to traditions and psychological barriers, the general attitude towards recycling in the building and construction industry is largely inhibitive to the utilization of recycled materials. Therefore, it is of great importance that recycled materials are officially certified so they can be accepted by all parties. It is recommended that considerable emphasis is placed on specifying the areas of utilization and quality standards for recycled materials. These must be in accordance with the local demand in order to improve confidence in the recycled materials and solve problems regarding the responsibility of using recycled materials.

5.4 Planning demolition projects

Optimal handling and recycling of concrete demolition waste depends on the materials being sorted in-situ and in co-ordination with the demolition process. It is therefore necessary to alter the traditional methods of demolition and introduce «selective demolition». This requires that before and during the demolition process an effective sorting of the different material categories is carried out, thereby preventing any mixture of materials leading to pollution of recyclable concrete with wood, paper, cardboard and plastics etc.

It is recommended that demolition projects should be planned and controlled in detail, in the same way as all other building and construction projects to ensure selective demolition and correct handling of the demolition waste.

5.5 Education and information

The improvement of cleaner technology and achievement of goals for recycling and waste minimization is inextricably tied to an understanding of the different problems in integrated waste management and recycling. Therefore, education and information within all parts of the industry on all levels are needed.

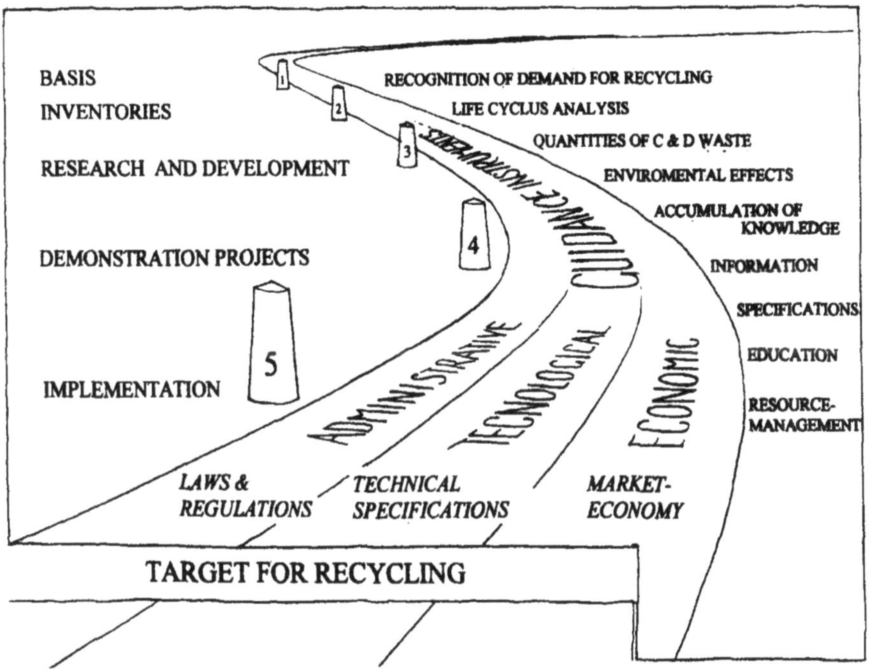

Fig 2. Economical-, technical- and administrative guidance instruments working jointly.

6 Conclusions

Recycling of concrete demolition waste is more feasible in some areas of the world than in others. With a well defined analysis where all items affecting the choice of concrete demolition waste treatment facilities are considered, it will be possible to find the optimal solution for each location. Many barriers have to be overcome during the implementation of a waste management plan, which requires long term action plans involving economical-, technical- and administrative guidance instruments jointly.

References

1. Symonds Travers Morgan/ARGUS (1995) *Construction and Demolition Waste Project in the framework of the Priority Waste Stream Programme of the European Commission.* Report of the project group to the European Commission.
2. Lauritzen, Erik K. (1993) *Demolition and Reuse of Concrete and Masonry*, Proceedings of Third International Symposium held by RILEM in Odense, Denmark, 24-27 October, E & FN Spon.
3. RILEM Recommendation 121-*DRG Guidance for Demolition and Reuse of Concrete and Masonry*, Materials and Structures, 1994, 27.
4. Lauritzen Erik K., Hansen Torben C. (1997) *Recycling of C&D Waste Research and Development 1986-1996*, Danish Environmental Protection Agency.

FUTURE NEEDS IN CONCRETE REPAIR TECHNOLOGY

G. HORRIGMOE
NORUT Teknologi AS, Narvik, Norway

Abstract

Increased requirements pertaining to safety and durability of concrete repair and rehabilitation are gradually developing. In the 21st century, demands on repair methods will be further emphasized by environmental considerations and shrinking public budgets. The present paper discusses the performance of current techniques for protection and repair of concrete structures and the need for new developments and innovations in this area.
Keywords: Concrete, damage, deterioration, repair, rehabilitation, strengthening, retrofit, performance, requirements.

1 Introduction

Concrete is the building material of choice and the dominating portion of the infrastructure is made of reinforced concrete. During the last two decades, deterioration of reinforced concrete structures has emerged as one of the most severe problems and, at the same time, one of the most demanding challenges facing the building and construction industry worldwide. The registered need for repair and rehabilitation and the associated, estimated costs is a major concern of private building owners, municipalities and public agencies. Already, public agencies are spending a significant portion of their annual construction budget on repair and rehabilitation. If the deterioration of vital infrastructure is allowed to continue, it is possible that implications for the future development and economic growth will surface in many countries. Growing public concern for lack of durability of concrete may be a threat to the future status and position of concrete as the leading engineering material.

The design and construction of durable concrete structures is an intricate problem which requires that all important factors influencing durability are considered simulta-

Concrete Technology for a Sustainable Development in the 21st Century edited by O.E. Gjørv and K. Sakai.
Published in 2000 by E & FN Spon, 11 New Fetter Lane, London EC4P 4EE, UK. ISBN: 0 419 25060 3

neously. The importance of considering durability as a holistic criterion has been excellently described in a recent paper by Metha [1]. It is obvious that a major portion of the deterioration and damage of concrete structures can be traced back to failure to comply with the holistic approach to concrete durability. Despite these and other shortcomings, the last decade has seen considerable progress in durability research and in the development of methods or systems to improve durability of reinforced concrete structures. Nevertheless, this does not seem to reduce the need for repair and rehabilitation since errors and bad judgements continue to be made in design and construction of concrete, leading to deterioration and damage of structural components and members. At the same time, the high costs of maintenance, repair and rehabilitation of concrete structures seem to cause delays and postponement of required actions, thereby further increasing the problems of deterioration and damage. The need for repair of concrete structures will therefore be with us for a long time, and we should consider the solution to the problems of deteriorating infrastructure as a part of the challenge of creating a sustainable development in the 21st century.

In the years to come, repair and rehabilitation of concrete structures will be subject to strict requirements with regard to environmental impacts and economical constraints. Shrinking budgets of governmental agencies and municipalities will be a critical factor for renewal of the infrastructure. It is therefore of utmost importance that concrete repairs are environmental friendly, cost efficient and reliable. As a minimum requirement, they should guarantee that the strength and durability of the structure are restored, so that its service life is unaltered. This may seem trivial, but most current repair methods fail to meet these conditions.

In the future, it is likely that we will se more of combined repair and strengthening of concrete structures. Globalization of the economy causes increased competition and needs for rapid adaptions and changes in industry. Reuse of production facilities and office buildings follows as a natural consequence of this. A similar situation will occur in the public sector as new objectives have to be met by government and municipalities. The need for strengthening of structures is also brought about by increased design loads, e.g. for bridges, and code revisions. Thus, it seems natural to assume that in the future, repair of concrete structures will more often be carried out in combination with structural strengthening and modification. The objectives of the combined repair and retrofit operations will be to increase the load-carrying capacity of the structure and to extend its service life, thereby further increasing the demands of concrete repair technology.

It seems clear that there is a need for significant improvements and innovations of existing repair techniques to meet future demands of costs and performance. Recently, fiber reinforced plastics (FRP) have been introduced as a means of retrofitting concrete structures, especially in seismic areas. This technique which has also been applied to concrete repair, is an example of innovative developments that may enable the concrete industry to meet some of the challenges with respect to repair and retrofitting of the infrastructure in the 21st century.

2 Repair of concrete structures

Repair and rehabilitation of concrete structures have evolved into a complex, inter- disciplinary science where it is necessary to master and combine knowledge from a number of different fields such as concrete technology, environmental loadings, transport mechanisms, electro-chemistry, structural mechanics, and composite materials, to name some of the most important.

Concrete repair is being conducted by various alternative methods, some of these are relatively new techniques developed during the last decade. Repairs are sometimes classified as either structural (i. e., load-carrying) or non-structural. Although, this distinction may represent an adequate description of the main purpose of the repair, it may also be confusing since durability and performance are generally related. Lack of durability will sooner or later influence mechanical behaviour or structural properties such as stiffness and ultimate load capacity. For this reason, it seems more appropriate to classify the various techniques applied to remedy damage or reduce deterioration of concrete structures in accordance with their actual effect, that is, what they really accomplish with the structure. Accordingly, existing methods may be grouped into two main categories:

- Protection methods
- Repair methods

Protection methods seek to limit the intrusion of moisture, chlorides, carbon dioxide, etc. by either surface treatments or through the use of electrochemical principles. To this category belongs various forms of sealers, membranes and overlays. Many of these products will delay the onset of future corrosion of the reinforcing steel. Cathodic protection systems seek to control corrosion by making the embedded reinforcing cathodic. Bridge piers were one of the earliest applications of cathodic protection and the method seems to have received increased interest in recent years. It is important to note that cathodic protection is only concerned with the control of future corrosion and does not replace corroded reinforcing bars. Cracked or spalled concrete will therefore have to be removed, corroding reinforcing bars should be cleaned, prefarably by sandblasting, before applying a cement-based mortar or concrete. Protection methods also include corrosion inhibitors, which improve the concrete's natural ability to protect embedded reinforcing bars by forming a passivating oxide layer on the steel. These products may be applied to the surface of existing structures (migrating corrosion inhibitors).

The objective of repair methods is to restore the reinforced concrete structure to its initial design conditions and to reduce or remove the actual cause of corrosion. The most widely used repair technique continues to be that of mechanical removal of carbonated and chloride contaminated concrete, followed by sandblasting of corroded reinforcing bars and application of a cement-based mortar or concrete. This procedure treats only the symptoms of corrosion and not its underlying causes. Unless the repair work is carefully executed and all contaminated concrete is removed, patch repair may under certain circumstances increase subsequent corrosion rather than prevent it. The removal of damaged concrete is limited by structural strength and safety requirements, hence one may be forced to accept less than optimum removal of carbonated and chloride contaminated concrete. Mechanical repair is sometimes denoted "structural repair". However, this is usually misleading since residual strength after completion of the repair is normally neither considered nor is it documented. Concrete repair may also be conducted by electrochemical treatments. In the chloride extraction method, chloride ions are removed under the influence of an electric field, whereas realkalization is a method that restores the high pH level in concrete. Spalled and cracked concrete must be removed and exposed reinforcement cleaned prior to application of these techniques. The major advantage of electrochemical treatments is that it deals with the actual cause of corrosion. Promising results with these methods have been reported, but unwanted side effects in the form of reduced bond strength, alkali-silica reactions and hydrogen embrittlement of high strength tendons may hamper their practical use.

Protection methods and repair methods should not be confused as alternative measures of repairing damaged concrete structures. On the contrary, they serve different purposes and can be combined to achieve more efficient concrete repair by extending the service life of the repaired structures.

There seem to be differences in opinion regarding the relative merits and actual performance of the various methods for protection and repair of concrete structures. One obvious reason for this is that systematic repair and rehabilitation efforts are relatively new and that long time experience with the performance of repairs therefore is limited. This fact should remind us that we are only in the initial stage of developing efficient and durable methods for concrete repair where the questions by far outnumber the answers. At this stage it is probably of particular importance to attempt an objective and critical survey of executed concrete repairs to map the actual performance of those repairs. Such a research project is currently being undertaken in Norway [2].

3 Structural consequences of concrete damage

Research concerning deterioration of concrete has been mainly focused on the non-structural aspects of damage, whereas only a minor effort has been devoted to develop methods to analyse the safety of damaged or repaired concrete members. This is somewhat surprising since the full integrity of the structure is implied after repair work has been completed and the structure is again subjected to the design loads. In addition, removal of chloride contaminated concrete over large areas and the associated debonding of steel reinforcement may significantly reduce the load-carrying capacity of the structure, which should be taken into account when selecting the sequence and amount of mechanical repair. Concrete repairs should not only meet the requirements of being cost-effective and durable, they should also be safe from a structural mechanics point of view. Documentation of structural stiffness and strength is necessary both for the conditions during repair when vital portions of the reinforcement are exposed, and after completion of the repair works when the contaminated concrete has been replaced by a suitable mortar.

Experimental investigations [3 - 6] of simple beams with exposed tensile reinforcement have demonstrated that the cracking behaviour and failure mechanisms may differ significantly from those of beams with normal composite action. In many cases, reduced strength was observed when the tensile reinforcement was exposed over a portion of the beam. These experiments did not fully simulate the conditions during mechanical repair. The beams in the experimental investigations were cast with various degrees of exposed reinforcement. Thus, they did not contain cracks due to service loads, nor was the tensile reinforcement reduced due to corrosion.

To achieve safe and efficient concrete repair it is necessary to have reliable methods for prediction of residual stiffness and strength of damaged and repaired concrete structures. Guidelines or recommendations for such calculations should be developed. The spatial distributions of strains and stresses in reinforced concrete structures are in many cases relatively complex, and predictions of the ultimate load carrying capacity can most conveniently be attained by nonlinear finite element analysis. For the numerical simulations to be accurate, it is necessary that constitutive and kinematical assumptions represent realistic approximations of the behaviour of damaged structures. Some of the most important ingredients in the modelling and numerical simulation of damaged and repaired concrete structures are:

- Failure or crushing criterion for concrete
- Crack model for concrete
- Bond slip model
- Modelling of damaged concrete
- Elasto-plastic model for reinforcing steel
- corrosion model
- Realistic representation of the loading and straining histories
- Reliable solution algorithms

Nonlinear finite element analysis of damaged and repaired concrete beams have been carried out by Horrigmoe and Tørlen [7,8] and verified against available experimental data. The main features of these finite element simulations can be explained from Fig. 1. First, the discretized model of the beam is loaded up to the full service load, where cracking occurs (Fig. 1a). In this condition, corrosion of the reinforcing steel takes place (Fig. 1b). This is accomplished by gradually reducing the area of the reinforcing bars. The structure is then taken out of service and the variable portion of the loading is removed, so that only the permanent load remains (Fig. 1c). Chloride contaminated concrete is removed, thereby exposing the tensile reinforcement over a major portion of the span (Fig. 1d). This is achieved by removing the associated finite elements from the mesh. The contaminated concrete is replaced by a cement-based mortar with constitutive properties that deviate from those of the original concrete by adding new elements to the model (Fig. 1e). The mortar is strain-free, whereas the surrounding concrete has been subjected to loading and partial unloading. Finally, the repaired beam is gradually loaded until failure occurs (Fig. 1f). The associated load-displacement diagram is illustrated in Fig. 2.

4 Strengthening and repair of concrete structures by externally bonded FRP sheets

The introduction of fiber reinforced plastics (FRP) as a means of strengthening concrete structures is one of the most significant innovations in concrete technology during the last decade. The major driving force behind this development has been the need for seismic retrofit of bridge structures in the United States and Japan. Thin carbon fabrics with unidirectional fibres used in conjunction with epoxy resins, can be easily wrapped on beams and columns to improve structural performance. Following the immediate success of FRP strengthening of structures subjected to earthquake loading, a large number of non-seismic applications have been reported, and the practical use of this technique is

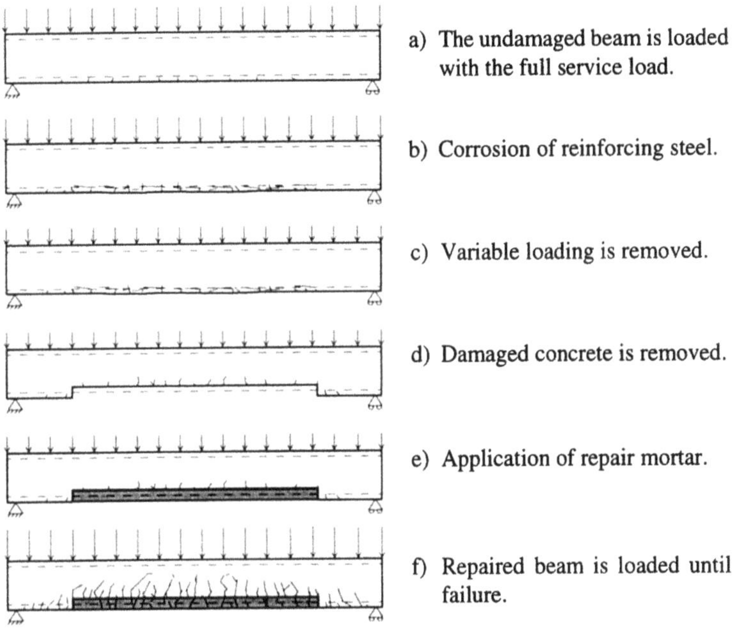

Fig. 1. Basic steps in nonlinear finite element analysis of damaged and repaired concrete beam.

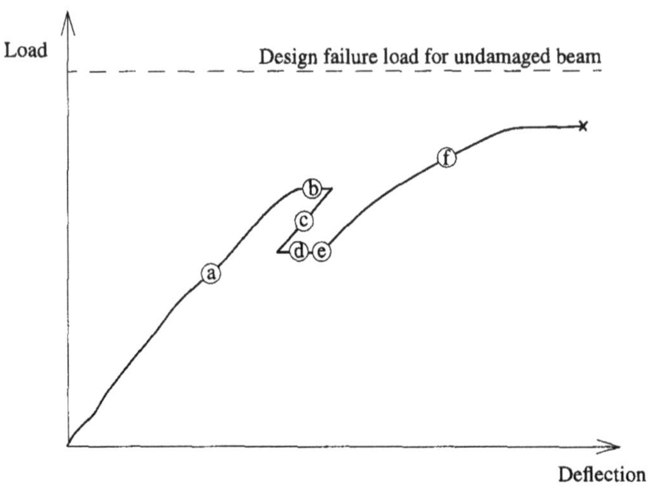

Fig. 2. Load-deflection curve for beam loaded and repaired according to Fig. 1.

growing rapidly.

The excellent durability of carbon fibres combined with low weight and low installation costs make externally bonded carbon fabrics attractive for repair of deteriorated concrete members. The main idea is the protection of the concrete surface provided by FRP together with the added strength. This is a very promising approach which, if it proves successful, may represent an important step forward for concrete repair technology. The major concern is durability. Since the effect of carbon fabric wrapping hinges on the bond between concrete and composite, long time effects on bond strength, especially in marine environment, need to be verified. Although some results have been reported in the literature [9-12], more intensified research is necessary.

Fibre reinforced composites are finding a wide range of applications which may enhance the durability of reinforced concrete structures. This includes the use of reinforcing bars and high strength tendons made of carbon fibre reinforced composites. In new designs of concrete beam structures composite material may replace tensile reinforcement or the entire web may be of composite material, thereby reducing the potential for tensile cracking of concrete as well some of the inherent durability problems in concrete construction.

The future of carbon fibre reinforced composites in concrete construction is very promising. These high performance materials provide opportunities for developing innovative solutions to serious problems of deterioration and repair. In the short term, the lack of design criteria and codes may hamper the practical use of composite materials for strengthening of concrete structures. It is important that these codes are specified in such a way that they are compatible with existing rules for design of reinforced concrete structures. Moreover, insufficient verification of the durability of carbon fibre reinforced plastics externally bonded to the concrete surface is another major concern regarding the practical use for repair of marine concrete structures or structures subjected to cycles of freezing and thawing.

In the longer term, the challenge is to prepare a new generation of architects and civil engineers to utilize the full potential of composite materials. This requires systematic education and training in the design and manufacture of fibre composites for the construction industry. A recent survey [13] of civil engineering and construction programs at universities in the United States, showed that the number of institutions that teach composite courses for civil engineers, architects and constructors is very small. It is reasonable to assume that the situation is similar in most countries. This should be taken as a warning and remind us that unless curricula are changed, practical use of carbon fibre reinforced composites will suffer.

5 Concluding remarks

Corrosion of imbedded reinforcing steel due to carbonation or chloride contamination continues to pose threats to the performance and integrity of reinforced concrete structures. Further degradation problems may be added by freezing and thawing actions. Deterioration of reinforced concrete infrastructure has become a problem of vital importance on an international scale. The need for repair and rehabilitation can be enormous.

Current methods for protection and repair of concrete structures leave much to be desired. These methods do little to prevent the intrusion of chlorides to the repaired areas and they are lacking in their ability to restore the stiffness and strength of the concrete members. In general, the structural effects of damage and repair have received relatively little attention and this topic should be given more serious considerations in the future. Increased focus on combined strengthening and repair of concrete structures will further emphasize the importance of performing structural analysis in conjunction with concrete repair.

In the 21st century, the concrete industry will have to comply with environmental restrictions and shrinking public budgets. In this context, it is important that concrete repair and rehabilitation are environmentally sound and that repair systems are cost-efficient, safe and durable. These objectives call for new developments and innovations in concrete repair technology. The use of externally bonded carbon fibre reinforced plastics for strengthening and repair is a promising new technique that may provide the concrete industry with a set of new tools to remedy some of the pressing problems of the deteriorating concrete infrastructure.

6 References

1. Metha, P. K. (1997) Durability - Critical issues for the future, *Concrete International*, Vol. 19, No. 7, pp. 27 - 33.
2. STRUCON - Structural Consequences of Deterioration and Repair of Concrete Members, *Strategic Research Program, Research Council of Norway*, NORUT Technology, 1998.
3. Minkarah, I. and Ringo, B. C. (1981) Behaviour and repair of deteriorated reinforced concrete beams, *Transport Research Record 821*, University of Cincinatti, Ohio, U.S.A.
4. Smith, D. N. and Wood L. A. (1989) The structural performance of damaged elements, *Proceedings of the International Seminar on The Life of Structures - The Role of Physical Testing*, Brighton, U. K.
5. Cairns, J. and Zhao, Z. (1993) Behaviour of concrete beams with exposed reinforcement, *Proceedings of The Institution of Civil Engineers, Journal of Structures and Buildings*, Vol. 99, pp. 141 - 154.
6. Raoof, M. and Lin, Z. (1997) Structural characteristics of RC beams with exposed main steel, *Proceedings of The Institution of Civil Engineers, Journal of Structures and Buildings*, Vol. 122, pp. 35 - 51.
7. G. Horrigmoe and A. Tørlen (1998) Load-carrying capacity of damaged and repaired concrete beams, *Concrete Under Severe Conditions 2* (eds. O.E Gjørv, K. Sakai and N. Banthia), E. & F.N. Spon, London, Vol. 2, pp 1075 - 1085.
8. A. Tørlen and G. Horrigmoe (1998) Analysis of repaired concrete beams with corroded reinforcing steel (in Norwegian), NORUT Technology, Report No. NTAS A98018.
9. R. J. Watson (1998) Case histories of structural rehabilitation utilizing composites in infrastructure and marine environments, *Fiber Composites in Infrastructure*, (eds. H. Saadatmanesh and M. R. Eshani), University of Arizona, Vol. 1, pp. 113 - 125.

10. L. K. Spainhour and I. Thompson (1998) Effect of carbon fiber jackets on reinforced concrete columns exposed to a simulated tidal zone, *Fiber Composites in Infrastructure*, (eds. H. Saadatmanesh and M. R. Eshani), University of Arizona, Vol. 1, pp. 426 - 439.
11. H. Toutjani and T. El-Korchi (1998) Tensile durability performance of cementitious composites externally wrapped with FRP sheets, *Fiber Composites in Infrastructure*, (eds. H. Saadatmanesh and M. R. Eshani), University of Arizona, Vol. 2, pp. 410 - 422.
12. K. M. Cetin, B. A. Shaw, C. Bakis, A. Nanni and T. Boothby (1998) Environmental degradation of repaired concrete structure, *Fiber Composites in Infrastructure*, (eds. H. Saadatmanesh and M. R. Eshani), University of Arizona, Vol. 2, pp. 488 - 500.
13. S. A. Chini and C. J. Kibert (1998) Fiber composites in civil engineering and construction curricula, *Fiber Composites in Infrastructure*, (eds. H. Saadatmanesh and M. R. Eshani), University of Arizona, Vol. 2, pp. 593 - 604.

FIBRE REINFORCED POLYMERS (FRP) AS AN ECONOMICAL ALTERNATIVE FOR CONSTRUCTION AND REHABILITATION

S.A. SHEIKH and A. BIDDAH
Department of Civil Engineering, University of Toronto, Canada

Abstract
As our infrastructures continue to deteriorate, the need for more economical and durable construction, strengthening and repair techniques has emerged. The use of Fibre Reinforced Polymers (FRP) promises to be one of the most attractive alternatives to traditional techniques for construction, strengthening and repair of concrete structures. The installation of these light-weight materials is less labour-and equipment-intensive. Fibre composites can also enhance durability as they prevent the passage of moisture, chlorides and other chemicals. Finally, fibre composites can considerably reduce the time for which the structures are taken out of service for retro-fitting. In many cases, this technique may result in considerable savings in comparison with conventional methods proposed for the same structure. This paper discusses the advantages of using fibre reinforced plastics for construction, strengthening and repair. The major applications in structural engineering are listed. Some of the research work in the field of application of FRP in structures is summarized. The FRP is studied from the technical as well as the economical points of view.

Keywords: Concrete structures, ductility, strengthening, fibre-reinforced polymers, repair, glass fibres, carbon fibres.

1 Introduction

Deterioration of concrete structures due to loss of material properties, inferior design, action of climate, substandard construction as a result of revision of design codes, or increase in traffic loads in the case of bridges, etc. often requires development of new

Concrete Technology for a Sustainable Development in the 21st Century edited by O.E. Gjørv and K. Sakai.
Published in 2000 by E & FN Spon, 11 New Fetter Lane, London EC4P 4EE, UK. ISBN: 0 419 25060 3

techniques which are more economical, durable and feasible than the existing ones. Construction of new structures as well as rehabilitation (strengthening or repair) of existing structures pose a challenge for engineers. Quite often this means consideration of new and innovative techniques such as the application of fibre reinforced polymers (FRP) in concrete structures. The development of comprehensive approaches for the use of FRP in structures can lower the cost and enhance the performance and durability of both types of structures. FRP products were first used to reinforce concrete structures in the mid 1950s [19]. Today, these FRP products take the form of bars, cables, 2-D and 3-D grids, strips, sheets, plates, etc. [2].

FRP are available in several forms. Available steel shapes and precast concrete components provide obvious examples for FRP applications. For use in new concrete construction, FRP rods and grids are the most common forms. Different FRP shapes are becoming available to be used as web and tension flange while in-situ concrete provides the compressive resistance in a flexural member. To simulate the function of steel plates, FRP sheets or fabric can be used to provide the reinforcement. The main difference being that FRP can apply forces only in the direction of fibres while a steel plate can exert force in any direction in its plane.

FRP shell system is suggested for new construction by Seible [25] for bridge columns to replace conventional external formwork and internal reinforcing steel. The FRP shell system represents a cost effective approach to bridge column design by taking advantage of the excellent mechanical properties of advanced composite materials, automated manufacturing processes and fast field implementation. The pre-manufactured FRP tubes eliminate the need for internal steel reinforcement and form-work removal. The observed excellent structural response of the concrete filled FRP shell columns led to the development of complete bridge systems.

Strengthening of existing structures is required for a variety of reasons including: a) aging of structural elements, b) original design assumptions that underestimated the actual loading on the members; c) construction errors that resulted in deficient members; d) demand for increasing the service loads on the member due to change in use; e) enhancing the structure's performance under certain load conditions (e.g. seismic load); and f) complying to new building code requirements that demonstrate the inadequate strength of the member as it was originally designed. Damage to structures can also occur due to several other actions such as earthquakes, overloading, winds and blasts.

Conventional repair techniques which involve traditional materials such as steel and concrete have been successfully employed for decades. However, in most cases these techniques are quite extensive and cumbersome. In situations where the conventional techniques may require traffic stoppage or closing of the entire damaged area of the structure, alternative repair techniques are needed which can allow regular activities to continue with minimal disruption. Results from recent work on FRP and its applications in new construction, strengthening and repair are very promising.

The FRP are obtained by the synthetic assembly of resin, fibres and selected filler. Depending on the fibres selected three families of composite materials are identified: namely glass fibre reinforced polymers (GFRP), carbon fibre reinforced polymers (CFRP), and aramid fibre reinforced polymers (AFRP). The resulting composite behaviour depends mainly on the fibre volume and mechanical properties of the constituents. Furthermore, procedures used to fabricate the composite also influence its behaviour. Each FRP composite has its own typical mechanical characteristics which make it suitable for a given structural application. Glass fibres are considerably cheaper than carbon fibres but some forms of this fibre tend to be very sensitive to the alkaline environments of concrete. Glass fibres also have a lower elastic modulus than carbon fibres. A comparison of mechanical properties of FRP with steel is provided in Table 1.

Table 1. Typical Comparative Properties of Fibre Reinforced Polymers and Steel (Mufti et al., 1991)

Material	Tensile Strength (MPa)	Modulus of Elasticity (MPa)
Glass-epoxy	1050	55 000
Carbon-epoxy	1500	180 000
Aramid-epoxy	1400	76 000
Steel	400-1000	200 000

FRP sheets could be a practical and economical method of strengthening. It has been reported that for a specified application of strengthening reinforced concrete beams, it was possible to replace 94 kg of steel by 4.5 kg of CFRP [14]. In early 1993, the Migros supermarket in Uzwil, Switzerland was expanded using CFRP laminates. Each laminate was 15.5 m long. If steel plates has been used, 120 kg of material would have been needed, compared to 3.5 kg of CFRP laminates actually used [13]. Although CFRP is normally considered an expensive material, the cost of the material in many applications is approximately only 20% of the total cost of the strengthening application, the remaining 80% being labour costs. Even a small reduction in labour cost can more than compensate for a large increase in the material cost.

2 Advantages of FRP

Although they are mainly used in aeronautical industries, composite materials are now selected for a variety of projects in the construction area. The use of FRP in construction industry and infrastructure-related applications has increased in recent years. They can be used in new construction, strengthening and repair of structures. FRP can provide a number of advantages over traditional techniques:

a) They have their high strength-to-weight ratio, which can reach 10 to 15 times that of steel. A high strength-to-weight ratio means a smaller proportion of the member's load carrying capacity is used to support its own dead load, thereby allowing for a larger live load and, hence, a more efficient use of available capacity. This feature becomes more attractive as spans increase, since larger proportions of capacity are then required to resist dead loads [15].

b) CFRP and AFRP have shown much better fatigue performance and life than structural metals while the fatigue strength of GFRP is lower than metals at a low stress ratio [23].
c) They are light weight and hence easy to transport and handle.
d) Due to their ease of installation, FRP reduces the construction time, the labour and equipment costs. The savings in labour costs can often outweigh the relatively high cost of the FRP material.
e) Their maintenance cost is low.
f) They are durable.
g) They have excellent corrosion resistance which enhances the durability of concrete as corrosion of steel in concrete structures is one of the main causes of concrete degradation.
h) They allow for regular activities to continue with minimal disruption during rehabilitation. This is particularly important in buildings that are occupied during the strengthening or repair processes and in case of bridges with heavy traffic. In many cases, strengthening and repairs can proceed without taking the structure out of service or halting industrial processes.
i) They cause no measurable loss of overhead or side clearance.

The attractiveness of using FRP in structures is thus quite clear. In many cases, the use of FRP can provide a more economical and technically superior solution to a problem compared with most other traditional methods. Table 2 summarizes the advantages of strengthening with CFRP and GFRP sheets in comparison with epoxy bonding of steel plates commonly used for strengthening of structures. The application process of the FRP strips requires minimum effort and in most cases does not interrupt the use of the structures during construction.

3 Applications in Structures

Advance composite materials have been used extensively in aeronautics, aircraft, aerospace, transportation and sporting goods industries for over a quarter of a century now. The first recorded application of glass fibre composite in aircraft industry dates as far as 1944 [8]. Since then, a variety of composites have been used in other industries such as ship building, chemical processing, medical, automotive, etc. Due to large number of applications, many types of composites have been produced in mass quantities in recent years. This has substantially reduced the production cost of these materials. In the last 3-4 decades, the cost of carbon fibre has dropped by about 80 to 90%. It is also projected that this cost will further decrease in the future. Therefore, the use of composites in structural engineering can be economically justified. Several field applications of externally bonded FRP reinforcement have been recently executed in Japan [16], Europe [12], Canada [3, 26, 28] and the United States [25]. Some advanced composite materials found their way into bracing systems, bridge girders [18] and decks, prestressing cables and other structural uses.

Table 2. Comparison between FRP sheets and steel plates strengthening systems

Criteria	Strengthening with CFRP sheets	Strengthening with GFRP sheets	Strengthening with steel plates
Own weight	Low (630 kg/m^3)	Low (740 kg/m^3)	High (7800 kg/m^3)
Tensile strength	Very high	Very high	High
Young's modulus	Intermediate	Low	High
Overall thickness	Very thin	Very thin	Small
Long-term behaviour	outstanding	adequate	adequate
Alkaline resistance	Very high	Low	High
Corrosion	None	None	Yes
Length of strips	Any	Any	Limited (6 to 8 m)
Handling	Flexible, easy	Flexible, easy	Difficult, rigid
Load bearing	In the fabric direction only	In the fabric direction only	In any direction
Laps	easy	easy	Complex
Fatigue behaviour	Outstanding	Adequate	Adequate
Material costs	High	Intermediate	Low
Installation costs	Low	Low	High
Application	Simple with minimal equipment		Require lifting equipment and clamping device

Examples of external strengthening of concrete structures with fibre composites are numerous. The main applications include beams, columns, slabs and walls. Several applications in tanks, pier caps, beam-column joints, elevator shafts, towers, etc. have also been reported in the literature.

3.1 Beams

In the past, concrete bridges were strengthened by additional beams, props, or external post-tensioning [20]. In the last several years, with the development of strong structural adhesives, plate-bonding techniques have been recognized to be an effective and convenient method of improving the performance of concrete structures under service loads, increasing their strength and stiffness. The field application of epoxy-bonded steel plates to concrete beams has been extensively used to strengthen buildings and bridges. The first documented application case was in 1964 in South Africa. The behaviour of concrete beams externally reinforced with steel plates bonded to their tension flanges were studied [10] and the results showed that significant improvements in performance could be achieved in terms of the ultimate load, crack control and stiffness. Long term experience with steel plates (20 years) has been good, and creep behaviour of strengthened members are as good as those for conventional reinforced concrete members [15]. However, the weight of the steel plates makes them difficult to handle and restricts the size of plate that can be used. There are also problems related to corrosion of the steel-adhesive interface. The simplicity and speed of application, the minimal additional self-weight, maintenance of original structural stiffness and the avoidance of head room problems have made FRP an attractive alternative to steel.

In case of FRP, it is necessary to remove the resin rich surface from CFRP laminate to achieve a correct adherent surface and treat the concrete surface by sandblasting. Pressure is required to bond the laminates to concrete and is achieved by using vacuum bags or lightweight clamps. It was reported that special attention must be paid to the development of shear cracks in concrete, which can lead to a premature peeling-off of the strengthening plates. Use of CFRP laminate also influences the flexural crack behaviour [14]. It reduces the total crack width and evenly distributes the cracks.

3.2 Columns

As a result of environmental and earthquake loads, columns in several structures have shown spectacular damage. Several columns supporting bridges show serious damage due to the corrosion of steel caused by chlorides from the de-icing salt. From the observations of damage from recent earthquakes, it was clear that column failures resulted in total collapse of several structures. Numerous columns sustained serious but repairable damage. Various design codes over the years have been revised to improve the seismic design of structures. As a result, structures designed and built in accordance with the older codes have substandard design details particularly in columns. Inadequate flexural strength and ductility, and shear strength are some of the main concerns in the design of those columns. Retrofitting of these columns is one of the most urgent needs for the safety of structures.

A rehabilitation or repair technique for columns, originally proposed in USA, consists of wrapping FRP sheets around existing reinforced concrete columns. This technique should improve, among other things, structural behaviour of columns under seismic loading. This approach is very similar to the reinforcement of tall chimneys with a combination of advance composite materials laminae and wires, and which has already been used in Japan [9]. Columns have been confined in the past with steel jackets around the potential plastic hinge region by encasing the column with steel sections, steel straps, welded wire fabric or additional spirals, hoops and longitudinal reinforcement. Most jackets are subsequently filled, injected or covered by concrete, shotcrete, or cement grout. It was shown that such reinforcement improves significantly the flexural and shear strength of the column and increase its ductility. However, installation is difficult and the steel jacket usually suffers from corrosion after exposure to environmental conditions. In addition, the stiffness of the column is increased substantially requiring modifications to other parts of the structure. Recent research has suggested that steel jackets could be replaced by FRP jackets, thus eliminating these problems [22, 27]. It should be noted that glass fibres degrade significantly and lose tensile strength when they are subjected to alkali penetration for a long period of time. FRP wraps containing glass fibres should not be used directly against the fresh grout. A plastic barrier between the glass fibre based FRP and the grout should work well [6, 27, 28].

3.3 Slabs

The improvement of one-way slab behaviour using FRP sheet is investigated by De Rose and Sheikh [5]. Three wall-slabs were tested under two line loads. One specimen was tested as a control specimen. The other two specimens loaded till reaching a crack width of approximately 0.4 mm then rehabilitated using CFRP and GFRP. FRP increased the wall-slab capacity and reduced its deflection. Sheikh and Biddah [29] tested full-scale reinforced concrete box culverts under concentrated load. Two culverts were tested to failure without strengthening and two culverts were tested after having been strengthened with CFRP and GFRP. The results showed that FRP significantly increased the slab flexural capacity. Malvar et al.[11] tested ten reinforced concrete slabs, simply supported on two edges. The slabs were subjected to a single patch load at mid span representing the loading from a crane outrigger float. Unidirectional CFRP sheets bonded to the bottom of two slabs demonstrated that both flexural and shear strengths could be significantly enhanced but the deflection was reduced. Orthogonal layers of CFRP bonded to the bottom of four slabs showed further strength increases as well as deformation comparable to that of the control slabs. The typical failure mode was punching shear. The CFRP sheets were able to enhance the punching shear resistance by providing increased in-plane constraint due to the added CFRP tension reinforcement.

3.4 Walls

Seismic retrofitting of lightly reinforced concrete walls can be achieved by epoxy bonding of the reinforcing material (composite fabric) to the exterior surface of the wall [6]. The fabrics can significantly increase the flexural and shear capacity of the member. The thin composite fabrics cause little change to the inertial loads and eliminating the need for strengthening of the footings. Large scale tests that were conducted by Schwegler [24] showed that existing shear walls can be efficiently strengthened by FRP. Six types of shear walls with top and bottom slabs were tested using either CFRP sheets or woven polyester fibre. Results showed that strengthening of shear walls by CFRP sheets increased the lateral load resistance by a factor of 4.3. An important factor in retrofitting the walls with FRP sheets is the anchoring of these sheets to the top and bottom concrete slabs. In the test reported, CFRP sheets did not fail despite of the large horizontal displacements of the shear wall. The sheets detached from the wall during higher deformation steps.

4 Research Work

Despite all the available information surrounding the use of fibre composites, many designers are still reluctant to recommend their use. New technologies raise liability concerns deterring practicing engineers from using them. As a result, the development of standards and design specifications are vital for the expansion of composites in civil engineering applications, which in turn require extensive research results from analytical as well as realistic experimental work. Several researchers (for example, Meier et al.[12], Saadatmanesh and Ehsani [21]) have studied the behaviour of concrete members strengthened with epoxy bonded FRP sheets or strips. Three experimental studies were carried out at the University of Toronto recently to investigate the behaviour of concrete members repaired and strengthened with FRP sheets. These studies are briefly discussed in the following.

In one experimental program, twelve reinforced concrete columns were tested [30]. All the columns were 356 mm in diameter and 1.473 m long. Each column was reinforced with 6-20M longitudinal reinforcing bars and spirals made of US #3 (77 mm^2) bars. The nominal concrete compressive strength was 40 MPa. The FRP tensile strength were 518 and 912 N/mm/layer for the 1.25 mm GFRP and the 1.00 mm CFRP, respectively. The columns were tested under cyclic flexure and shear loads while simultaneously subjected to a constant axial load. The variables in the test program were the amount of lateral steel, level of axial load, number of FRP layers and type of FRP (carbon or glass). Four specimens were tested without wrapping (control specimens), two specimens were tested to a certain level of damage, then repaired under load and tested again and six specimens were tested after strengthening with FRP. Table 3 shows the details of the specimens.

Figures 1a to 1d show the moment-curvature relationships of a select group of these specimens (S-1NT, S-3NT, ST-2NT and ST-3NT). The displacement ductility factors @ 0.8 P_{max} are 1.9, 1.3, 4.2 and 3.0 and the curvature ductility factors @ 0.8 M_{max} are 5.4, 2.6, 8.9 and 7.7 for the specimens S-1NT, S-3NT, ST-2NT and ST-3NT, respectively.

A comparison of specimens S-1NT and S-3NT shows the effect of amount and spacing of spiral steel on the behaviour of unwrapped columns. An increase in spiral pitch and the resulting decrease in the amount of lateral steel greatly reduces the

Table 3. Comparison of composite systems with steel plate system

	Carbon fibre	Glass fibre
Nominal thickness of fabric layer (mm)	1.04	1.24
ultimate tensile force per unit width (N/mm/layer)	912	518
Weight of fabric (grams/m^2/layer)	658	923
Weight of sheet (fabric + epoxy) (grams/m^2/layer)	1660	2500
Equivalent steel plate thickness (mm)	3.04	1.73
Weight of steel plate (grams/m^2)	23707	13494
Weight of steel plate + epoxy (grams/m^2)	24750	14590
Relative weight (fabric/ steel)	0.07	0.17
Price per m^2	US$ 91.4	US$37.6
Price per m^2 of equivalent steel plate	US$32.5	US$25.4
Relative price (fabric/ steel)	2.8	1.48

ductility and energy absorption of the column. Whereas the ductility numbers clearly show the effect of confinement provided by spiral steel, the energy index based on the area enclosed by the loops show the effect more dramatically. The relative values for energy indices for Specimens S-1NT and S-3NT are 69 and 5, respectively. The spiral steel in Specimen S-1NT was designed according to the seismic provisions of design codes [1, 4].

The effect of using FRP on the behaviour of column can be studied by comparing specimens ST-2NT and ST-3NT with S-3NT. All the specimens had 0.3% lateral spiral steel and were tested under an axial load of 0.54 P_o. Wrapping of columns with GFRP (ST-2NT) and CFRP (ST-3NT) improved their behaviour remarkably. Two layers of GFRP wraps resulted in the improvement of column behaviour that was very similar to that achieved by one layer of CFRP. The relative energy indices of

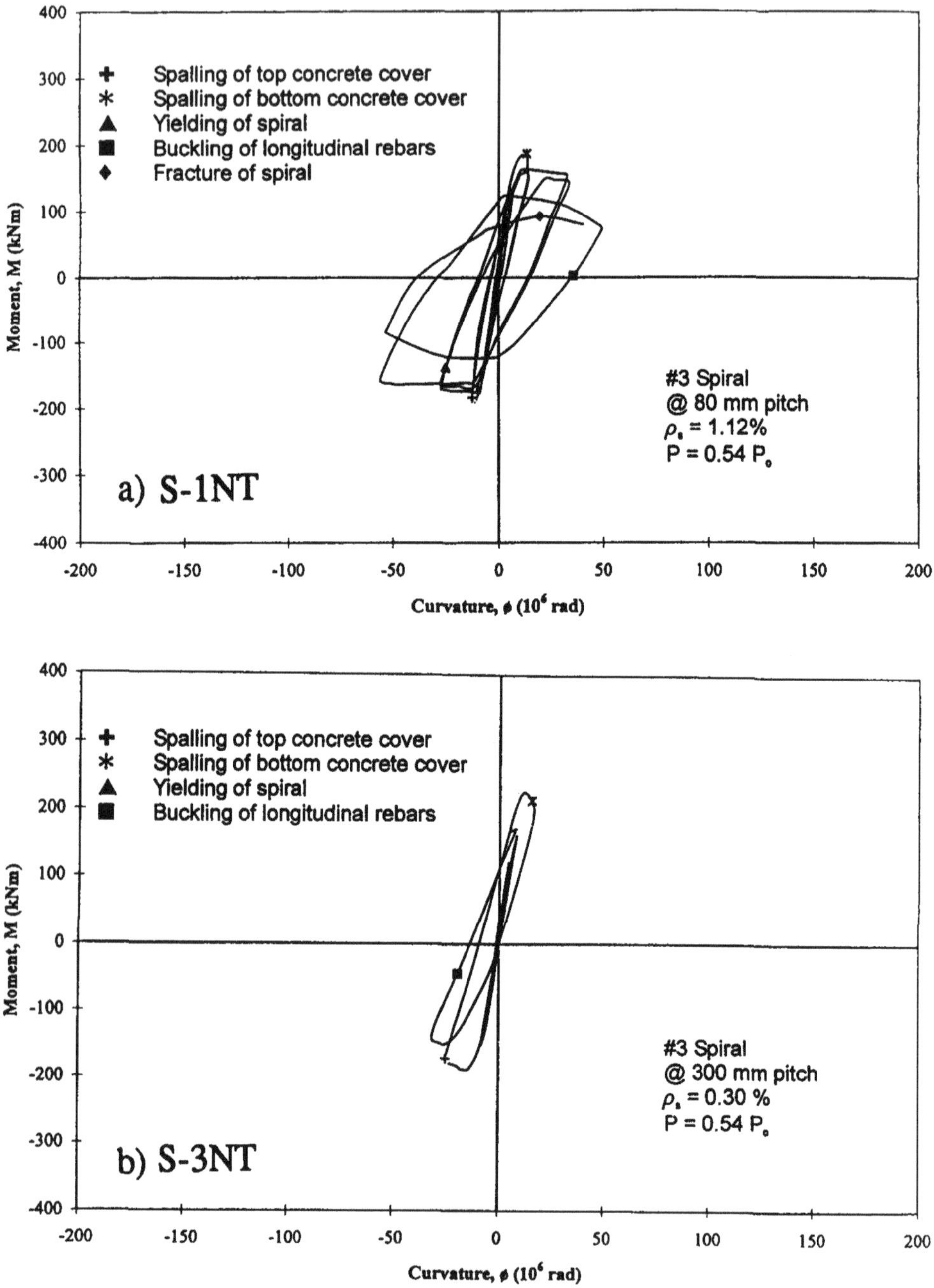

Figure 1 Moment-curvature behaviour of column specimens

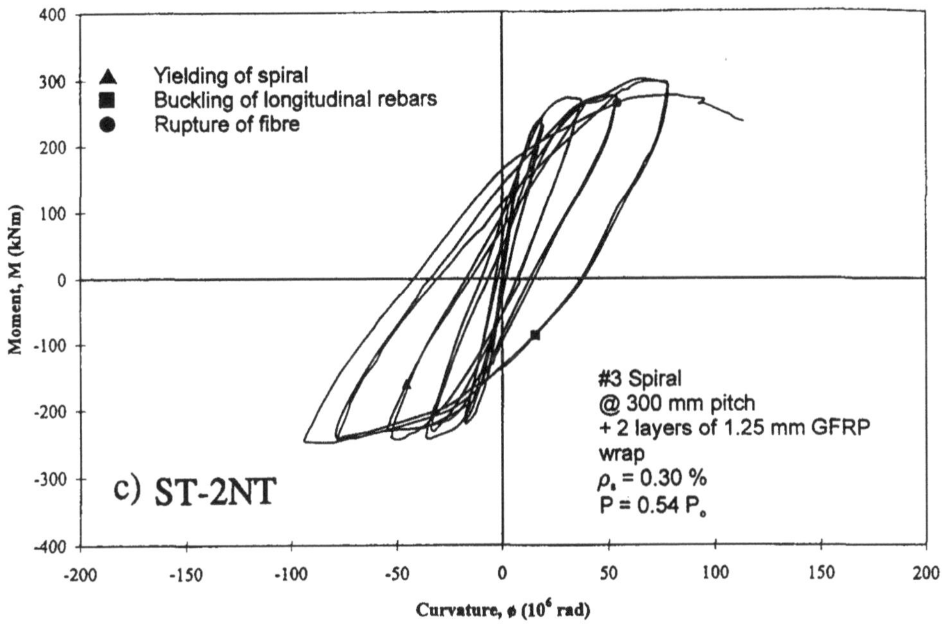

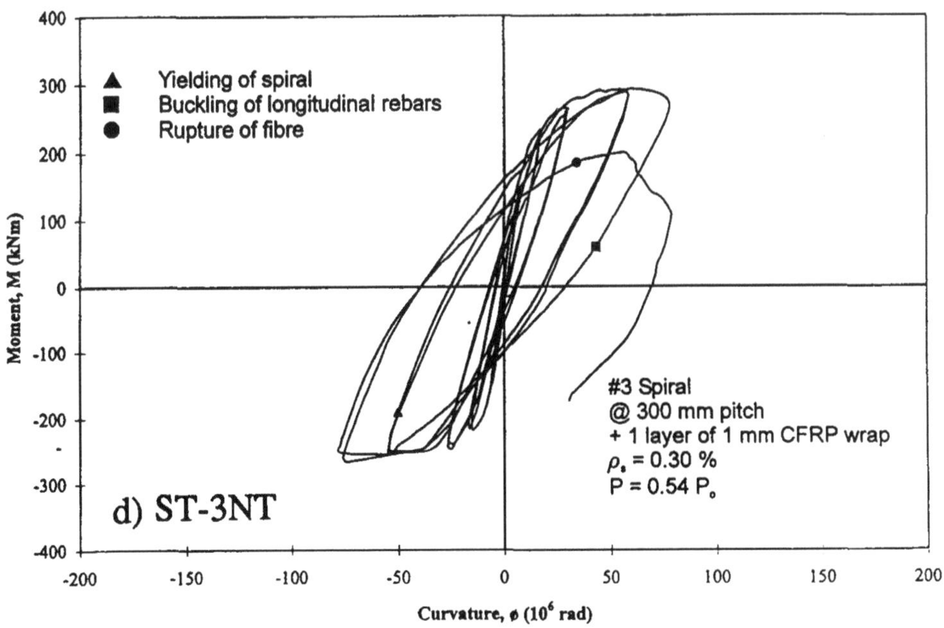

Figure 1 Moment-curvature behaviour of column specimens (continue)

Specimens S-3NT, ST-2NT and ST-3NT are 5, 181 and 202, respectively. It should be noted that the behaviour of wrapped specimens was superior to that of Specimens S-1NT which was designed to meet the seismic requirements for confinement steel.

In the second experimental program [27], tests on five half-scale models of field columns were conducted. All the column specimens were 406 mm in diameter and 1.37 m long. Each column was reinforced with 6-20M longitudinal reinforcing bars and 10M spiral at 75 mm spacing. Clear concrete cover was 20 mm. The 28 day strength of the concrete was 26 MPa. Four specimens were subjected to accelerated corrosion to produce damage similar to the observed field damage. Specimen 5 was not corroded while specimen 4 was corroded, but not repaired. For the three repaired specimens, the details are given as follows:

Specimen 1: Patching by a rheoplastic mortar (EMACO) followed by a protective epoxy coating and two layers of GFRP wrap.

Specimen 2: Type K shrinkage compensating cement grout in a formwork made of a stiff plastic sheet over which two layers of GFRP were applied. The plastic sheet stayed in place to provide a barrier between new grout and GFRP.

Specimen 3: Same as specimen 2 except that expansive cement replaced type K cement.

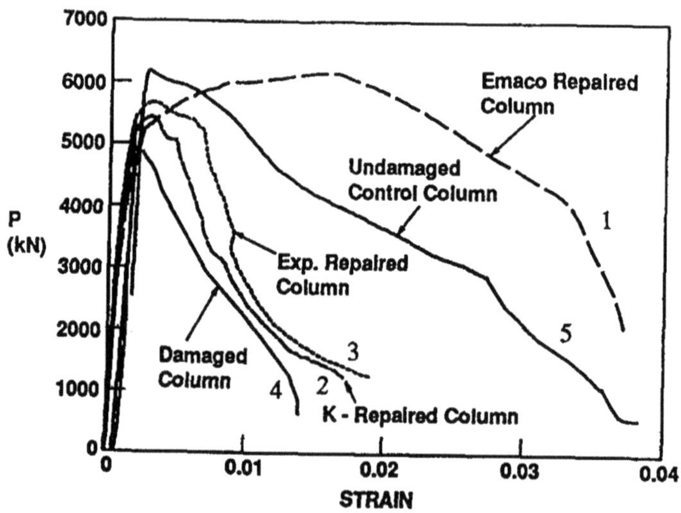

Figure 2 Behaviour of column specimens under axial load

Figure 2 shows the axial load-axial strain behaviour of the five column specimens. All the repair methods improved the behaviour of the specimens above that of the damaged unrepaired specimen. The repair methods, in which plastic sheet formwork was used (specimens 2 and 3), did no display as good a performance as the one (specimen 1) without the formwork. At large strain the plastic sheet opened out and engaged the fibre wrap creating large local strain and premature failure of wraps and the columns. The lower strength of specimen 2 compared with specimen 3 could be attributed to the lower expansion of cement and higher steel loss due to corrosion in that column. As a result of corrosion specimen 4 lost most of the spiral steel at several locations and substantial amount of cover. It failed in a very brittle mode. Compared to specimen 5, its strength is 20% lower. The effectiveness of spiral in providing ductility is also obvious. Efficient confinement, provided by spiral steel, and low-strength concrete, used for casting the specimens, contributed to the ductile behaviour of the undamaged specimen. Specimen 1 wrapped with 2 layers of GFRP without plastic sheets showed very ductile behaviour and exceeded the performance of undamaged control specimen.

In the third experimental program [5], two tests on full scale models of field beams were conducted. The beams were 1000 mm in depth, 550 mm in width and 3.85 m long. Each beam was reinforced with 6-30M bottom longitudinal reinforcing bars, 5-25M top longitudinal reinforcing bars and deformed US #3 (77 mm^2) stirrups at 400 mm spacing. The nominal concrete strength was 45 MPa. One beam was tested as a control specimen. The second beam was loaded until the distress resembled that in the field and then repaired with 1 mm thick carbon FRP wrap under load. The fabric was applied in the transverse direction to resist shear. A fabric overlap of 200 mm was

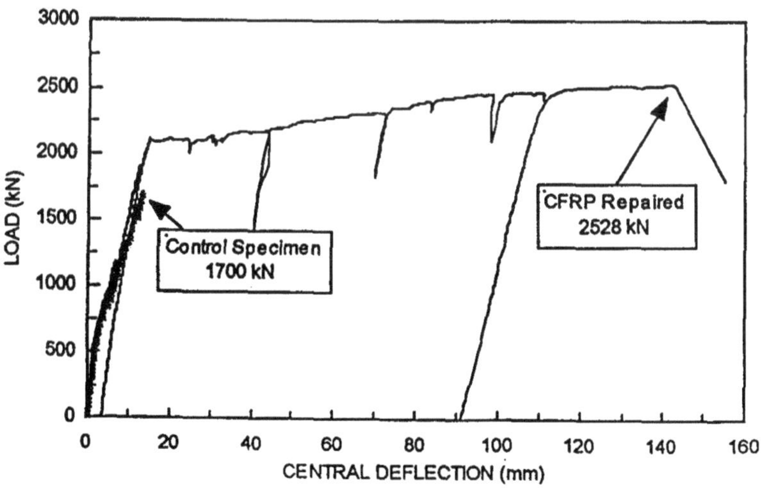

Figure 3 Load-deflection behaviour of beam specimens

used at the top of the specimen. Figure 3 shows the load-centre point deflection of the beams. The control specimen failed in a brittle shear mode. The ultimate failure of the rehabilitated specimen was by rupture of the fabric at the edge of the beam after the flexure steel was past the strain hardening stage. The sharpness of the beam edges was somewhat reduced by slight grinding. A well-rounded edge perhaps would have eliminated this type of fabric failure. The use of carbon FRP wrap resulted in a substantial increase in the ultimate capacity of the beam specimen. More importantly, premature shear failure was eliminated as the beam failed in ductile flexural mode. In fact, the maximum centre point deflection at ultimate failure was increased from 14 mm in the control specimen to 143 mm in the repaired specimen. In addition, the response of the repaired specimen resulted in much larger energy dissipation as can be seen from comparing the area below the curves in Figure 3.

5 Economic Considerations

Many advantages have been discussed above for the use of FRP in construction, strengthening and repair of concrete structures. At first glance, the main drawback seems to be the higher material cost of FRP over traditional materials. The material costs of fibre and resin in advanced composites are greater than their equivalent in steel. For example, the approximate pricing for the components of one composite system are as follows: carbon fibre US$ 75.3/m^2, glass fibre US$ 21.5/m^2 and epoxy US$ 16.1/m^2 The price of steel plate (yield strength 300 MPa) is US$ 0.70.

Based on these prices, a comparison is shown in Table 4. For the same in-plane force in 1 m^2 of reinforcement, the weight of steel plate is more than 14 times that of CFRP. The material cost of CFRP is about 2.8 times that of steel. A similar analysis shows that steel is 6 times heavier than GFRP but it costs 1.5 times more than steel. It should be noted that the cost comparison can not be made directly between GFRP and DFRP from the table since the forces in both the materials are different. The thickness of the fabric was chosen based on the availability of the material. For the same force in the two composite materials, CFRP will cost about 90% more and weight 67% less than GFRP.

The materials cost alone should not influence the decision regarding the use of FRP. The costs of manufacture, fabrication and erection of the complete system need to be considered. Other factors, such as the potential permanent loss of leasable space and the short-term costs associated with loss of productivity and revenue generation by the occupants if the facility is taken out of service, are also major issues to be considered. One of the main considerations may be to ensure ongoing operation of the facility during strengthening or repair. Light weight materials can be installed quicker and more cheaply than heavy steel. With all these factors considered, FRP may provide an economical alternative to traditional materials.

CFRP sheets were used in strengthening of reinforced concrete bridge girders located south of Edmonton, Alberta, Canada. The bridge had three 18 m long spans. Each girder was simply supported. All of the ten south span girders (1/3 of the bridge) were strengthened in shear by using CFRP sheets. The total duration of the project was 16

Table 4. Details of test specimens (Yau and Sheikh 1998)

Specimen	Spiral		GFRP	CFRP	Treatment	Axial Load P/P_o
	Spacing (mm)	ρs (%)	Layers*			
S-1NT	80	1.12	--	--	Control	0.54
S-2NT	80	1.12	--	--	Control	0.27
S-3NT	300	0.30	--	--	Control	0.54
S-4NT	300	0.30	--	--	Control	0.27
R-1NT	160	0.56	2	--	Test to a certain damage level, repair under load and test to failure	0.54
R-2NT	160	0.56	--	1	Test to a certain damage level, repair under load and test to failure	0.54
ST-1NT	300	0.30	1	--	Strengthen with FRP and test to failure	0.54
ST-2NT	300	0.30	2	--	Strengthen with FRP and test to failure	0.54
ST-3NT	300	0.30	--	1	Strengthen with FRP and test to failure	0.54
ST-4NT	300	0.30	--	1**	Strengthen with FRP and test to failure	0.27
ST-5NT	300	0.30	1	--	Strengthen with FRP and test to failure	0.27
ST-6NT	300	0.30	--	1	Strengthen with FRP and test to failure	0.27

* Each GFRP layer is of 1.25 mm thick fabric and each CFRP layer is of 1.00 mm thick fabric

** This CFRP layer is of 0.50 mm thick fabric

working days with a total of 37.5 man work days. As for the major part of the project, there were only two workers on site. The surface preparation including grinding, priming and puttying took almost 60% of the total project time without affecting the traffic. The total cost estimated if the entire bridge was strengthened [3] is as follows:

Man power	$22 500
CFRP material	$33 000
Traffic control	$ 7 000
Miscellaneous	$ 8 000
Total	$70 500

The total cost corresponds to a unit cost of about $428/ m^2. An alternative to the CFRP method is using external steel stirrups. A rough estimate would put the cost at about $100 000. This would require bridge closure for at least one month, removing the bridge deck, coring through the girder flanges and replacing the bridge deck.

After Northridge earthquake of January 1994, one of the industrial complexes, that was damaged, was repaired using FRP. The damage was concentrated in the tilt-up wall panels in two bays of the building [6]. A few different alternatives were evaluated and the merits of each system including cost, construction time and disruption to daily operations of the complex were carefully evaluated. It was decided that the best solution would be to strengthen both faces of the walls with FRP. The repair activities were carried out without interrupting the operation of the place. The cost of the project was about 70% of the lowest cost of alternative techniques and the construction was completed in two months.

In a recently completed project in Toronto area, Canada, involving repair of eight 690 mm square columns in an industrial structure, the total cost using GFRP system was approximately $20 000. The repair was carried out in three days without closing the operation of the industry. An alternative traditional method of repair would have cost approximately 25% less but the plant would have been closed for 4 to 5 working days resulting in a loss of hundred of thousands of dollars.

El-Mikawi and Mosallam [7] have developed a methodology for evaluating the use of fibre composites in civil engineering applications. The model incudes not only primary costs but also secondary or social costs. Primary costs include material costs, installation costs and maintenance requirements. Although composites have a higher material cost, they also have a lower installation cost and maintenance requirements. Easier installation reduces requirements for labour and heavy equipment. The durability of composites also increases their economic life and reduces maintenance costs. Secondary costs are often overlooked leading to inappropriate decisions. For instance, the availability of the structure for its intended use should be considered. If industrial process must be halted, loss of revenue may result. In the case of transportation projects, traffic congestion or detours may lead to social costs such as lost time, increased fuel consumption and an increase in fuel emissions. Many of these costs can be minimized if the work can be carried out without too much obstruction to traffic. The model is capable not only of comparing economic factors but also a variety of criteria including material performance, construction duration, architectural aspects. codes and regulations and material availability. As a result, the model provides decision makers with a tool to select an optimal structural material for structure rehabilitation, repair and construction.

6 Conclusions

A summary of various applications of FRP has been presented in this paper. Test results from a few research studies are discussed which show that strengthening and repair of concrete structures with FRP are very effective. Field application of use of FRP in new construction, strengthening of existing structures and repair of buildings and bridges that were damaged by earthquakes or other loads shows that this technique may, in many cases, be a superior alternative to traditional techniques both from technical and economical perspectives. One of the most important advantages of using FRP may be that the facility will not be taken out of service. FRP has much to offer the industry in the future which has the potential of radically changing the construction and rehabilitation methods. Considering the lower total cost, the ease of application, the no down time and the improved structural behaviour that will result, the FRP provides a viable alternative for new construction, strengthening and repair of structures.

7 Acknowledgements

The financial support of the Natural Sciences and Engineering Research Council of Canada is gratefully acknowledged. Additional financial and technical assistance from Hexcel Fyfe of Del Mar California, R.J. Watson of E. Amherst New York, Road Savers of Toronto and Petro Canada of Toronto are gratefully acknowledged.

8 References

1. ACI Committee 318 (ACI 318-95/ACI 318R-95), Building Code Requirements for Reinforced Concrete, American Concrete Institute, Detroit, 369 p.
2. ACI 440R-96 (1997) State-of-the-Art report on fibre reinforced plastics (FRP) reinforcement for concrete structures, ACI manual of concrete practice, part 5, American Concrete Institute, Farmington Hills, MI.
3. Alexander, J.G.S. and Cheng J.J.R. (1996) Filed application and studies on using CFRP sheets to strengthen concrete bridge girder, Proceedings of the Advanced Composite Materials in Bridges and Structures ACMBS II Conference, Canadian Society for Civil Engineering, Montreal, Quebec, pp. 465-472.
4. Code for Design of Concrete Structures for Buildings (CAN3-A23.3-94) (1994) Canadian Standards Association, Rexdale, Ontario, 199 p.
5. De Rose, D. and Sheikh, S.A. (1997) The rehabilitation of a concrete structure using fibre reinforced plastics, Research report, Department of Civil Engineering, University of Toronto, 170 p.

6. Ehsani, M.R. and Saadatmanesh, H. (1997) Fibre Composites: An Economical Alternative for Retrofitting Earthquake-Damaged Precast-Concrete Walls, Earthquake Spectra, Vol. 13, No. 2, pp. 225-241.
7. El-Mikawi, M. and Mosallam, S.A. (1996) A methodology for evaluation of the use of advances composites in structural civil engineering applications, Composites Journal, Part B.
8. Hoskin, C.B. and Baker, A.A. (1986) Composite materials for aircraft structures, AIAA education series, American Institute of Aeronautic and Astronautic Inc., New York, NY.
9. Labossiere, P. (1993) Current and Future Applications of Advanced Composite Materials in Structural Engineering, Canadian Society for Civil Engineering, 43p.
10. Lloyd, G.D. and Calder, A.J.J. (1982) Microstructure of Epoxy Bonded Steel-to-Concrete Joints, Supplementary Report No. 705, Transport and Road Research Laboratory, Structural Department, Bridge Division, Crowthorne, Berkshire, England, pp. 119-127.
11. Malvar, L.J., Warren, G.E. and Inaba, C.M. (1996) Large scale tests on navy reinforced concrete pier decks strengthened with CFRP sheets, Proceedings of the Advanced Composite Materials in Bridges and Structures ACMBS II Conference, Canadian Society for Civil Engineering, Montreal, Quebec, pp. 497-504.
12. Meier, U., Deuring, M., Meier, H. and Schwegler, G. (1993) Strengthening of Structures With Advanced Composites, Alternative Materials for the Reinforcement and Prestressing of Concrete, edited by J.L. Clarke, Blackie Academic and Professional, UK, pp. 153-171.
13. Meier, U. and Erki, M-A. (1997) Advantages of composite materials in the post strengthening technique for developing countries, Proceedings of the Sixth International Colloquium on Concrete in Developing Countries, Lahore, Pakistan, pp. 1-13.
14. Meier, U. and Kaiser, H. (1991) Strengthening of Structures with CFRP Laminates, Advanced Composite Materials in Civil Engineering Structures, Proceedings of the Speciality Conference, Las Vegas, Nevada, ASCE, pp. 288-301.
15. Mufti, A.A., Erki, M.A. and Jaeger, L.G. (1991) Advanced Composite Materials with Application to Bridges, Canadian Society for Civil engineering, 297p.
16. Nanni, A. (1995) Concrete Repair With Externally Bonded FRP Reinforcement-Examples from Japan, Concrete International, Vol. 17, No. 6, pp. 22-26.
17. Norris, M.S. and Nanni, A. (1993) Construction and Repair of Concrete Members with FRP Reinforcement, Construction Technology Update, pp. 2-3.
18. Rizkalla, S.H. and Tadros, G. (1994) A smart highway Bridge in Canada, Concrete International, Vol. 16, No. 6, pp. 42-44.
19. Rubinsky, I.A. and Rubinsky, A. (1954) An investigation into the use of fibre-glass for prestressed concrete, Magazine of Concrete Research, Vol. 6.
20. Saadatmanesh, H., Albrecht, P. and Ayyub, B.M. (1989) Experimental Study of Prestressed Composite Beams, Journal of the Structural Division, ASCE, Vol. 115, No. 9, pp. 2349-2364.

21. Saadatmanesh, H. and Ehsani, M.R. (1991) RC Beams Strengthened With GFRP Plates, I: Experimental Study, Journal of Structural Engineering, ASCE, Vol. 117, No. 11, pp. 3417-3433.
22. Saadatmanesh, H., Ehsani, M.R. and Limin, J. (1996) Seismic Strengthening of Circular Bridge Pier Models with Fibre Composites, ACI Structural Journal, Vol. 93, No. 6, pp. 639-647.
23. Schwarz, M.M. (1992) Composite Materials Handbook, McGraw-Hill, Inc., New York.
24. Schwegler G. (1994) Masonry construction strengthened with fibre composites in seismically endanger zones, Tenth European Conference on Earthquake Engineering, Vienna, Austria.
25. Seible F. (1996) Advanced composites materials for bridges in the 21st Century, Proceedings of the Advanced Composite Materials in Bridges and Structures ACMBS II Conference, Canadian Society for Civil Engineering, Montreal, Quebec, pp. 17-30.
26. Shehata, E., Rizkalla, S., Adams, A. and Stewart, D. (1997) Strengthening of concrete roof using CFRP strips, Annual Conference of the Canadian Society for Civil Engineering, Sherbrooke, Quebec, Vol. 6, pp. 1-10.
27. Sheikh, S.A., Pantazopoulou, S.J., Bonacci, J.F., Thomas, M.D. and Hearn, N. (1996) Repair of delaminated circular pier columns by ACM, Final report to the Ministry of Transportation of Ontario.
28. Sheikh, S.A., Agarwal, A.C. and Pianca, F. (1997) Repair of the Leslie street bridge columns using glass fibre wraps, Proceedings of the Sixth International Colloquium on Concrete in Developing Countries, Lahore, Pakistan, pp. 379-388.
29. Sheikh, S.A. and Biddah, A. (1998) Behaviour of concrete box culverts and their strengthening by fibre reinforced plastics, Ontario Joint Transportation Research Program Report, Department of Civil Engineering, University of Toronto, 41 p.
30. Yau, G. and Sheikh, S.A. (1998) Repair and strengthening of columns with fibre reinforced composites, Research report, Department of Civil Engineering, University of Toronto, 144 p.

HIGH QUALITY CONCRETE MADE MORE RESISTANT AND BETTER ECO MANAGED IN AGGRESSIVE ENVIRONMENT BY COATING

O. G. TJUGUM
Rescon AS, Sagstua, Norway

Abstracts

This article describes practical cases applied to concrete offshore constructions, bridge pillars in marine environment and mention industrial floor, silos and towers where the technique with coating just after pouring is used, laboraratory tests at Rescon and at the Technical University of Trondheim and the Public Road Authorities are also treated..

The author will point out future development possibilities for such applications to give concrete longer service life and achieve optimum eco-management of concrete in special aggressive marine environment.

1 Introduction

Concrete is a very complex material. It contains three phases; gas, liquid and solid, which of each may have varying structure. The material contains a capillary and pore system which let gas and liquid slowly through. 10-20% of the volume is water or air. Further, the material is complicated by the embedded steel reinforcement surrounded by the electrolytic concrete. In the concrete electrolyte the electricity is transported by different water soluble ions from the cement gel or polluted ions as Cl^- and Na^+. These are mobile charge carriers which is at disposal in the pore water. The electric conductivity varies with moist content, type and amount of ions and temperature.

A closer look at the dimensions of capillary and micropores and cracks in concrete, it is obvious that the Cl^- ion may penetrate even high quality concrete. To put the Cl^--ion into the "concrete dimension", concrete pores and cracks with chloride iones are

Concrete Technology for a Sustainable Development in the 21st Century edited by O.E. Gjørv and K. Sakai.
Published in 2000 by E & FN Spon, 11 New Fetter Lane, London EC4P 4EE, UK. ISBN: 0 419 25060 3

measured in diameter as dimention in Table 1 below, but given in mm to give an association.

Table 1 Dimensions

Cl-ion	0,00000018 mm
Water molecule:	0,00000030 mm
Gel pore:	0,00000200 mm, i.e. 10 times bigger than Cl-ion
Capillary pores:	0,00000200-0,00500000 mm, i.e. 10-25000 times bigger than the Cl-ion
Macro pores:	0,00500000-0,50000000 mm, i.e. approx. 250000-250000000 times bigger than the Cl-ion
Crack example:	
	0,18000000 mm, i.e. approx. 1.000.000 times bigger than the Cl-ion.

The pores or the capillaries in even the best practical concrete will not make any effective physical barrier against the steel corrosion initiating Cl^--iones, which is abundantly existing both in air, as aerosoles, and in sea water around marine concrete constructions. The local micro environment on the concrete surface is determinating whether the chloride iones penetrate to the reinforcement and may initiate very destructive reinforcement corrosion. When the porous concrete works as a "salt extraction" material and the bulk concrete is moist enough so that the chloride iones may be driven inwards by natural concentration motive power, the chloride iones migrate inwards even in good concrete. Remember, the chloride iones cannot "fly" in the concrete, they move only by Brownian movements in the liquid. (In free air the Cl^--iones flies in aerosoles, tiny water drops).

Around the world, chloride corrosion on reinforced concrete constructions is a very huge technical and economical problem.

Within many areas, like industry, purification plants etc., a protective coating of thermosetting plastic is an accepted necessity to protect and increase the concrete's lifetime in areas exposed to concrete aggressive chemicals. It is special plastics tightness, i.e. as opposed to the concretes throughgoing pore-capillary system (10-20%), these does not have a connected pore system which may be less than 1% of the volume. Of course the curing mechanism, durability and strength must also be considered.

My idea - point of view, is that a high quality concrete in marine environment, as constructed today, may be made more durable by surface treatment, especially at location where the micro environment means considerable salt extraction on the concrete surface. This is places on the construction where water evaporate and sea salt remains on the concrete surface without being washed away. A strategy by treating the above mentioned areas, will demand a much more intimate knowledge of the micro environment the concrete is exposed to than usually today..

The strategy will practically applied give a sustainable use of the construction material concrete, i.e. a more proper eco-managed use of the material.

The salt extraction follows simple physical laws and should be simulated in computers based on registered metroligical data at the side of the constructions. Important factors will be the construction geometry, relative humidity, temperature, wind direction, wind speed, salt concentration etc.

A strategy is also that "airtight" coatings may reduce oxygen ingress, and thereby reduce the reinforcement corrosion. An example is given in Appendix 1.

Resources are saved when concrete which shall be coated is coated few hours after casting. Extra curing treatment, special cleaning as sand blasting and new scaffolding for treatment is not necessary. This may also be an environmental important advantage.

A main problem by using the above strategies and philosophies is a well-established conservative attitude among practical concrete experts. Concrete is good enough and tight enough against chlorides in marine environment. When you look at the extent of the chloride damages, the cost to the community today, one may ask: How long shall one "experiment" with concrete without taking into use elementary barrier coating material technology where it is needed? The concrete will have a "spongy" effect on chlorides in an unfortunate micro environment, which we will find in many places, for instance on our coastal bridges. The connected pore and capillary geometry and water quantity in the concrete, decides how easy the chlorides "swim" inwards by a definite concentration motive power.

2 Environmental effect

According to Mehta [1] we have limestone and clay, the basic rawmaterials for making portland cement are plentiful throughout the world. The major bottlenecks in increasing the portland cement production by almost five hundred million tons annually by the year 2005 appear to lie in the high energy requirements and the high rate of CO_2 emision. Every ton PC release one ton of CO_2. This will give an unacceptable "green house effect". An environmental disaster would be unavoidable if China, India, and developing countries of South and Central America start consuming as much energy and materials as the West did in it's march to industrialization.

From references [2], [3]and [11] we can set up a comparative table (Table 1), which compares the materials energy consumption and some pollution by production. This is of course not absolute correct values, but it gives clear indications that durable coatings on concrete can be a sustainable use of concrete in aggressive environment.

Table 2: Total energy consumption and some pollution per m^2 of concrete by two different thicknesses, coated and reinforced or not.

No	1	2	3	4	5	6	7	8
Concrete wall Thickness reinforcement epoxy coating per m^2	30 cm not reinforced concrete	30 cm not reinforced concrete + 2 kg/m^2 epoxy coating	As no. 2 + 20 kg/m^2 reinforcement	As no. 2 + 50 kg/m^2 reinforcement	50 cm not reinforced concrete	50 cm not reinforced concrete + 2 kg/m^2 epoxy	As no. 6 + 20 kg/m^2 reinforcement	As no. 6 + 50 kg/m^2 reinforcement
Pollution Energy per m^2								
CO_2 kg/m^2	91	97	117	147	155	161	181	211
NO_2 kg/m^2	0,42	0,44	0,47	0,52	0,70	0,74	0,77	0,82
SO_2 kg/m^2	0,16	0,17	0,23	0,32	0,27	0,28	0,34	0,43
Total energy MJ/m^2	575	859	1059	1359	958	1242	1442	1742

From Table 2 one may do simple calculations on energy consumption and pollution per year on the assumption that the epoxy coating increases the lifetime in aggressive environment.

Table 3: Total energy consumption and some pollution per m^2 and per year by different assumed lifetimes.

Pollution Energy per m^2 and year	30 cm reinforced concrete 50 kg reinforcement/m^2	30 cm reinforced concrete 50 kg reinforcement/m^2 + 2 kg epoxy/m^2		
Lifetime	20 years	25 years	30 years	40 years
CO_2 kg/m^2 and year	7,1	5,9	4,9	3,7
NO_2 kg/m^2 and year	0,025	0,021	0,017	0,013
SO_2 kg/m^2 and year	0,016	0,013	0,011	0,008
Total energy MJ/m^2 and year	53,8	54,4	45,3	34,0

The example in Table 3 shows a clear positive environmental effect by coating of concrete constructions assumed that the lifetime increases.

Figure 1 shows pollution or energy consumption per year, and as 100% for untreated reinforced concrete from Table 3 as y-axis and % longer lifetime with coating as the x-axis.

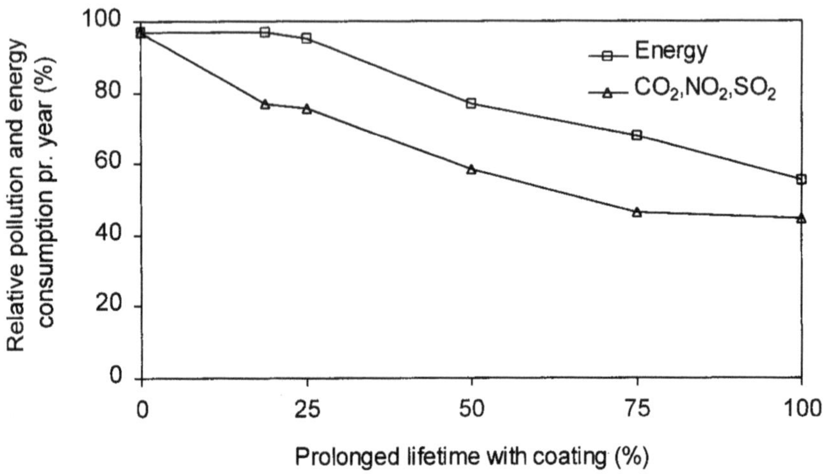

Fig. 1 Relative pollution and energy consumption per year

3 Coating, experience and laboratory experiments

In the following I will discuss at the laboratory experiments at Rescon/NTNU and SVV, field experiments and practical use with relative long time of function, mixed with different opinions based on approx. 30 years of experience with concrete and thermosetting plastics.

3.1 Bridges

I will disregard the thermosetting plastic coatings/impregnations of bridge decks, parapets and edges and repair with coating in areas with too low reinforcement covering. Two interesting experiments are treated.

Field tests our company has carried out together with the Public Road Authorities in a larger concrete rehabilitation development project; Gimsøystraumen bridge (approx. 20 years old bridge), applied different types of thicker coatings in 1993, 1994 and 1995 (ref. [7]). In addition a considerable concrete slab program was carried out where the slabs were surface treated and placed outdoor for exposion at a typical Norwegian coast bridge not far away from Svolvær.

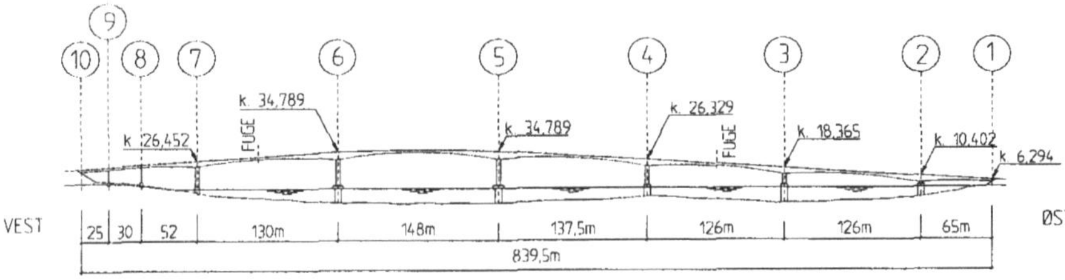

Fig. 2: Gimsøystraumen Bridge 1994

Table 4: Relevant coating data from field tests

	1993	1994	1995
Elastic polymer cement grout 3-4 kg/m^2 diffusion open	227 m^2	157 m^2	953 m^2
Flexibel epoxy 2,5 kg/m^2 diffusion tight	0	159 m^2	
Cement epoxy slurry 3-4 kg/m^2 diffusion open	0	0	197 m^2

The two first coatings may build over cracks which opens and closes 0,2 mm by ±20°C. Non-treated concrete acts as reference and the test areas are instrumented to monitor the corrosion development, potentials, LPR, oxygen transport, electric resistance and moist in the concrete ar registered.

After 3-5 years it is to early to draw positive or negative conclusions on the corrosion development in the saline coated concrete, but it will be followed up over the years to come. It is an indication of a change in chloride distribution under the elastic polymer cement grout.

In 1993, two pillars on Stokksund bridge were totally coated with reinforced epoxy till approx. ½ m below low water (ref. [8]). The bridge on Figure 2 was 10 years old and the pillars hade a chloride content of 0,1 - 0,2% of the concrete weight. The reinforcement was in the beginning of corrosion. The aim was to reduce the reinforcement corrosion activity by a diffusion tight coating, i.e. reduce the oxygen ingress to stop the corrosion (cathodic process). One of the pillars were coated with corrosion inhibitor before coating. Further, the experiment should demonstrate practical coating under water. Yet it is to early to say if the corrosion process has stopped, but it is measured very low oxygen transport which makes the results very interesting. No inihibitor effect has been observed after approx. 5 years. The coating does not show any frost damages.

Totally this will be a cost effective rehabilitation method.
The concrete is instrumented to follow the corrosion situation in the future. Concrete detorioration caused by corrosion is a long time process, therefore the results will com in the future.

3.2 Concrete platform

Inspection of coated and uncoated concrete structures exposed in the North Sea, clearly indicates a difference in environmental impacts, mainly chloride uptake and resulting reinforcement corrosion - inspection of epoxy coated concrete in the splash zone indicates that the coating after 15 years exposure still has a good bond to concrete (2-3 Mpa) and that the coating is intact, except in areas where it has been exposed to mechanical wear [4]. Epoxy coating effectively protects the concrete from chloride uptake.

On most of the concrete platforms manufactured before 1980, the splash zone and the top of the shafts were treated with epoxy coatings. Frigg TCP2, Beryl A, Brent B, Brent D, Statfjord A, B and C. With improved concrete quality as the main argument for excluding the coating in the splash zone for later constructions. However, epoxy coating were applied in certain areas to compensate for insufficient concrete cover and to seal cracks, Gullfax A, B and C, Oseberg A, Sleipner A, Draugen, Heidrun, Troll A.

The above information is based on information from Norwegian Contractors about coatings applied on their constructions. Results from inspections on Statfjord and Gullfax and Oseberg A:

Table 5: Results from inspections on Statfjord, Gullfax and Oseberg A

Coating thickness	3,2 mm
Bond to concrete	1,8 Mpa
Chloride penetration	insignificant
Carbonation	not detected
Microcracks under coated concrete	not observed
Microcracks without coating	5-30 mm deep

A chloride profile of uncoated concrete is given in Figure 3.

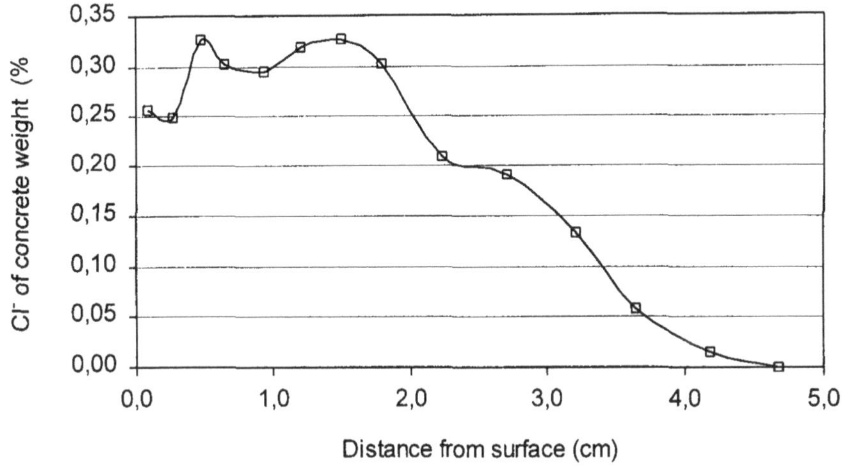

Fig. 3 Chloride profile

When the first concrete platform that was coated, in the beginning of 1970, there was no talking about protection of concrete against chlorides in the splash zone. At the time, the concrete was durable. The purpose of the coating explained to me as a material technichian, and responsible for mix design of the coating: long lasting curing membrane and flexible coating which could build over some cracks so that the crew in case of crisis should have some more time to get out of the concrete shafts.

Direct epoxy coating of the concrete constructions like on the concrete platforms as the concrete appears under the slipform, we have used with great success in chemical industry by great concrete (prelltårn) works, great silos in purfication plants etc. The technique by early coating is also used in industry, for instance electrolysis halls where the floor is renewed during running of production.

3.3 Frost resistance laboratory tests of coated concrete

It is performed freeze-thaw tests [12] on several surface treatments, the modified SS137244 (Borås method). There are three main differences: dimension of the test piece (cylinder 10 and 4 cm high), coating on concrete and initial moist conditions are more extreme (RF up to 100%). Tests are also made with access to water on the under side of the sample. Cracks are made in the coating to simulate the effect of errors in the surface coat on the frost peeling. The coatings are applied 5 hours and 24 hours after casting of the concrete. Two concretes after NS 3420 NA and MA-concrete are tested as a reference and with coating, see Tables 6 and 7.

Table 6 Concrete mixes

	NA-concrete V/C=0,59	MA-concrete V/(C+25)=0,4
Cement Norcem Standard	300	375
Sand 0-8 mm	980	934
Aggregate 8-16 mm	904	828
Silica		18
Lignosulphonate 40%	2,4	2,4
Modified melamine 25%		3,6
Tenside		0,1
Water	176	166

Adhesion is measured both before and after the frost test. Samples of NA and MA-concrete shows that NA-concrete is far from frost proof, while the MA-concrete may be characterized as acceptable frost proof.

Table 7 Coating specifications

Surface treatment	
Elastic polymer cement grout with emulsion primer	3,5 kg/m^2
Solvent free epoxy primer Cement epoxy screed Solvent free epoxy coating	3 kg/m^2
Solvent free epoxy primer Solvent free epoxy concrete glue (Low shrinkage designed for power transmission)	2,5 kg/m^2

After 112 frost cycles +20^{0}C tol - 20^{0}C in 24 hours, it was practically no frost peeling of the epoxy samples, but the concrete was considerably damaged in cases where open cracks was made in the samples. The elastic polymer cement grout had some peeling after 70 cycles. The MA-concrete was in every case better than the NA-concrete. On the epoxy samples it was taken adhesion tests both before and after the frost cycles. Results are given in Table 8.

Table 8 Adhesion tests

	Applied 5 hours after casting				Applied 24 hours after casting			
	Before frost cycles		After 112 frost cycles		Before frost cycles		After 112 frost cycles	
Concrete	NA	MA	NA	MA	NA	MA	NA	MA
Solvent free epoxy primer Cement-epoxy Solvent free epoxy coating	3	2,5	1,9	2,8	2,4	2,8	1,9	3,0
Solvent free epoxy primer Solvent free epoxy concrete glue	3,2	3,0	1,6	1,2	2,0	3,2	2,3	2,4

The tests show that all surface treatments give reduced peeling compared to untreated samples of both NA and MA-concrete according to SS 137244. The results show that the surface treatment of concrete may reduce the frost/salt peeling of concrete considerably. Damages in the coating have a negative effect on the frost resistance, especially on stiff epoxy systems. The difference between concrete treated directly, i.e. 5 hours after casting, and later treatment seems little. One may therefore conclude that the surface treatment of fresh concrete is secure as regards adhesion.

The tests show that surface treatment which is exposed to moist and frost must be porefree without damage in the coating to give optimal protection. This is analog to the demands to a surface treatment which shall protect the concrete against chemical aggressives.

Early treatment of concrete with thicker coating based on cement polymer, or specially adjusted epoxy coating, will be an economical and environmental advantage, as it gives the concrete good curing conditions. The working procedure with curing of the concrete with, for instance, curing membrane, is not necessary, and the same applies for later cleaning of the concrete surface before coating. From experience in industry in aggressive environment, the pores and damagefree coating of marine concrete constructions will be possible in most areas, if not exposed to mechanical strain.

3.4 Laboratory test with concrete pillars

At the Rescon Centre we have the following tests going on with coated and non coated steel reinforced concrete pillars:

1. Shaping
   5 pillars with diameter 30 cm and height 150 cm. Concrete cover 20 mm. The

reinforcement is made as rings of 10 mm reinforcement steel. The rings were welded to 6 vertical 10 mm reinforcement steel with a distance of 13 cm. The pillars are casted with NS 3420 MA-concrete with v/c 0,45 and 5% silica D_{max} = 8 mm.
Pillar no.:

1. Reference with conventional admixture, lignosulphonate and melamine.
2. As 1, which is coated immediately by removal of forms with light solvent free epoxy.
3. As 1, but coated after 3 months curing.
4. Reference with superplasticizing and inhibitating admixture. Modified melamin (HP-SF).
5. As 4, but extra addition of inhibitor (GPI).
6. As 1, but only coated with epoxy in the splash zone.
   Temporarily from august 1996 till 1998 the pillars has been kept in sea water in laboratory, but the plan is to put them out in a fjord.

Instrumentation
The reinforcement in the pillars was instrumented with regard to potential, corrosion rate, oxygen transport and electric resistance.
This was performed by mounting reference electrodes and probes in 4 heigths from the lowest reinforcement (15, 45, 75 and 118 cm). Probes for corrosion rate measurements were mounted only in the two lowest. The system is connected to a computer and network at the Rescon Centre.
The placing of the pillars:

* one measure point under water
* one just above the water surface
* one well above the water surface
* one in the top of the pillar

2. Conclusions
After nearly two years it is difficult to draw any conclusion, but no negative effects have been observed.

4 References

1. Mehta PK, Bringing concrete industry into a new era of sustainable development, Italy 1997.
2. M. Vold and A. Rønning. LCA of Cement and concrete - Main report. November 1995.
3. I Boustead. ECO-profiles of the European plastics industry. Report 12. Liquid Epoxy Resins. Brussels November 1977.
4. R. Aarstein, O. E. Rendarøy, O. Liodden and Bjarne Jensen, Experience of inspection of coated and uncoated concrete of Offshore structures exposed to the North Sea for 2 - 20 years. OFU-Repair Conference, Svolvær 1997, Norway.
5. UK Cement Manufacture and the Environment. British Cement Association.

6. G. Tjugum - Repcon rapport. Tradisjonell rehabilitering. Elektrokjemi. Hoveddrivkraften ved tradisjonell rehabilitering?
7. Publikasjon nr. 84. Prøvereparasjon og Produktutvikling. Statens Vegvesen, Veglaboratoriet.
8. G. Tjugum - Repcon rapport, Stokksund bru.
9. S. Wehrte. Einsatzmöglichkeiten und Grensen für Riss überbüchende Beschichtungen auf Betongflächen in Parkhäusern, Institutt für Bautenschutz, Fellbach, Deutschland.
10. M. Homann. Entwicklung Riss überbüchender Beschictung für Industriboden, Universität Dortmund, Deutschland.
11. Sverre Fossdal, Energi og miljøregnskap for bygg. Norges byggforskningsinstitutt, Rapport 1995.
12. Hovedoppgave NTNU 1996 Franing og Sundel. Faglærer Proff Sellevoll. Samarbeide SVV (Kompen) Rescon (Kristiansen) og Norsk Byggforskningsinstitutt (S. Jackobsen) og videre arbeide i 1997 og 1998.

Appendix 1

Epoxy film resistance to oxygen penetration and possible contribution to reduce reinforcement corrosion rate.
This is a calculation based on literature values.

Water saturated concrete:

w/c = 0,40, thickness = 20 mm, O_2-flux = $6{,}5\times10^{-13}$ mol/cm^{2}s
Gjørv and Vennesland (1986)

Epoxy coal tar:

thickness: 0,1 mm, O_2-flux = 15.10^{-13} mol/cm^{2}s

Guruviah (1967/70)

I.e.: <u>0,23 mm</u> epoxy lets through the same oxygen amount as <u>20 mm</u> water saturated concrete (w/c = 0,40).

Basis:

Insignificant corr.rate.: $I_{corr} \leq 0{,}1\ \mu A/cm^2$ High corr.rate.: $I_{corr} > 1\ \mu A/cm^2$ (in concrete)

Assume that 1 meter long reinforcement (Ø = 12 mm) consuming the oxygen via 20 cm x 100 cm concrete surface:

100 cm

Surface area concrete: A = 2000 cm^2
Surface area reinforcement: a = $2\pi r$. L = 377 cm^2

Assume small local corrosion areas on the surface are: totally 10% of the surface: 37,7 cm^2 (377-37,7=339,3 cm^2 is cathode for O_2-reduction).
In these small areas the corrosion rate is quite high: 1 μA/cm^2. Totalt 1x37,7=<u>37,7μ A</u>

The oxygen is consumed (corresponding to 37,7 μA) on the rest of the steel surface (339,3 cm^2).

The oxygen is supplied through 2000 cm² concrete surface. Number of mol O_2 which is reduced on the cathode surface (339,3 cm² pr. second) is given by:

$n_{o2} = 1 \times 37{,}7 \times 10^{-6}/4 \times 96487 = 9{,}8 \times 10^{-11}$ mol/s

The oxygen is supplied from outside the concrete surface, and must penetrate the 2000 cm² concrete surface, i.e.:

$9{,}8 \times 10^{-11}$ mol/s
or $4{,}9 \times 10^{-14}$ mol/cm² . s
or $4{,}9 \times 10^{-14}$ 32 . 1000/(1/3,154x10^7) = 0,05 mg/cm² . year

The literature shows permability approx. 1 - 2 mg O_2/cm² x year through 0,1 mm epoxy.

We must down to (or below) 0,05 mg/cm² . year.

i.e. at least $\frac{1\ .\ 0{,}1\ \text{mm}}{0{,}05} = 2\ \text{mm}$

The diffusion in the concrete under the coating is not evaluated.

RETROFITTING: AN ANSWER FOR SUSTAINABLE DEVELOPMENT IN THE 21st CENTURY

Retrofitting for sustainable development

T. UEDA
Division of Structural and Geotechnical Engineering, Hokkaido University, Sapporo, Japan

Abstract

A retrofitting design system which consists of investigation the performance of existing structures, selection of retrofitting method and examination of retrofitted structures is outlined. The design system which uses the performance-based design concept and integrates structural and durability design deals with both general and seismic retrofit. This is the first time ever in Japan to unify design methods for major retrofitting methods, namely external cable method, wrapping (or jacketing) method and concrete wrapping (or jacketing) method.
Keywords: External cable, jacketing, performance-based design, retrofit design, wrapping

1 Introduction

It can be said that the 20th century is the century in which infrastructure has been constructed and stocked. The 21st century, however, is the century in which we will be asked how to hand the infrastructure as the property of human being constructed in the 20th century over to the next generation. In the 21st century aiming at "Sustainable Development", we, civil engineers, will be faced with a new task of showing what is the most appropriate infrastructure in the closed environment on the earth and how to realize it instead of the old task of supplying infrastructures of excellent quality. From this point of view we can see a clear picture of "Retrofitting" as not only a measure for seismic retrofitting after the Great Hanshin Earthquake or to cope with the change in the design loads for highway bridges in Japan, but also as a measure to substitute demolition and reconstruction for all types of structures. Retrofitting seems to be the best solution for handing the infrastructures constructed

Concrete Technology for a Sustainable Development in the 21st Century edited by O.E. Gjørv and K. Sakai.
Published in 2000 by E & FN Spon, 11 New Fetter Lane, London EC4P 4EE, UK. ISBN: 0 419 25060 3

during the 20th century, whose service lives will be ended in the 21st century, to the next generation. It can be easily seen that energy and resources necessary for retrofitting are much less than those for reconstruction.

Design concept for retrofitting is essentially the same as that for constructing a new structure. The difference between retrofitting an existing structure and constructing a new structure is the great variety of retrofitting methods. The variation can be seen not only in structural types for retrofitting but also material types. As a result there are various design details for retrofitting. The major reason for the variety is restrictions at construction site. If the existing structures to be retrofitted are in service, it is preferable to continue the service during retrofitting works. Portions to be retrofitted are generally only a part of a whole structure. These facts impose severe restrictions in space as well as time for retrofitting works. The space available for retrofitting works is limited by the existing structures and the service space, and the available time limited by the service time.

Variety in the retrofitting methods is the major reason why no rational unified design method for retrofitting has been established yet. Generally the size of each retrofitting work is not large enough to prepare a design manual for retrofitting. As retrofitting becomes more significant in the 21st century, there is an urgent need for a rational retrofitting design method together with a chart for selection of an appropriate retrofitting method.

In this paper the performance-based design system for retrofitting proposed by Sub-Committee on Retrofitting Design of JSCE Concrete Committee [1], which may be a model for retrofitting design in the coming century, is introduced. Then necessary measures to make retrofitting design methods more rational are discussed.

2 Performance-Based Design System for Retrofitting Proposed by JSCE [1]

2.1 Outline

The retrofitting design system proposed by JSCE in 1998 covers major retrofitting methods currently applied in Japan, namely external cable method (both steel and continuous fiber cable), wrapping (or jacketing) method (both steel and continuous fiber reinforcement wrap/jacket) and concrete wrapping (or jacketing) method. Types of retrofitting considered in the JSCE design system are both general and seismic retrofitting.

The JSCE retrofitting design system uses the performance-based design concept which is considered to be the next generation of the design method for structures. The flow chart given in the retrofitting design system is illustrated in Fig. 1. The retrofitting design system contains 1) investigation of existing structures, 2) examination of performances of existing structures, 3) selection of retrofitting method, and 4) examination of performances in retrofitted structures. In order to examine performances of a structure, required performances should be known. The required performances for an existing structure may be different from those for the same structure to be retrofitted. For example aesthetics which was not considered when the structure was constructed may be considered when the structure is retrofitted. The service life of the structure when it was constructed could be longer than that when it is retrofitted, so that the requirement for the structure to

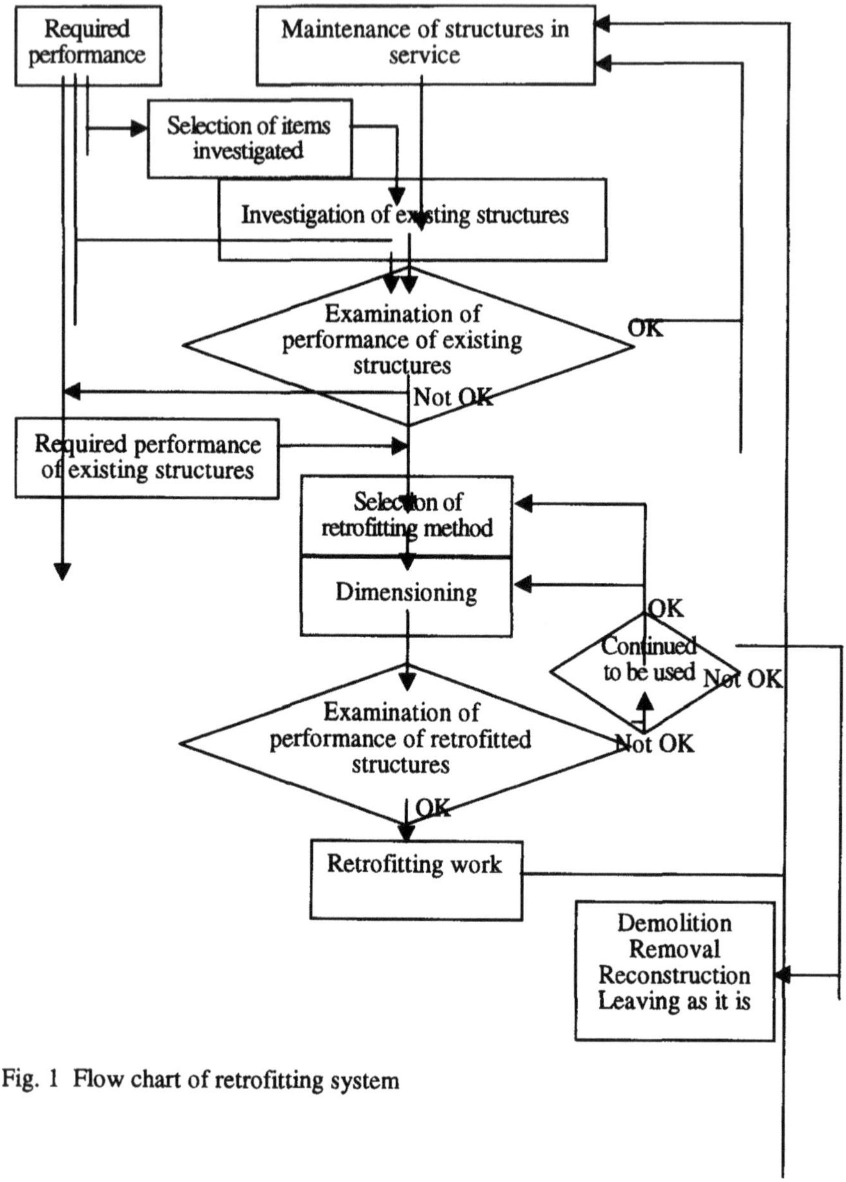

Fig. 1 Flow chart of retrofitting system

maintain its performances may be eased at the time of retrofitting. The required performances for retrofitted structures, therefore, are newly provided as shown in the flow chart (see Fig. 1).

2.2 Concept of performance-based design

2.2.1 Required performance

Required performances are classified into the following categories:

- social and environmental friendliness
  ability to contribute to healthy social, economical and cultural activities and to minimize harmful effects to surrounding social and natural environment (aesthetics at long distance can be included)
- safety
  ability to avoid casualty due to structural failure and collapse under any actions
- serviceability
  ability to make users of and people around structures feel comfortable with the structures and to provide functions such as water tightness
- restorability
  ability to restore lessened performances with economically and technically feasible way
- constructability
  ability to assure safety and reliability during construction
- easy maintenance
  ability to make maintenance easy
- easy demolition and recycling
  ability to make demolition and recycling easy

In Table 1 items of the above major categories are listed. In the above categories, durability or seismic performance is not seen, since the required performances are not classified according to types of actions. It is considered that durability and seismic performance are description of the required performance under environmental and seismic actions respectively.

2.2.2 Actions

Actions considered in the retrofitting design system are as follows:

- normal loads (permanent loads, variable loads and accidental loads)
- seismic actions (level 1 and level 2 seismic actions)
- environmental and chemical actions

Fatigue loads are included in variable loads. Seismic actions are a kind of accidental loads. However, seismic actions are treated differently from other normal loads since effects of seismic actions are different from those of other normal loads and the most significant for most of structures in Japan. Level 1 seismic actions corresponds with earthquakes which likely happen more than once during service life of a structure, while level 2 seismic actions mean practical maximum earthquakes which may happen at the site of a structure. Environmental and

Table 1. Performance index

Required performance			Performance index I_P		Actions[1)]
Classification	Major items	Items	Precise	Simple	
Social and environmental friendliness	Contribution to social, economical and cultural activities	No items	Selection of appropriate I_P		Normal loads
	Harmful effect to social and natural environment	Loads to environment			
Safety		Member failure (normal loads)	Failure analysis by FEM	Axial force, bending moment, shear force and torsional moment	Normal loads
		Member failure (seismic actions)		Axial force, bending moment, shear force, torsional moment and ultimate deformation	Level 2 seismic actions
		Stability	Stability analysis by FEM	Buckling force, overturning moment and sliding force	Normal loads
Serviceability	Comfortable feeling	Comfortable ride/walk	Acceleration	Deformation, step, gap, angle and type of pavement	Normal loads, level 1 seismic actions
		Anti-vibration	Vibration level	Type of pavement, step, gap, stiffness, mass	
		Anti-noise	Noise level		
		Soundproof	Leaking noise level	Shape, thickness, height and type of soud-proof wall	
		Heat-insulation	Leaking and/or reflecting heat level	Thickness of heat insulation member and heat reflection characteristics of member surface	

1) Environmental actions are considered in any case.

Table 1. Performance index (continued)

Required performance			Performance index I_P		Actions[1)]
Classification	Major items	Items	Precise	Simple	
Serviceability	Comfortable feeling	Anti-odor	Density of substance with odor	Amount of substance with smell	Normal loads, level 1 seismic actions
		Anti-humidity	Humidity	Water content of concrete	
		Aesthetics (short distance)	Crack width, crack density, size and density of dirty spot		
		Visual safety	Deformation, crack width and crack density		
	Functioning	Water-tightness	Amount of leaking water	Crack width	
		Air-tightness	Amount of leaking air	Crack width	
		Energy and substance insulation	Amount of energy and substance leaking	Crack width and member thickness	
Construct-ability	Safety	Danger to worker / surroundings	Working space and heavy instrument	Construction manual	Loads during construction
	Reliability	Quality of material	Difficulty of work		
Restorability			Difficulty of restoring work	Maximum deformation and remaining deformation	Level 2 seismic actions
Easy maintenance		Routine investigation	Whether accessible and visible or not		Normal loads
		Investigation of material	Whether monitoring system is available or not		
Easy demolition and recycling	Easy demolition		Difficulty of demolition		Normal loads during demolition
	Easy disposal and recycling	Disposal	Energy needed for disposal and product after disposal		None
		Recycling	Whether recyclable or not		

chemical actions are causes of material deterioration.

2.2.3 Examination of performance

Performances are quantified by performance indexes, I_P. Examination of the performances is conducted by verifying the following condition:

$$I_{P,req} > I_{P,act} \qquad (1)$$

where $I_{P,req}$: performance index representing required performance

$I_{P,act}$: maximum performance index caused by actions

For the examination of safety one of I_P is bending moment. $I_{P,req}$ and $I_{P,act}$ are the bending capacity of a section and the maximum bending moment caused by actions. I_P for aesthetics which is one of serviceability may be crack width at concrete surface. In this case $I_{P,req}$ represents the maximum crack width which does not damage the aesthetics and $I_{P,act}$ is the maximum crack width under the considered actions. Performance indexes for each item of required performances are given in Table 1.

Seismic actions which are major actions in seismic regions affect significantly safety of structures. The examination of safety under seismic actions is generally different from that under normal loads. A typical performance index for seismic actions is the maximum deformation during earthquake or the remaining deformation after earthquake, while performance indexes for normal loads are axial force, bending moment, shear force and torsional moment (see Table 1). Fatigue loads which are a kind of normal loads (variable loads) also affect safety. In this case performance indexes may be load or stress ranges.

For serviceability there are many items to be considered as shown in Table 1. Comfortable ride/walk, anti-vibration, anti-noise, soundproof, heat insulation, anti-odor, anti-humidity, aesthetics (at short distance), visual safety are items for comfortable feeling for users of and people around structures, while water tightness, air tightness and insulation of energy and substance are items for functioning of structures. Appropriate performance indexes for those items should be chosen. For example the precise performance index for comfortable ride/walk may be acceleration which passengers/pedestrians feel. Vertical and horizontal gaps, angles of road surface as well as type of pavement can be chosen as simple performance index. Precise examination for serviceability has not been conducted in practical design. More research is needed for identifying appropriate performance indexes and ways to quantify those performance indexes.

The concept of restorability comes from seismic design. Through the experience of major earthquakes it is obvious that avoiding any loss of required performances for serviceability is practically impossible. Instead another requirement was introduced, so that rather quick restoration of the lost performances is made possible. A typical performance index is damage of structures, such as the maximum deformation during earthquake or the remaining deformation after earthquake.

Common concept for three other categories, namely social and environmental friendliness, easy maintenance and easy demolition and recycling, has not been established. Further study is necessary.

Examination for ultimate limit states and serviceability limit states in the current limit state design is a good example of the examination for safety and serviceability.

2.2.4 Durability aspect

Durability is not considered to be one of the required performances in this performance-based design system for retrofitting. Instead all the performances are examined at every moment during construction, service life and demolition, considering the chronological change in performances induced by material deterioration and so on. For example the bending capacity may be reduced due to corrosion of steel reinforcement, while the bending moment acting in the considered section may be changed due to stiffness reduction caused by material property change such as concrete shrinkage. Therefore the performance index in Eq.(1) is time-dependent and the following condition is derived:

$$I_{P,\,req}(t) > I_{P,\,act}(t) \quad \text{for } t_{start} \le t \le t_{end} \tag{2}$$

where t: time

t_{start}: time at the beginning of examination period (usually the beginning of construction)

t_{end}: time at the end of examination period (usually the end of demolition)

As a rule the above condition should be satisfied at any time during construction, service life and demolition. However, if it is found that the above condition is satisfied at the end of service life, it can be considered satisfied at any time during service life in many cases. Because many $I_{P,\,req}(t)$ decrease with time and many $I_{P,\,act}(t)$ increase with time. For example the bending capacity as $I_{P,\,req}(t)$ decreases with time, and the crack width as $I_{P,\,act}(t)$ increases with time. The sole examination at the end of service life can simplify the examination process significantly.

Similarly fatigue phenomena is taken into account in the examination of performance when the fatigue effect is considerable. The bending capacity under the fatigue effect can be reduced. The crack width under the same magnitude of loads may increase due to the fatigue effects.

For precise approach physical and chemical models to express chronological change in material properties and structural conditions under environmental and chemical actions are necessary. However, simplified models can be introduced considering material type, structural type, construction condition and environmental actions and loads. Proper detailing, such as minimum concrete cover, may further simplify the process to consider the chronological change in material properties. For example if a thick cover is provided, corrosion of steel reinforcement may not be considered at all.

2.3 Retrofitting design system

2.3.1 Investigation of existing structures

There are preliminary investigation and site investigation. Items to be investigated should be carefully selected so that the necessary information needed to examine the required performances can be obtained. Tables 2 and 3 show the items for preliminary and site investigation and required performances related to those items.

Table 2. Preliminary investigation

Items for investigation	Required performances[7]					
	A[1]	B[2]	C[3]	D[4]	E[5]	F[6]
Time for design and construction	✓	✓	✓	✓	✓	✓

Applied code	✓	✓	✓	✓	✓	✓
Type of structure	✓	✓	✓	✓	✓	✓
Design condition	✓	✓	✓	✓	✓	✓
Ground condition		✓	✓	✓	✓	
Others[8)]	✓	✓	✓	✓	✓	✓

1) Social and environment friendliness, 2) Safety, 3) Serviceability, 4) Restorability
5) Easy maintenance and 6) Easy demolishing and recycling
7) with ✓: related, without ✓: not related
8) Whether retrofitted or not / Location of attachment

Table 3. Site investigation

Type of investigation	Investigated items	Required performances[7)]					
		A[1)]	B[2)]	C[3)]	D[4)]	E[5)]	F[6)]
Harmonization with preliminary examination		✓	✓	✓	✓	✓	✓
Environmental condition	Site condition	✓			✓	✓	✓
	Traffic condition	✓	✓	✓	✓	✓	✓
Damage	Cracking		✓	✓	✓		✓
	Loss of section			✓			✓
	Deformation						
	External appearance						✓
	Vibration/noise	✓	✓	✓	✓	✓	✓
Internal investigation	Mechanical properties of material		✓	✓	✓		✓
	Other material properties						✓
	Reinforcement size and arrangement		✓	✓	✓		✓
	Concrete cover			✓			✓
	Internal defect			✓			✓
	Reinforcement corrosion		✓	✓	✓		✓
	Amount of chemical substance						✓
	Depth of environmentally affected zone						✓
Others	Ground condition				✓		✓
	Location of attachment				✓	✓	

1) Social and environment friendliness, 2) Safety, 3) Serviceability (crack width as performance index), 4) Restorability, 5) Easy maintenance and 6) Easy demolishing and recycling
7) with ✓: related, with ✓: related slightly, without ✓: not related

Table 4. Retrofitting method and required performance

Retrofitting method		Required performances[7)]					
Major category	Type	A[1)]	B[2)]	C[3)]	D[4)]	E[5)]	F[6)]
External cable method	External cable method (steel cable)	✓	✓	✓	✓	✓	✓
	External cable method (continuous fiber)	✓	✓	✓	✓	✓	✓
Wrapping or jacketing method	Steel plate wrapping	✓	✓	✓			✓

	FRP plate wrapping	✓	✓	✓			✓
	FRP sheet wrapping	✓	✓	✓			✓
	Steel plate jacketing	✓	✓	✓	✓		✓
	FRP sheet jacketing	✓	✓	✓	✓		✓
	FRP plate jacketing	✓	✓	✓	✓		✓
Concrete wrapping and jacketing	Concrete wrapping on top surface	✓	✓	✓			
	Concrete wrapping on bottom surface	✓	✓	✓			
	Mortar spraying on bottom surface	✓	✓	✓			
	Reinforced concrete jacketing	✓	✓	✓	✓		
	Prestressed concrete jacketing	✓	✓	✓	✓		
	Mortar spraying	✓	✓	✓	✓		
	Precast panel jacketing	✓	✓	✓			✓

1) Social and environment friendliness, 2) Safety, 3) Serviceability (crack width as performance index), 4) Restorability, 5) Easy maintenance and 6) Easy demolishing and recycling
7) with ✓: effective, with ✓: effective slightly, without ✓: not effective

2.3.2 Examination of performances with existing structures

Performances with existing structures to be examined are social and environmental friendliness, safety, serviceability, restorability, easy maintenance, and easy demolition and recycling. Constructability is not necessary because existing structures have already been constructed. The examination is conducted according to what is described in Sec. 2.2.3.

2.3.3 Selection of retrofitting method

An appropriate retrofitting method is selected among external cable method (both steel and continuous fiber cable), wrapping (or jacketing) method (both steel and continuous fiber reinforcement wrap/jacket) and concrete wrapping (or jacketing) method. The required performances to be retrofitted should be considered for the selection. Effects of retrofitting methods to required performances are indicated in Table 4.

2.3.4 Examination of performances with retrofitted structures

Social and environmental friendliness, safety, serviceability, restorability, constructability, easy maintenance, and easy demolition and recycling are the performances to be examined. Constructability is for retrofitting works. The examination is conducted in a similar way to that for existing structures according to what is explained in Sec. 2.2.3.

3 Needs for future study

Through preparation of the retrofitting design system the followings are identified for further study:

- Improvement of shear strength prediction of members retrofitted with external cable
- Establishment of seismic design of members retrofitted with external cable
- Improvement of shear strength and ductility prediction of members retrofitted with wrapping
- Establishment of shear design of beams retrofitted with wrapping
- Quantification of effects of creep and shrinkage in members retrofitted with concrete wrapping
- Improvement of ductility prediction of members retrofitted with concrete jacketing
- Establishment of fatigue design for all the retrofitting methods
- Establishment of durability design of all the retrofitted methods
- Establishment of maintenance method for members retrofitted with wrapping (jacketing)

4 Concluding remarks

A retrofitting design system with the performance-based design which covers major retrofitting methods applied for both general and seismic retrofit in Japan is outlined. It represents an example of the retrofitting design in the 21st century. Further study is still needed especially on how to predict the time dependency of structural performances.

5 Acknowledgement

The author would like to express his sincere gratitude to the members of JSCE 307 Committee for their valuable discussion which is used as a basis for preparing this paper.

6 References

1. Sub-Committee on Retrofitting Design, Concrete Committee (1998) Design and Construction of Retrofitting Concrete Structures in Future - Tentative Draft for Performance-Based Design for Retrofitting -, *Concrete Engineering Series*, JSCE, No.28 (in Japanese).

Key Word Index

This index is compiled from the keywords provided by the authors of these papers, edited and extended as appropriate. The numbers refer to the first page of the relevant papers.

For Product Safety Concerns and Information please contact our EU representative GPSR@taylorandfrancis.com
Taylor & Francis Verlag GmbH, Kaufingerstraße 24, 80331 München, Germany

www.ingramcontent.com/pod-product-compliance
Ingram Content Group UK Ltd.
Pitfield, Milton Keynes, MK11 3LW, UK
UKHW020022200926
475702UK00014B/2020